Mathematical Foundations of Quantum Mechanics - in memoriam Paul Busch

Journal of Physics: Conference Series Volume 1638

York, United Kingdom
18 – 19 June 2019

Editors:

Roger Colbeck
Leon Loveridge
Stefan Weigert

ISBN: 978-1-7138-3914-9
ISSN: 1742-6588

Printed from e-media with permission by:

Curran Associates, Inc.
57 Morehouse Lane
Red Hook, NY 12571

Some format issues inherent in the e-media version may also appear in this print version.

Additional copies of this publication are available from:

Curran Associates, Inc.
57 Morehouse Lane
Red Hook, NY 12571 USA
Phone: 845-758-0400
Fax: 845-758-2633
Email: curran@proceedings.com
Web: www.proceedings.com

TABLE OF CONTENTS

Journal of Physics: Conference Series **1638** (2020) 011001 doi:10.1088/1742-6596/1638/1/011001

Editorial

Roger Colbeck[1], Leon Loveridge[2] and Stefan Weigert[1]

[1] Department of Mathematics, University of York, York YO5 10DD, United Kingdom
[2] Quantum Technology Group, Department of Science and Industry Systems, University of South-Eastern Norway, 3616 Kongsberg, Norway

E-mail: roger.colbeck@york.ac.uk, leon.d.loveridge@usn.no, stefan.weigert@york.ac.uk

In June 2018, our friend and colleague Paul Busch [1–3] passed away unexpectedly. Paul's death put an abrupt end to a particularly productive period during which many of his ideas—after a long incubation period—came to fruition. There would have been more to come.

To remember Paul and to celebrate his achievements a memorial event was held in York a year after his untimely death. On the 18th and 19th June 2019, the workshop "Mathematical Foundations of Quantum Mechanics—*in memoriam* Paul Busch" took place at King's Manor, a historic building dating back to the 15th century. Around 35 people attended the event: colleagues old and new, current and former students, collaborators, friends, admirers and family members. The meeting was kindly supported by the Department of Mathematics at the University of York, Paul's academic home since 2005, and by the Institute of Physics, of which Paul was an Elected Fellow.

The twelve invited talks presented at the meeting mirrored Paul's interests, which elegantly straddled quantum foundations, quantum information, mathematical physics and the philosophy of science. *Uncertainty*, *complementarity* and *incompatibility* played a major role in the presentations given by Pekka Lahti (Turku, Finland), Guido Bacciagaluppi (Utrecht, NL), Teiko Heinosaari (Turku, Finland), Jukka Kiukas (Aberystwyth, UK), and Oliver Reardon-Smith (York, UK). *Philosophical questions of realism* related to Paul's work were discussed by Brigitte Falkenburg (Dortmund, Germany) and Dennis Dieks (Utrecht, NL).

The other half of the contributions touched upon a spread of technical issues related to the *foundations of quantum mechanics* and to *quantum measurements*: Victoria J Wright (York, UK) presented a refinement of Gleason's theorem for unsharp observables and Roberto Beneduci (Calabria, Italy) discussed localization observables for photons. Stan Gudder (Denver, USA) proposed a method to quantify entanglement, David B Pearson (Hull, UK) spoke on a new technique to approach problems in spectral theory and Heinz-Jürgen Schmidt (Osnabrück, Germany) reflected upon the properties of measurements in quantum thermodynamics.

This volume of *proceedings* contains thirteen refereed papers and a bibliography of Paul's scientific work. Seven of the papers are based on talks presented at the meeting; the other six papers are written by authors unable to speak at the time. The contributions contain a mix of original and review material. In addition, historical and personal remarks can be found in them, as well as asides relating to the development of the ideas presented. Often, Paul's work is still pertinent to the papers presented, so much so in some that you can almost hear him comment.

This is also true for the contributions to a special issue [4] of the journal *Foundations of Physics*, of which Paul was a long-standing editor.

The first five papers of the proceedings directly connect to Paul's work on complementarity (Lahti), incompatibility (Heinosaari), joint measurability (Kiukas), coexistence (Stulpe) and time-energy uncertainty (Roberts and Butterfield). The following four papers belong to the quantum theory of measurement, understood in a broad sense: a technical discussion of Naimark's theorem (Beneduci) and a study of the role of sequential measurements in quantum thermodynamics (Schmidt and Gemmer) are given, followed by two investigations into quantum reference frames in quantum measurement (Miyadera and Loveridge; Loveridge). The subsequent pair of contributions ponder the question of unsharp relativistic localisation observables (Jaeger) and critically comment on the concept of unsharp reality (Falkenburg), respectively. The series of papers concludes with a novel quantification of entanglement (Gudder) and a method to generate entanglement in spin chains (Estarellas, Spiller and D'Amico).

At the end of this volume the reader is presented with a bibliography listing, we believe, all scientific publications which Paul wrote over nearly four decades. This body of work is testament to his desire to resolve fundamental and important problems posed by quantum theory, without prejudice to difficulty or popular trends. We anticipate that Paul's contributions and the insights they contain will continue to be of value to future generations of quantum physicists.

References

[1] Gudder S, Lahti P and Loveridge L 2018 Paul Busch 1955–2018 *Found. Phys.* **48** 1128–30
[2] Weigert S 2019 Paul Busch 1955–2018 *January Newsletter* (London: London Mathematical Society) p 47
[3] Gudder S and Lahti P 2019 Paul Busch: at the heart of quantum mechanics *Found. Phys.* **49** 457–9
[4] P Lahti and S Gudder (eds) 2019 Paul Busch: at the heart of quantum mechanics *Found. Phys.* **49** 457–692

Journal of Physics: Conference Series **1638** (2020) 011002 doi:10.1088/1742-6596/1638/1/011002

Peer review declaration

All papers published in this volume of Journal of Physics: Conference Series have been peer reviewed through processes administered by the Editors. Reviews were conducted by expert referees to the professional and scientific standards expected of a proceedings journal published by IOP Publishing.

- **Type of peer review: Single-blind / Double-blind / Triple-blind / Open / Other (please describe) Single-blind**
- **Conference submission management system: Email to Editors**
- **Number of submissions received: 13**
- **Number of submissions sent for review: 13**
- **Number of submissions accepted: 13**
- **Acceptance Rate (Number of Submissions Accepted / Number of Submissions Received X 100): 100**
- **Average number of reviews per paper: 14/13=1.08**
- **Total number of reviewers involved: 14**
- **Any additional info on review process: the bibliography of Paul Busch's publications has not been sent out for review as it does not contain research.**
- **Contact person for queries: Stefan Weigert – stefan.weigert@york.ac.uk**

Uncertainty and complementarity: two key notions in Paul Busch's studies at the heart of quantum mechanics

Pekka Lahti

Department of Physics and Astronomy, University of Turku, Turku, Finland

E-mail: pekka.lahti@utu.fi

Abstract. The problem of approximate joint measurement of complementary observables, like position and momentum, and the relevance of the uncertainty relations to that question were at the heart of the investigations of Paul Busch into the foundations of quantum mechanics. A good third of his published work dealt with this and closely related questions. This paper is an attempt to survey some of the steps taken in that research starting with Paul's first papers on the subject matter in the early 1980's and reaching its height in recent years.

1. Introduction

The problem of approximate joint measurement of complementary observables and the relevance of the uncertainty relations to that question were at the heart of Paul Busch's investigations into the foundations of quantum mechanics.

The intuitive background of these investigations was defined in the early writings of Niels Bohr, Wolfgang Pauli, and Werner Heisenberg and may be summarised as follows: on one hand, according to Bohr [1, 2] and Pauli [3], position and momentum are complementary observables in the sense that all the experimental arrangements allowing their unambiguous operational definitions and measurements are mutually exclusive. Since mutually exclusive arrangements cannot be applied jointly, such observables cannot be operationally defined and measured together. On the other hand, according to Heisenberg [4], complementary observables, like position and momentum, can be defined and measured jointly if sufficient ambiguities are allowed in their definitions. For the necessary defining ambiguities or measurement inaccuracies $\delta q, \delta p$ for position and momentum Heisenberg gave his famous relation $\delta q \cdot \delta p \sim h$.

The problem arising from these heuristics read and still reads: how to express and possibly confirm or reject these intuitive ideas in (the Hilbert space formulation of) quantum mechanics?

With this contribution I try to survey some steps taken in that research starting with Paul's first papers on the subject matter and reaching its height in recent years. I have been fortunate to follow and participate in much of that work since 1981. A lot of our common work, with many collaborators, is summarised in the monographs *The Quantum Theory of Measurement* [5], *Operational Quantum Physics* [6], and *Quantum Measurement* [7]. I will use freely the standard notions of quantum mechanics, with its minimal interpretation, as developed, for instance, in our last book [7]. Part of the present contribution is based on a paper [8] appearing in the special issue of Foundations of Physics, entitled *Paul Busch: At the Heart of Quantum Mechanics.*

2. First steps

Though important conceptual and mathematical clarifications and extensions of Heisenberg's groundbreaking paper [4] of 1927 appeared almost immediately in the subsequent writings of Kennard [9], Weyl [10], and Robertson [11], the interpretation and meaning of Heisenberg's uncertainty relations, notably their relevance to the joint measurability of the involved observables, remained controversial over several decades to come. A clear modern account of this development is given in [12].

A reason for that controversy was, at least partly, in the adopted mathematical theory of quantum mechanics which contained the result, due to John von Neumann [13], according to which observables, as self-adjoint operators, have a joint observable if and only if they commute, and for commuting observables there are no (nontrivial) uncertainty relations. Another reason, already clearly pointed out by Kennard [9] and further emphasized, for instance, by Popper [14], is that quantum probabilities, measurement outcome distributions, do not reflect as such any measurement errors or inaccuracies.

The possibility to resolve this controversy was slowly emerging in the late 1960s with the works of Ludwig [15], Davies and Lewis [16], and others, where it was recognised that the probabilistic or statistical structure of quantum mechanics is compatible with the assumption that observables are given not only as spectral measures, associated with self-adjoint operators, but, more generally, with semispectral measures, normalised positive operator measures, an operator representation of an observable being just an idealized special case. This extension was soon utilized to describe fuzzy observables and approximate measurements, to mention only the papers of Ali, Emch, and Prugovečki [17, 18] and the monograph of Davies [19]. It is in this context that Paul defended his doctoral thesis "Unbestimmtheitsrelation und simultane Messungen in der Quantentheorie" in Cologne 1982 [20].

This thesis can be seen as a first systematic attempt to distinguish between preparation (statistical) and measurement (individualistic) uncertainty relations, the latter being discussed in terms of fuzzy position and fuzzy momentum observables and exemplified through an elaboration and extension of the Arthurs-Kelly model [21] for an approximate joint measurement of position and momentum. Since the emphasis of the thesis is on an individualistic/realist interpretation of quantum mechanics, it contains also a detailed study of necessary ideality conditions of the involved measurement processes needed in such an interpretation. Leaving aside the question of a realist interpretation, the thesis argues, even only within a model, that the two aspects of the uncertainty relations are closely related with each other, preparation uncertainty relations appearing as a part of the measurement uncertainty relations, reflecting the intuitive idea that the possibilities for measurements cannot overtake the possibilities for preparations. A generic notion of a measurement error adequate for quantum mechanics still waited to be found. Perhaps, a first clear formulation of this task is given in the 1983 monograph of Ludwig [22, pp 197-8].

3. Complementarity as a lack of joint tests

In spite of the numerous essays Bohr wrote on the topic, he never gave an explicit definition of the notion of complementarity, nor wrote an extensive treatise on the subject matter, a fact which may explain the abundance of the secondary literature on this theme. Leaving that aside, let us note that complementarity can be, and has been expressed in quantum mechanics (as well as in more general probabilistic theories) in several alternative ways in terms of measurement outcome probabilities, observables, instruments, or measurement schemes, see, e.g. [6, 7, 22]. In [23, 24] a formulation of the *complementarity* of observables as a *lack of joint tests* was advanced. Here it will be expressed directly on the level of effects constituting the observables, as further developed more recently in [8].

For any two effects $E, F \in \mathcal{E}(\mathcal{H}) = \{A \in \mathcal{L}(\mathcal{H}) \,|\, 0 \leq A \leq I\}$, the yes-outcome of a yes-no

Journal of Physics: Conference Series **1638** (2020) 012001 doi:10.1088/1742-6596/1638/1/012001

measurement of a (nontrivial) dichotomic observable $\{0, A, I - A, I\}$, with $A \leq E, A \leq F$, gives probabilistic information on both E and F. Such a measurement is a *joint test* of E and F. The bigger A the better joint test though, unless E or F is a projection, the set of their common lower bounds

$$\text{l.b.}\{E, F\} = \{A \in \mathcal{E}(\mathcal{H}) \,|\, A \leq E, A \leq F\}$$

need not have a greatest element. No such tests exist exactly when E and F are disjoint:

$$\text{l.b.}\{E, F\} = \{0\}, \text{ that is, } E \wedge F = 0.$$

For any two observables $\mathsf{E} : \mathcal{A} \to \mathcal{L}(\mathcal{H})$ and $\mathsf{F} : \mathcal{B} \to \mathcal{L}(\mathcal{H})$ their lack of joint tests may vary between the two extremes,

$$\text{l.b.}\{\mathsf{E}(X), \mathsf{F}(Y)\} = \{0\} \text{ for all } X, Y,\ \mathsf{E}(X) \neq I \neq \mathsf{F}(Y),$$
$$\text{l.b.}\{\mathsf{E}(X), \mathsf{F}(Y)\} \neq \{0\} \text{ for all } X, Y,\ \mathsf{E}(X) \neq 0 \neq \mathsf{F}(Y),$$

giving rise to the question of where to put the definition of complementarity? To decide on that, recall that for an operational definition of an observable $\mathsf{E} : \mathcal{A} \to \mathcal{L}(\mathcal{H})$ it suffices to know the effects $\mathsf{E}(X)$ for a semiring $\mathcal{R}$ which *generates* the σ-algebra $\mathcal{A} \subset 2^{\Omega}$ and *covers* the value set Ω in the sense of countable (disjoint) union. Hence, we define:

> Observables E and F are *complementary* if $\mathsf{E}(X) \wedge \mathsf{F}(Y) = 0$ for all $X \in \mathcal{R}, Y \in \mathcal{S}$ (such that $\mathsf{E}(X) \neq I \neq \mathsf{F}(Y)$) for some generating and covering semirings $\mathcal{R} \subset \mathcal{A}$ and $\mathcal{S} \subset \mathcal{B}$.

As an immediate observation we note that complementary observables have no joint measurements. In fact, if M is a joint measurement of E and F, that is, $\mathsf{E}(X) = \mathsf{M}(X, \Xi), \mathsf{F}(Y) = \mathsf{M}(\Omega, Y)$ for all $X \in \mathcal{A}, Y \in \mathcal{B}$, then $\mathsf{M}(X, Y)$ is a common lower bound of $\mathsf{E}(X)$ and $\mathsf{F}(Y)$ for all X, Y; thus, if E and F are complementary, then $I = \mathsf{M}(\Omega, \Xi) = \mathsf{M}(\cup X_i, \cup Y_j) = \sum \mathsf{M}(X_i, Y_j) = 0$, for some disjoint covers $(X_i) \subset \mathcal{R}$ and $(Y_j) \subset \mathcal{S}$ of the value spaces Ω and Ξ, respectively. This is a contradiction.

The study of the complementarity of observables reduces to the study of the set of the lower bounds of their effects. Let us recall that if one of the effects E and F is a projection, then the set l.b.$\{E, F\}$ contains the greatest elements $E \wedge F$ (in the order of $\mathcal{E}(\mathcal{H})$ [25]) and if both are projections then $E \wedge F$ is equal to their greatest lower bound in the projection lattice, the projection onto the closed subspace $E(\mathcal{H}) \cap F(\mathcal{H})$, the intersection of the ranges of E and F. In general, the range $E(\mathcal{H}) = \text{ran}(E)$ of an effect E need not be closed. If P_E denotes the support projection of E, that is, the projection onto $\mathcal{H}_E = (\ker E)^{\perp} = \overline{\text{ran}(E)}$, we have $E \leq P_E$, so that, in general, the set l.b.$\{E, F\}$ is bounded from above by the projection $P_E \wedge P_F$.

4. The case of Q and P

The canonical position-momentum pair (Q, P), as studied in the thesis [20], is the prototype example of complementary observables, their complementarity can be derived from their Fourier equivalence and is manifested in the relation:

$$\mathsf{Q}(X) \wedge \mathsf{P}(Y) = 0, \tag{4.1}$$

for all bounded $X, Y \in \mathcal{B}(\mathbb{R})$, independently of the Lebesgue measures $\ell(X), \ell(Y)$ of the sets. Their Fourier equivalence gives equally well $\mathsf{Q}(X) \wedge \mathsf{P}(Y)^{\perp} = \mathsf{Q}(X)^{\perp} \wedge \mathsf{P}(Y) = 0$ for those sets, underlying together with (4.1) the non-Boolean structure of the projection lattice $\mathcal{P}(\mathcal{H})$. For the complements of such sets one instead has

$$\mathsf{Q}(\mathbb{R} \setminus X) \wedge \mathsf{P}(\mathbb{R} \setminus Y) \neq 0, \tag{4.2}$$

Journal of Physics: Conference Series **1638** (2020) 012001 doi:10.1088/1742-6596/1638/1/012001

a result going back to [26,27]. In addition, Q and P have both totally noncommutative spectral projections (associated with half-lines) as well as mutually commutative spectral projections (associated with appropriate periodic sets), cases of interest also to Busch *et al.* [28]. In view of the importance of these results in the quantum probability calculus as developed in [29], they were collectively referred to as the Jauch theorem in [30].

Many more examples of complementary observables are known and extensively discussed in [8], to mention number N and canonical phase Φ, any two (rotated) quadratures $\mathsf{Q}_\alpha, \mathsf{Q}_\beta$, as well as the pairs $(\mathsf{Q}_\alpha, \mathsf{H})$, where H is an energy observable, with the operator $H = \frac{1}{2m}P^2 + V(Q)$, where V is any function such that H has a purely discrete spectrum.

The equal importance of the complementary observables for the full description of (the state of) the system was underlined in some of Bohr's writings. Hence, it is worth recalling that, due to the informational incompleteness of the pair (Q, P), as shown by Bargmann and reported in [31, p 92], some complementary information is hereby lacking for a full state description. Any triple $(\mathsf{Q}, \mathsf{P}, \mathsf{H})$, with a discrete energy H, is pairwise complementary but none is known to be informationally complete. However, any set of rotated quadratures $\{\mathsf{Q}_\theta \,|\, \theta \in J \subset [0, \pi)\}$ is pairwise complementary and informationally complete if J is a dense subset of $[0, \pi)$ [32].

5. Breaking complementarity of Q and P: generalized Jauch theorem

The fuzzy position and momentum observables studied in [20] were of the convolution form $\mu * \mathsf{Q} = \mathsf{Q}_\mu, \nu * \mathsf{P} = \mathsf{P}_\nu$, with the effects

$$\mathsf{Q}_\mu(X) = \int (\mu * \chi_X)(q)\, d\mathsf{Q}(q), \tag{5.1}$$

$$\mathsf{P}_\nu(Y) = \int (\nu * \chi_Y)(p)\, d\mathsf{P}(p), \tag{5.2}$$

and with the probability measures μ and ν having the densities $f = |\varphi|^2$ and $g = |\psi|^2$ (defined by unit vectors φ, ψ in $L^2(\mathbb{R})$). It was further assumed that the standard deviations of μ and ν, $\Delta(\mu), \Delta(\nu)$, describe the measurement inaccuracies. From [18,19] it was known that if ψ is the Fourier transform of φ, that is, $\psi = \hat{\varphi}$, then the fuzzy observables Q_μ and P_ν can be obtained as the marginal observables of a (covariant) phase space observable G^T, $T = |\varphi\rangle\langle\varphi|$, with the effects

$$\mathsf{G}^T(X \times Y) = \frac{1}{2\pi} \int_{X \times Y} W_{q,p} T W_{q,p}^*\, dqdp, \tag{5.3}$$

where $W_{q,p}$ are the Weyl operators. In this case

$$0 \neq \mathsf{G}^T(X \times Y) \in \text{l.b.}\{\mathsf{Q}_\mu(X), \mathsf{P}_\nu(Y)\},$$

for all X, Y, meaning that the thus introduced fuzziness has broken the complementarity of Q and P. Moreover, now $\Delta(\mu)\Delta(\nu) = \Delta(\mathsf{Q}, \varphi)\Delta(\mathsf{P}, \varphi) \geq \frac{1}{2}\hbar$, indicating that the assumed measurement inaccuracies are related to the preparation of the joint measurement modelled by the phase space observable G^T.

This model led to the study of the question of how much inaccuracy in the form of convolutions is to be introduced in position and momentum to break their complementarity and possibly allow their joint measurements [24,30].

The first question is answered in what Paul called generalised Jauch theorem, [30, Theorem $(\widetilde{J})$], a result generalising the equivalence of the statements $\text{l.b.}\{\mathsf{Q}(X), \mathsf{P}(Y)\} \neq \{0\}$ and $\text{ran}(\mathsf{Q}(X)) \cap \text{ran}(\mathsf{P}(Y)) \neq \{0\}$, to the effects $\mathsf{Q}_\mu(X)$ and $\mathsf{P}_\nu(Y)$:

$$\text{l.b.}\{\mathsf{Q}_\mu(X), \mathsf{P}_\nu(Y)\} \neq \{0\} \iff \text{ran}(\sqrt{\mathsf{Q}_\mu(X)}) \cap \text{ran}(\sqrt{\mathsf{P}_\nu(Y)}) \neq \{0\}. \tag{5.4}$$

A partial answer to the second question came from the observation that

$$\mathsf{Q}_\mu(X) \leq \mathsf{Q}(\mathrm{supp}(\mu * \chi_X)), \ \mathsf{P}_\nu(Y) \leq \mathsf{P}(\mathrm{supp}(\nu * \chi_Y)), \tag{5.5}$$

showing that if the spectral projections $\mathsf{Q}(\mathrm{supp}(\mu * \chi_X))$ and $\mathsf{P}(\mathrm{supp}(\nu * \chi_Y))$ are disjoint then also the effects $\mathsf{Q}_\mu(X)$ and $\mathsf{P}_\nu(Y)$ are disjoint, independently of the standard deviations of μ and ν. The converse question whether $\mathsf{Q}_\mu(X) \wedge \mathsf{P}_\nu(Y) = 0$ also implies $\mathsf{Q}(\mathrm{supp}(\mu * \chi_X)) \wedge \mathsf{P}(\mathrm{supp}(\nu * \chi_Y)) = 0$ was left open in [30] (but was to be answered in the negative, see below).

Result (5.4) extends to arbitrary effects. Indeed, [33, Theorem 3] characterises the range of the square root of an effect in terms of the rank-1 operators contained in it: for any effect $E \in \mathcal{E}(\mathcal{H})$ and for any rank-1 operator $|\varphi\rangle\langle\varphi|$, $\|\varphi\| \leq 1$, one has

$$\varphi \in \mathrm{ran}\sqrt{E} \Longleftrightarrow \exists\lambda > 0 \text{ s.t. } \lambda|\varphi\rangle\langle\varphi| \leq E. \tag{5.6}$$

This then shows that for any $E, F \in \mathcal{E}(\mathcal{H})$,

$$\mathrm{l.b.}\{E, F\} \neq \{0\} \Longleftrightarrow \mathrm{ran}\sqrt{E} \cap \mathrm{ran}\sqrt{F} \neq \{0\}. \tag{5.7}$$

A direct study of either of these equivalent conditions may be difficult in concrete cases, as already exemplified by the study of the various cases of $\mathsf{Q}(X)$ and $\mathsf{P}(Y)$, and a further complication may arise since the range (of the square root) of an effect need not be closed. To study complementarity, it is thus useful to have some upper bound estimates for (5.7). From $\mathrm{ran}\sqrt{E} \subset \overline{\mathrm{ran}\sqrt{E}} = \mathcal{H}_E$, one immediately gets one: if $\mathcal{H}_E \cap \mathcal{H}_F = \{0\}$, that is, $P_E \wedge P_F = 0$, then also $E \wedge F = 0$. However, the converse implication does not hold: there are effects E, F which are disjoint though $P_E \wedge P_F \neq 0$ [8, Proposition 7].

The fuzzy position and momentum effects are of a special form of being (nontrivial) functions of a spectral measure. In general, if $E = \int h\, d\mathsf{A}$ for some spectral measure $\mathsf{A} : \mathcal{B}(\mathbb{R}) \to \mathcal{L}(\mathcal{H})$ and a Borel measurable function $h : \mathbb{R} \to [0,1]$, then a direct calculation shows that $P_E \leq \mathsf{A}(\mathrm{supp}(h))$, extending the observation (5.5). Again, there are (even nontrivial) examples showing that this inequality may be strict [8, Lemma 4].

For any two effects $E, F \in \mathcal{E}(\mathcal{H})$ of the above form $E = \int h d\mathsf{A}, F = \int k d\mathsf{B}$, one thus has the equivalent conditions:

$$\begin{aligned} \mathrm{l.b.}\{E, F\} &\subseteq \{D \in \mathcal{E}(\mathcal{H}) \,|\, D \leq P_E \wedge P_F\} \\ &\subseteq \{D \in \mathcal{E}(\mathcal{H}) \,|\, D \leq \mathsf{A}(\mathrm{supp}(h)) \wedge \mathsf{B}(\mathrm{supp}(k))\}, \\ \mathrm{ran}\sqrt{E} \cap \mathrm{ran}\sqrt{F} &\subseteq \mathcal{H}_E \cap \mathcal{H}_F \subseteq \mathsf{A}(\mathrm{supp}(h))(\mathcal{H}) \cap \mathsf{B}(\mathrm{supp}(k))(\mathcal{H}), \end{aligned}$$

where all the inclusions may be proper.

To close this section let us return briefly to the fuzzy position and fuzzy momentum studied in [20]. By now it is known that any observable which shares the symmetry properties of position (translation covariance and boost invariance) is of the form $\mu * \mathsf{Q}$ for some probability measure μ. Similarly, any observable which shares the symmetry properties of momentum (boost covariance, translation invariance) is of the form $\nu * \mathsf{P}$ for some ν [34]. Moreover, any such pair $(\mu * \mathsf{Q}, \nu * \mathsf{P})$ is jointly measurable exactly when they can be obtained as the margins of some covariant phase space observable G^T as defined in (5.3) [35]. In addition, any observable M which is covariant under the phase space translations (translations and boosts) is of that form for some $T \geq 0, \mathrm{tr}\, T = 1$, a result going back to Holevo [36] and Werner [37] (for alternative revised proofs, see [38,39]). Further, any such an observable can be realised by a modified Arthurs-Kelly model (eight-port homodyne detection scheme) of an approximate joint measurement of position and momentum [40].

For any unsharp position $\mu * \mathsf{Q}$ and momentum $\nu * \mathsf{P}$ we now have

$$\text{l.b.}\{\mathsf{Q}_\mu(X), \mathsf{P}_\nu(Y)\} \neq \{0\} \;\Leftrightarrow \text{ran}\left(\sqrt{\mathsf{Q}_\mu(X)}\right) \cap \text{ran}\left(\sqrt{\mathsf{P}\nu(Y)}\right) \neq \{0\}.$$

Since, for instance, $\text{supp}(\chi_X * \mu) \subset \overline{X + \text{supp}(\mu)}$, one observes that $\mu * \mathsf{Q}$ and $\nu * \mathsf{P}$ remain complementary if the unsharpness measures μ and ν have bounded supports, independently of the sizes of $\Delta(\mu)$ and $\Delta(\nu)$. On the other hand, the pair $(\mu * \mathsf{Q}, \nu * \mathsf{P})$ is jointly measurable if and only if they can be obtained as the margins of a phase space observable G^T, in which case $\Delta(\mu)\Delta(\nu) = \Delta(\mathsf{Q}, T)\Delta(\mathsf{P}, T) \geq \frac{\hbar}{2}$. Still, the question of the possible relation of μ, or $\Delta(\mu)$, to the actual measurement accuracy is not explicitly addressed. Finally, the question left open in [30] is answered in the following [8, Proposition 13]:

> For any bounded intervals $X, Y \subset \mathbb{R}$ with lengths d_X, d_Y satisfying $d_X d_Y \leq \pi/2$, there exist probability measures μ, ν with finite variance, such that the effects $\mathsf{Q}_\mu(X)$ and $\mathsf{P}_\nu(Y)$ are complementary, but $\mathsf{Q}(\text{supp}(\mu * \chi_X)) \wedge \mathsf{P}(\text{supp}(\nu * \chi_Y)) \neq 0$.

6. Approximate measurements and measurement uncertainty

The notion of a fuzzy or approximate observable, like $\mu * \mathsf{Q}$, has been used in the above discussed investigations to model an approximate measurement of the given observable, like Q, with the idea that the probability measure μ accounts for the approximation. This is well justified since the convolving measure simply adds state independent noise in the measurement outcome statistics, like $(\mu * \mathsf{Q})_\rho = \mu * \mathsf{Q}_\rho$. Still, no systematic theory of an approximate measurement of an observable is given here.

With the revised operational tools of quantum measurement theory, the study of the measurement uncertainty relations, including noise and disturbance, got a new boost in the turn of the millennium. Part of this new wave of interest was triggered by the work of Ozawa [41, 42] on the error-disturbance relations. The proposed notions of Ozawa were independently criticised in the papers of Busch *et al.* [43] and Werner [44].

Common to [43, 44] was a search for operationally meaningful measures of measurement error, noise, and disturbance, with the idea that such notions should be built on comparing the outcomes and statistics of the target observable, the observable one intends to measure, and the approximating observable, the observable that is actually implemented by the measurement scheme – an idea advanced already by Ludwig [22, pp 197-8] in 1983 but missing in Ozawa's work. Following this idea several measures of error (noise and disturbance) have been proposed and developed, among them are the notions of metric error, calibration error, and error bar width, studied also in many writings of Paul and his collaborators [45–50], and [7, Chpt. 13].

A breakthrough step was taken in the 2004 paper of Werner [44] where a metric error $\Delta_1(\widetilde{\mathsf{E}}, \mathsf{E})$ in measuring the approximator $\widetilde{\mathsf{E}}$ in place of the target E was defined as the worst-case (state independent) limit of the Monge distance d_1 between the distributions $\widetilde{\mathsf{E}}_\rho$ and E_ρ:

$$\Delta_1(\widetilde{\mathsf{E}}, \mathsf{E}) = \sup_\rho d_1(\widetilde{\mathsf{E}}_\rho, \mathsf{E}_\rho).$$

In subsequent work [47, 49, 50] this was extended to use the Wasserstein distance of order α, $1 \leq \alpha < \infty$, that is, replacing d_1 with

$$d_\alpha(\widetilde{\mathsf{E}}_\rho, \mathsf{E}_\rho) = \inf_\gamma \Big(\int |x - y|^\alpha \, d\gamma(x, y)\Big)^{\frac{1}{\alpha}},$$

where the infimum (actually minimum) is taken over all the couplings γ of the (real) probability measures $\widetilde{\mathsf{E}}_\rho$ and E_ρ. If, for instance, E is a spectral measure and $\widetilde{\mathsf{E}} = \mu * \mathsf{E}$, then

$$\Delta_\alpha(\widetilde{\mathsf{E}}, \mathsf{E}) = d_\alpha(\mu, \delta_0) = \Big(\int |x|^\alpha \, d\mu(x)\Big)^{\frac{1}{\alpha}},$$

where δ_0 is the point measure at 0. For $\alpha = 2$, this gives $\Delta_2(\widetilde{\mathsf{E}}, \mathsf{E}) \geq \Delta(\mu)$, with equality whenever $\int x \, d\mu(x) = 0$.

The paper [44] also developed an averaging method to construct, under appropriate conditions, a covariant observable from a given observable. For approximate joint measurements of position and momentum this all led to the following fundamental result [47, 49], see also [7, Chpt. 15], a result which can be seen as a completion of the study of the problem faced in [20], now in the frame of the minimal interpretation of quantum mechanics.

Theorem 1. Let M be any phase space observable (covariant or not), with the marginal observables $\mathsf{M}_1, \mathsf{M}_2$, and $1 \leq \alpha, \beta < \infty$. If the measurement errors $\Delta_\alpha(\mathsf{M}_1, \mathsf{Q})$ and $\Delta_\beta(\mathsf{M}_2, \mathsf{P})$ both are finite then there is a covariant phase space observable G^T such that $\Delta_\alpha(\mathsf{G}^T_1, \mathsf{Q}) \leq \Delta_\alpha(\mathsf{M}_1, \mathsf{Q})$ and $\Delta_\beta(\mathsf{G}^T_2, \mathsf{P}) \leq \Delta_\beta(\mathsf{M}_2, \mathsf{P})$, and thus

$$\begin{aligned} \Delta_\alpha(\mathsf{M}_1, \mathsf{Q})\Delta_\beta(\mathsf{M}_2, \mathsf{P}) &\geq \Delta_\alpha(\mathsf{G}^T_1, \mathsf{Q})\Delta_\beta(\mathsf{G}^T_2, \mathsf{P}) \\ &= d_\alpha(\mathsf{Q}_{\Pi T \Pi, \delta_0}) d_\beta(\mathsf{P}_{\Pi T \Pi, \delta_0}) \geq c_{\alpha\beta}\hbar, \end{aligned} \tag{6.1}$$

where Π is the parity operator, and the constant $c_{\alpha\beta}$ is connected to the ground state energy $g_{\alpha\beta}$ of the Hamiltonian $H_{\alpha\beta}$, the closure of the essentially selfadjoint operator $|Q|^\alpha + |P|^\beta$, by the equation

$$c_{\alpha\beta} = \alpha^{\frac{1}{\beta}} \beta^{\frac{1}{\alpha}} \Big(\frac{g_{\alpha\beta}}{\alpha + \beta}\Big)^{\frac{1}{\alpha} + \frac{1}{\beta}}.$$

The latter inequality in (6.1) is an equality exactly when $\Pi T \Pi$ arise from the ground state of $H_{\alpha\beta}$ by phase space translation and dilation.

For $\alpha = \beta = 2$, $H_{\alpha\beta}$ is twice the harmonic oscillator Hamiltonian with the ground state energy $g_{22} = 1$, giving rise to the familiar lower bound $\hbar/2$. In the case of finite measurement errors, the errors are bounded from below by the preparation uncertainties characterizing the measurement scheme implementing the approximate joint measurement G^T; for $\alpha = \beta = 2$, one has $\Delta_2(\mathsf{M}_1, \mathsf{Q}) \geq \Delta(\mathsf{Q}_{\Pi T \Pi})$ and $\Delta_2(\mathsf{M}_2, \mathsf{P}) \geq \Delta(\mathsf{P}_{\Pi T \Pi})$.

There is now an increasing flow of papers analysing in one or another form something like a measurement uncertainty region for two (or more) incompatible observables E_i, with the value spaces $(\Omega_i, \mathcal{A}_i)$,

$$\mathsf{MU}(\Omega_1, \Omega_2) = \{(\Delta_1(\mathsf{M}_1, \mathsf{E}_1), \Delta_2(\mathsf{M}_2, \mathsf{E}_2)) \mid \mathsf{M} : \mathcal{A}_1 \otimes \mathcal{A}_2 \to \mathcal{L}(\mathcal{H})\}, \tag{6.2}$$

and with various measures Δ_1, Δ_2 of uncertainty. During the past five years also Paul was highly active in this research, as illustrated, e.g., by the papers [51, 52].

7. Fragility of complementarity

For any two effects E, F the set of their common lower bounds is characterized by the intersection of the ranges of their square roots. This set can easily be modified with a small perturbation of one of the effects. Indeed, we have [8, Proposition 11]:

For any effect E and for any $\lambda, p \in (0,1)$ define $E_{\lambda,p} = \lambda E + (1-\lambda)pI$. Then ran $\sqrt{E_{\lambda,p}} = \mathcal{H}$.

Journal of Physics: Conference Series **1638** (2020) 012001 doi:10.1088/1742-6596/1638/1/012001

This shows that the complementarity of any two observables $\mathsf{E}_1, \mathsf{E}_2$ is immediately broken by adding trivial noise in one of the observables, say, $\mathsf{E}_1 \to \widetilde{\mathsf{E}}_1 = \lambda \mathsf{E}_1 + (1-\lambda)\mathsf{T}_1$, with the trivial observable $X \mapsto \mathsf{T}_1(X) = \mu_1(X)I$ defined by the probability measure μ_1.

In fact, any two observables E_1 and E_2 can even be made compatible by mixing them with trivial noise,

$$\begin{aligned}\widetilde{\mathsf{E}}_1 &= \lambda \mathsf{E}_1 + (1-\lambda)\mathsf{T}_1,\\ \widetilde{\mathsf{E}}_2 &= \gamma \mathsf{E}_2 + (1-\gamma)\mathsf{T}_2,\end{aligned}$$

and choosing the weights $0 < \lambda, \gamma < 1$ appropriately, for instance, $\gamma = 1 - \lambda$. See, for instance, [53, 54].

In studying position-momentum uncertainty, this way of breaking complementarity and making them compatible is, however, not very useful since the errors grow without bound. This can indirectly be concluded from [44, 49] together with [54] but can also be seen directly [55], as reported below.

Lemma 1. Let $\widetilde{\mathsf{Q}} = \lambda \mathsf{Q} + (1-\lambda)\mathsf{T}$ be an approximate position obtained by mixing Q with a trivial observable T with a weight $0 < \lambda < 1$. Then $\Delta_\alpha(\widetilde{\mathsf{Q}}, \mathsf{Q}) = \infty$ for all $1 \leq \alpha < \infty$.

Proof. Let μ be the probability measure defining T. We use the Kantorovich duality theorem [56, Theorem 5.10] (in the form [49, Lemma 5]) to compute the distance $\Delta_\alpha(\widetilde{\mathsf{Q}}, \mathsf{Q})^\alpha$. To that end, fix any pair of positive continuous functions $f, g : \mathbb{R} \to \mathbb{R}$ with compact supports satisfying the condition

$$f(y) - g(x) \leq |x - y|^\alpha \text{ for all } x, y. \tag{7.1}$$

Assume, in addition, that $f(0) > 0$ (if necessary, replace both functions with the translates $f(y+y_0)$ and $g(x+y_0)$ where $f(y_0) > 0$). Fix a compact set K containing the supports of f and g. Then, for any $r > 0$, also the scaled functions $f_r(y) = rf(r^{-1/\alpha}y)$ and $g_r(x) = rg(r^{-1/\alpha}x)$ satisfy (7.1) and their supports are contained in the set $r^{1/\alpha}K$ extending along r. For each $r > 0$ choose a unit vector $\phi_r \in L^2(\mathbb{R})$ such that its support is outside the set $r^{1/\alpha}K$. Thus, $\langle \phi_r | \mathsf{Q}[f_r] \phi_r \rangle = \int f_r \, d\mathsf{Q}_{\phi_r, \phi_r} = 0$ and $\left\langle \phi_r | \widetilde{\mathsf{Q}}[f_r] \phi_r \right\rangle = (1-\lambda)\mu[f_r]$ for each r. Thus for each $r > 0$

$$\Delta_\alpha(\widetilde{\mathsf{Q}}, \mathsf{Q})^\alpha \geq \left\langle \phi_r | (\widetilde{\mathsf{Q}}[f_r] - \mathsf{Q}[g_r]) \phi_r \right\rangle = (1-\lambda)\mu[f_r] = r(1-\lambda) \int f(r^{-1/\alpha}y) d\mu(y).$$

Since the continuous bounded function $y \mapsto f(r^{-1/\alpha}y)$ converges pointwise to the constant function $f(0)$, the dominated convergence theorem implies that $\lim_{r\to\infty} \int f(r^{-1/\alpha}y) d\mu(y) = \int f(0) d\mu(y) = f(0) > 0$, showing that the right hand side of the above inequality approaches infinity along with $r \to \infty$. The Kantorovich duality theorem thus gives $\Delta_\alpha(\widetilde{\mathsf{Q}}, \mathsf{Q}) = \infty$. □

This result shows that though the trivial perturbation may be arbitrarily small on the level of effects, it is not at all small in the sense of the Wasserstein distance of observables.

8. Where do we stand?

The problem of approximate joint measurability of complementary observables and the relevance of the uncertainty relations to that question was a major subject in Paul's investigations into the foundations of quantum mechanics. Perhaps, the first poetic formulation of the problem is due to Wolfgang Pauli in a letter to Werner Heisenberg, dated 19 October 1926 [57]:

> *Man kann die Welt mit dem p-Auge und man kann sie mit dem q-Auge ansehen, aber wenn man beide Augen zugleich aufmachen will*[,] *dann wird man irre.*

Journal of Physics: Conference Series **1638** (2020) 012001 doi:10.1088/1742-6596/1638/1/012001

Paul's equally poetic answer to Pauli is given in the Epilogue of [6]:

> *We hope to have demonstrated that one can safely open a pair of complementary 'eyes' simultaneously. He who does so may even 'see more' than with one eye only. The means of observation being part of the physical world, Nature Herself protects him from seeing too much and at the same time protects Herself from being questioned too closely: quantum reality, as it emerges under physical observation, is intrinsically unsharp. It can be forced to assume sharp contours – real properties – by performing repeatable measurements. But sometimes unsharp measurements will be both, less invasive and more informative.*

The notion of joint test of a pair of effects, which is behind our formulation of the notion of complementarity of observables is a weaker notion than the notion of their joint measurability (coexistence). Even the pairwise joint measurability of all the effects constituting the observables does not guarantee their joint measurability (see, e.g. [58]). Hence, it is natural that one may break the complementarity of observables without allowing their approximate joint measurement.

The complementarity of position and momentum is manifested in the relation $\mathsf{Q}(X)\wedge\mathsf{P}(Y)=0$, equivalently, $\mathrm{ran}(\mathsf{Q}(X))\cap\mathrm{ran}(\mathsf{P}(Y))=\{0\}$, for all bounded sets X,Y. To break this relation one needs to introduce inaccuracies (unsharpness, fuzziness) in the measurements of the observables to get jointly measurable or at least jointly testable approximators $\widetilde{\mathsf{Q}}$ and $\widetilde{\mathsf{P}}$ with $\mathrm{ran}\sqrt{\widetilde{\mathsf{Q}}(X)}\cap\mathrm{ran}\sqrt{\widetilde{\mathsf{P}}(Y)}\neq\{0\}$ (possibly) for all X,Y. This is the content of Paul's generalized Jauch theorem.

There are natural measurement schemes which lead to jointly testable, though possibly incompatible approximators $\widetilde{\mathsf{Q}}=\mathsf{Q}_\mu$ and $\widetilde{\mathsf{P}}=\mathsf{P}_\nu$. The quality of these approximations can be quantified using, for instance, the Wasserstein 2-distances $\Delta_2(\mathsf{Q}_\mu,\mathsf{Q})\geq\Delta(\mu)$ and $\Delta_2(\mathsf{P}_\nu,\mathsf{P})\geq\Delta(\nu)$, but there need be no correlation between the errors $\Delta(\mu)$ and $\Delta(\nu)$. Such approximators are jointly measurable exactly when the error measures μ and ν are Fourier related, in which case $\Delta(\mu)\Delta(\nu)\geq\frac{\hbar}{2}$. Moreover, in this case the errors can be traced back to the preparation uncertainties of the position and momentum of the probe (measuring apparatus). This is essentially the result which led Paul to formulate his answer to Pauli.

A simple way to break the complementarity of any two observables is mixing trivial noise in the statistics of one of them. But, for instance, in the case of position or momentum such an approximation is always extremely bad. Indeed, for instance, $\Delta_2(\mathsf{Q}_p,\mathsf{Q})=\infty$ for any $\mathsf{Q}_p=p\mathsf{Q}+(1-p)\mathsf{T}_1$, $p<1$.

As concerns position and momentum, the problem formulated in the opening sentence of this paper is now solved by the following result. Consider any biobservable B, that is, an observable with two independent outcomes, as an approximate joint observable of Q and P. If the degrees of approximations $\Delta_2(\mathsf{B}_1,\mathsf{Q})$ and $\Delta_2(\mathsf{B}_2,\mathsf{P})$ are finite then there is also a covariant phase space observable G^T which serves no worse joint approximator, with

$$\Delta_2(\mathsf{B}_1,\mathsf{Q})\Delta_2(\mathsf{B}_2,\mathsf{P})\geq\Delta_2(\mathsf{G}_1^T,\mathsf{Q})\Delta_2(\mathsf{G}_2^T,\mathsf{P})\geq\Delta(\mathsf{Q}_{\Pi T\Pi})\Delta(\mathsf{P}_{\Pi T\Pi})\geq\frac{\hbar}{2}.$$

This is just a reformulation of Theorem 1.

The notion of joint measurability of two (or more) observables is well understood and it has various equivalent characterisations, see, for instance, [7, Theorem 11.1]. However, as noted above, there are obvious weakenings of this notion based on the joint measurability and joint testability of pairs of the effects constituting the observables. A further, perhaps extreme weakening of this concept is the joint measurability of two (or more) observables understood as any measurement the statistics of which allows one to reconstruct the statistics of the observables in question. Then any two observables can be measured together using tomographic methods (state reconstruction). In the case of Q and P there are even non-tomographic methods (applying informationally incomplete measurements) to measure Q and P together in the sense of "statistical postprocessing" [59].

There is no end in sight in this story.

References

[1] Bohr N 1928 The quantum postulate and the recent development of atomic theory *Nature* **121** 580–90
[2] Bohr N 1935 Can quantum-mechanical description of physical reality be considered complete? *Phys. Rev.* **48** 696–702
[3] Pauli W 1980 *General Principles of Quantum Mechanics* (Berlin: Springer)
[4] Heisenberg W 1927 Über den anschaulichen Inhalt der quantentheoretischen Kinematik und Mechanik *Z. Phys.* **43** 172–98
[5] Busch P, Lahti P and Mittelstaedt P 1991 *The Quantum Theory of Measurement* (Berlin: Springer) 1996 2nd edition
[6] Busch P, Grabowski M and Lahti P 1995 *Operational Quantum Physics* (Berlin: Springer) 1997 2nd corrected edition
[7] Busch P, Lahti P, Pellonpää J-P and Ylinen K 2016 *Quantum Measurement* (Berlin: Springer)
[8] Kiukas J, Lahti P, Pellonpää J-P and Ylinen K 2019 Complementary observables in quantum mechanics *Found. Phys.* **49** 506–31
[9] Kennard E 1927 Zur Quantenmechanik einfacher Bewegungstypen *Z. Phys.* **44** 326-52
[10] Weyl H 1928 *Gruppentheorie und Quantenmechanik* (Leipzig: Hirzel) English translation 1931 *Group Theory and Quantum Mechanics* (London: Methuen)
[11] Robertson H 1929 The uncertainty principle *Phys. Rev.* **34** 163–4
[12] Werner R F and Farrelly T 2019 Uncertainty from Heisenberg to today *Found. Phys.* **49** 460–91
[13] Von Neumann J 1932 *Mathematische Grundlagen der Quantenmechanik* (Berlin: Springer)
[14] Popper K R 1980 *The Logic of Scientific Discovery* (London: Hutchinson); *Logik der Forschung* 1934 (Wien: Springer)
[15] Ludwig G 1967 Attempt of an axiomatic foundation of quantum mechanics and more general theories II *Commun. Math. Phys.* **4** 331–48
[16] Davies E B and Lewis J T 1970 An operational approach to quantum probability *Commun. Math. Phys.* **17** 239–60
[17] Ali S T and Emch G G 1974 Fuzzy observables in quantum mechanics *J. Math. Phys.* **15** 176–82
[18] Ali S T and Prugovečki E 1977 Systems of imprimitivity and representations of quantum mechanics on fuzzy phase spaces *J. Math. Phys.* **18** 219–28
[19] Davies E B 1976 *Quantum Theory of Open Systems* (London: Academic Press)
[20] Busch P 1982 *Unbestimmtheitsrelation und simultane Messungen in der Quantentheorie* (PhD thesis, Cologne) English summary 1985 Indeterminacy relations and simultaneous measurements in quantum theory *Int. J. Theor. Phys* **24** 63-92
[21] Arthurs E and Kelly J L 1965 On the simultaneous measurements of a pair of conjugate observables *Bell Syst. Tech. J.* **44** 725–9
[22] Ludwig G 1983 *Foundations of Quantum Mechanics I* (Berlin: Springer)
[23] Bugajski S and Lahti P 1985 Fundamental principles of quantum mechanics II - From a convexity scheme to the DHB theory *Int. J. Theor. Phys.* **24** 1051–80
[24] Busch P and Lahti P 1984 On various joint measurements of position and momentum observables in quantum theory *Phys. Rev.* D **29** 1634–46
[25] Du H, Deng C and Li Q 2006 On the infimum problem of Hilbert space effects *Sci. Chin. Ser. A* **49** 545–56
[26] Lenard A 1972 The numerical range of a pair of projections *J. Funct. Anal.* **10** 410–23
[27] Berthier A M and Jauch J M 1976 A theorem on the support of functions in $L^2(\mathbb{R})$ and of their Fourier transforms *Lett. Math. Phys.* **1** 93–7
[28] Busch P and Lahti P 1986 To what extent do position and momentum commute? *Phys. Lett.* **115** 259–64
[29] Jauch J M 1974 The quantum probability calculus *Synthese* **29** 131–54
[30] Busch P 1984 On joint lower bounds of position and momentum observables in quantum mechanics *J. Math. Phys.* **25** 1794–7
[31] Reichenbach H 1944 *Philosophic Foundations of Quantum Mechanics* (Berkeley, CA: University of California Press)
[32] Kiukas J, Lahti P and Pellopää J-P 2008 A proof for the informational completeness of the rotated quadrature observables *J. Phys. A: Math. Theor.* **41** 175206

[33] Busch P and Gudder S P 1999 Effects as functions on projective Hilbert space *Lett. Math. Phys.* **47** 329–37
[34] Carmeli C, Heinosaari T and Toigo A 2004 Position and momentum observables on $\mathbb{R}$ and on $\mathbb{R}^3$ *J. Math. Phys.* **45** 2526–39
[35] Carmeli C, Heinosaari T and Toigo A 2005 On the coexistence of position and momentum observables *J. Phys. A: Math. Gen.* **38** 5253–66
[36] Holevo A S 1979 Covariant measurements and uncertainty relations *Rep. Math. Phys.* **16** 385–400
[37] Werner R F 1984 Quantum harmonic analysis on phase space *J. Math. Phys.* **25** 1404–11
[38] Cassinelli G, De Vito E and Toigo A 2003 Positive operator valued measures covariant with respect to an irreducible representation *J. Math. Phys.* **44** 4768–75
[39] Kiukas J, Lahti P and Ylinen K 2006 Normal covariant quantization maps *J. Math. Anal. Appl.* **319** 783–801
[40] Lahti P, Pellonpää J-P and Schultz J 2010 Realistic eight-port homodyne detection and covariant phase space observables *J. Mod. Opt.* **57** 1171–9
[41] Ozawa M 2002 Position measuring interactions and the Heisenberg uncertainty principle *Phys. Lett.* A **299** 1–7
[42] Ozawa M 2003 Universally valid reformulation of the Heisenberg uncertainty principle on noise and disturbance in measurement *Phys. Rev.* A **67** 042105
[43] Busch P, Heinonen T and Lahti P 2004 Noise and disturbance in quantum measurement *Phys. Lett.* A **320** 261–70
[44] Werner R F 2004 The uncertainty relation for joint measurement of position and momentum *Quant. Inf. Comput.* **4** 546–62
[45] Busch P and Pearson D B 2007 Universal joint-measurement uncertainty relation for error bars *J. Math. Phys.* **48** 082103
[46] Busch P, Heinonen T and Lahti P 2007 Heisenberg's uncertainty principle *Phys. Rep.* **452** 155–76
[47] Busch P, Lahti P and Werner R F 2013 Proof of Heisenberg's error-disturbance relation *Phys. Rev. Lett.* **111** 160405
[48] Busch P, Lahti P and Werner R F 2014 Heisenberg uncertainty for qubit measurements *Phys. Rev.* A **89** 012129
[49] Busch P, Lahti P and Werner R F 2014 Measurement uncertainty relations *J. Math. Phys.* **5** 042111
[50] Busch P, Lahti P and Werner R F 2014 Quantum root-mean-square error and measurement uncertainty relations *Rev. Mod. Phys.* **86** 1261–82
[51] Busch P and Stevens N 2015 Direct tests of measurement uncertainty relations: what it takes *Phys. Rev. Lett.* **114** 070402
[52] Bullock T and Busch P 2018 Measurement uncertainty relations: characterising optimal error bounds for qubits *J. Phys. A: Math. Theor.* **51** 283001
[53] Busch P, Heinosaari T, Schultz T and Stevens N 2013 Comparing the degrees of incompatibility inherent in probabilistic physical theories *EPL* **103** 10002
[54] Heinosaari T, Schultz J, Toigo A and Ziman M 2014 Maximally incompatible quantum observables *Phys. Lett.* A **378** 1695–9
[55] Kiukas J (private communication)
[56] Villani C 2009 *Optimal Transport - Old and New* (Berlin: Springer)
[57] Pauli W 1926 (see `cds.cern.ch/record/84535`)
[58] Reeb D, Reitzner D and Wolf M M 2013 Coexistence does not imply joint measurability *J. Phys. A: Math. Theor.* **46** 462002
[59] Busch P, Kiukas J and Lahti P 2008 Measuring position and momentum together *Phys. Lett.* A **372** 4379–80

Journal of Physics: Conference Series **1638** (2020) 012002 doi:10.1088/1742-6596/1638/1/012002

Quantum incompatibility from the viewpoint of entanglement theory

Teiko Heinosaari

QTF Centre of Excellence, Turku Centre for Quantum Physics, Department of Physics and Astronomy, University of Turku, FI-20014 Turku, Finland

E-mail: teiko.heinosaari@utu.fi

Abstract. In this essay I discuss certain analogies between quantum incompatibility and entanglement. It is an expanded version of my talk presented at *Mathematical Foundations of Quantum Mechanics in memoriam Paul Busch*, held in York in June 2019.

1. Introduction

The basics of entanglement theory belong to any first course on quantum information, while incompatibility is not necessarily covered at all, mentioned just briefly, or perhaps falsely identified with non-commutativity. For this reason, in the following I will use some elementary well-known facts about entanglement to guide the presentation on quantum incompatibility, through some analogies. I hope this could be helpful for a reader who has not encountered incompatibility so much. For a more expert reader, the highlighted analogy between entanglement and incompatibility may give thoughts about additional, yet unexplored, connections. My intention has been to keep the presentation light, simple and concrete, and therefore I am using only elementary mathematical machinery and not e.g. the resource theoretic framework where some of the discussed features could be treated in a unified manner.

Personally, the analogy between entanglement and incompatibility became visible to me when I visited Paul at the Perimeter Institute in February 2007, and we were continuously discussing incompatibility of quantum observables. Before that, I had studied incompatibility of certain pairs of quantum observables, such as noisy position and noisy momentum, and I had also worked with Paul on those things. During that visit we started to talk with Paul about a goal of having a 'theory of incompatibility', something similar to entanglement theory. This basically meant that we wanted to study general features of incompatibility, not only some physically interesting examples. That is what we then both devoted time to in the following years, together and separately. We still agreed that physically relevant examples are important to find new mathematical techniques and to demonstrate the domain of incompatibility, so that research line was not forgotten. But since that visit the general features of incompatibility have been one of my main research interests and the visit had, in fact, long-lasting implications on my research. In all of the subtopics discussed in this paper Paul has either directly worked with me, boosted my research with discussions on those topics, or encouraged me to investigate some questions. It is clear to me that I would not have involved myself so deeply in incompatibility questions without his encouragement and support.

The general goal of 'incompatibility theory' was made public in one form in an article of Paul and Heinz-Jürgen Schmidt a couple of years after my visit at the Perimeter Institute. In the introduction of [1] they wrote:

> *The present paper is a contribution to the emerging programme of investigating the structure of the set of observables, which should complement current studies of the dual structure of the set of quantum states.*

That sentence does not mention 'incompatibility' so it can refer even to a more general programme, but the article is, in fact, about incompatibility and I believe that our discussions at the Perimeter partly motivated that sentence. Two other public 'announcements' of the goal can been seen as the two workshops devoted solely to incompatibility. These were held at the Technical University of Munich in 2013 and at Maria Laach Abbey in 2017. The workshops were in large part organized by our friends and colleagues, but Paul had an important role for the events to happen.

I need to make two beginning remarks. Firstly, this paper is not meant to be a review paper. I have chosen rather arbitrarily some topics that have been of my personal interest and there are many more that are not mentioned here at all. A more systematic invitation to incompatibility is given in [2], although many important results have been found after that article appeared. Secondly, I have written some personal comments (like those above) on the background of some of the research results. They are subjective memories and someone else can remember the events differently.

2. Definition of incompatibility

Entanglement is a property of bipartite (or multipartite) states, while incompatibility is a property of pairs (or collections) of observables. The first similarity between entanglement and incompatibility is that they are defined by saying what they are not. Namely, an entangled state is a composite state that is not separable, and a separable state ω is, by definition, a convex mixture of product states,

$$\omega = \sum_i t_i \, \varrho_i^A \otimes \varrho_i^B \,. \tag{1}$$

Similarly, an incompatible collection of observables is a collection that is not compatible, and I will shortly recall the definition of compatibility. In conclusion, both separability and compatibility have clear operational definitions, but entanglement and incompatibility get their meaning in this negative way. In the following I will concentrate on pairs of observables, similarly as one would start entanglement theory with bipartite states. I make some comments on the general case in Section 9.

A quantum observable is mathematically described as a positive operator valued measure (POVM). For simplicity, in this paper all observables are assumed to have a finite number of outcomes. An observable A is a hence a map $x \mapsto \mathsf{A}(x)$ from a finite set of outcomes $[m] \equiv \{1, \ldots, m\}$ to the algebra $\mathcal{L}(\mathcal{H})$ of bounded linear operators on a Hilbert space $\mathcal{H}$, such that each $\mathsf{A}(x)$ is a positive operator (i.e., $\langle \psi \,|\, \mathsf{A}(x)\psi \,\rangle \geq 0$ for all $\psi \in \mathcal{H}$) and $\sum_x \mathsf{A}(x) = \mathbb{1}$. The physical interpretation is that if the system is prepared in a state ϱ and a measurement of A is performed, then the outcome x is obtained with probability $\mathrm{tr}\,[\varrho \mathsf{A}(x)]$. Many of the definitions and results are equally valid for finite and countably infinite dimensional Hilbert spaces. However, I will assume that Hilbert spaces are finite dimensional as this is the usual setting where one learns entanglement theory and I want to keep the discussion simple.

Two observables A and B, with outcome sets $[m]$ and $[n]$, are *compatible* if there exists a third observable C, with an outcome set $[k]$, and functions $f : [k] \to [m]$ and $g : [k] \to [n]$ such that

$$\mathsf{A}(x) = \sum_{z:f(z)=x} \mathsf{C}(z) \,, \quad \mathsf{B}(y) = \sum_{z:g(z)=y} \mathsf{C}(z) \,. \tag{2}$$

Journal of Physics: Conference Series **1638** (2020) 012002 doi:10.1088/1742-6596/1638/1/012002

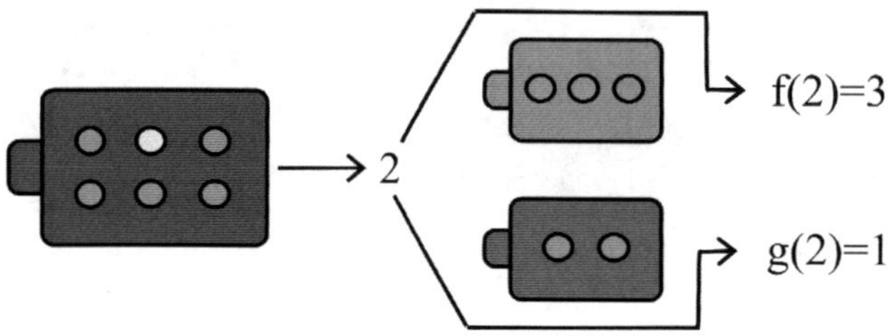

Figure 1: Two observables (green and red) are compatible if their measurements can be simulated with a single observable (blue) by relabeling the obtained measurement outcomes.

The functions f and g are simply relabeling the outcomes of C. The intuitive idea is depicted in Fig. 1. If A and B are not compatible, then they are *incompatible*. The power of stating the definitions of compatibility and incompatibility in this way is that they have direct generalizations to any number of observables, even an infinite number.

There is a useful equivalent definition when we are considering the incompatibility of a finite number of observables. If A and B are compatible and C, f, g are such that (2) holds, then we can define a new observable G on the product outcome set $[m] \times [n]$ as

$$\mathsf{G}(x,y) = \sum_{z:f(z)=x \wedge g(z)=y} \mathsf{C}(z), \tag{3}$$

and then (2) implies that

$$\mathsf{A}(x) = \sum_y \mathsf{G}(x,y), \quad \mathsf{B}(y) = \sum_x \mathsf{G}(x,y). \tag{4}$$

This is a special case of (2), where the relabeling functions are taken to be the projections $f(x,y) = x$ and $g(x,y) = y$. Any observable G satisfying (4) is called a *joint observable* of A and B. We conclude that one can take the existence of a joint observable as an equivalent definition of compatibility for two (or a finite number of) observables. Further, it can be shown that even if relabeling functions would be stochastic, it is still possible to construct a joint observable [3]. This means that by allowing relabeling functions to take different values with some fixed probabilities (hence being Markov kernels), the definition of compatibility is still the same.

3. The basic case

When investigating some mathematically defined property, it is very useful to have a simple 'test class' of objects where that property is easy to check. The simplest case in entanglement theory is the composite state space of two qubits. The entanglement in this class of states has a neat analytic solution: a bipartite qubit state is entangled if and only if its partial transpose has a negative eigenvalue [4, 5].

The simplest class of pairs of observables is the class of two unbiased dichotomic qubit observables. This class has turned out to be very useful in testing (in)compatibility conditions, quantifications and many other developments. An unbiased dichotomic qubit observable is determined by a vector $\mathbf{a} \in \mathbb{R}^3$, $\|\mathbf{a}\| \leq 1$, via

$$\mathsf{M}^{\mathbf{a}}(\pm) = \tfrac{1}{2}(\mathbb{1} \pm \mathbf{a} \cdot \boldsymbol{\sigma}), \tag{5}$$

where $\mathbf{a}\cdot\boldsymbol{\sigma} = a_1\sigma_1 + a_2\sigma_2 + a_3\sigma_3$ and σ_i are the Pauli matrices. (It is customary and convenient to label the outcomes $\pm$ instead of $1,2$.) The necessary and sufficient incompatibility condition was found by Paul in his seminal work [6]: $\mathsf{M}^{\mathbf{a}}$ and $\mathsf{M}^{\mathbf{b}}$ are compatible if and only if

$$\|\mathbf{a}+\mathbf{b}\| + \|\mathbf{a}-\mathbf{b}\| \leq 2\,. \tag{6}$$

The physical interpretation of (6) has been analyzed in detail by Paul and his coworkers in [7–9].

A general (i.e., not necessarily unbiased) dichotomic qubit observable has an additional parameter $\alpha \in [0,1]$ and has the form

$$\mathsf{M}^{\alpha,\mathbf{a}}(\pm) = \tfrac{1}{2}((1\pm\alpha)\mathbb{1} \pm \mathbf{a}\cdot\boldsymbol{\sigma})\,, \tag{7}$$

with the parameters satisfying $|\alpha| \leq 1 - \|\mathbf{a}\|$. We did consider this class of observables with Paul in [10] and found some partial results on their compatibility, but it took some time to digest the ideas and find the final answer. The necessary and sufficient condition for $\mathsf{M}^{\alpha,\mathbf{a}}$ and $\mathsf{M}^{\beta,\mathbf{b}}$ being compatible was found independently (in different but equivalent forms) in [1,11,12]. Paul and I knew that we were working on the same problem, but our approaches and mathematical techniques were so different that we decided to keep the works separate. We coordinated the arXiv submissions and they appeared with the same title and on the same day in February 2008. Admittedly, the form proven in [12] is particularly concise; $\mathsf{M}^{\alpha,\mathbf{a}}$ and $\mathsf{M}^{\beta,\mathbf{b}}$ are compatible if and only if

$$\left(1 - s(\alpha,a)^2 - s(\beta,b)^2\right)\left(1 - \frac{\alpha^2}{s(\alpha,a)^2} - \frac{\beta^2}{s(\beta,b)^2}\right) \leq (\mathbf{a}\cdot\mathbf{b} - \alpha\beta)^2\,, \tag{8}$$

where $a = \|\mathbf{a}\|$, $b = \|\mathbf{b}\|$ and s is a function defined as

$$s(\alpha,a) = \tfrac{1}{2}\left(\sqrt{(1+\alpha)^2 - a^2} + \sqrt{(1-\alpha)^2 - a^2}\right)\,. \tag{9}$$

The meaning of s (or more precisely $1-4s^2$) as a sharpness measure was discussed by Paul in [13].

A simple but important observation of Paul in [6], preceding the proof of (6), was that a joint observable of any pair of dichotomic observables (i.e., not only qubit) is determined by a single operator. In fact, dichotomic observables A and B are compatible if and only if there exists $G \in \mathcal{L}(\mathcal{H})$ such that the following four operators are positive:

(i) G
(ii) $\mathsf{A}(1) - G$
(iii) $\mathsf{B}(1) - G$
(iv) $\mathbb{1} + G - \mathsf{A}(1) - \mathsf{B}(1)$

The equivalence is straightforward to verify: the operators (i)–(iv) are just the operators $\mathsf{G}(x,y)$, $x,y=\pm$, for a joint observable G. This characterization gives a starting point for any attempt to determine the compatibility of two dichotomic observables belonging to some specific class, e.g. when the operators belong to the von Neumann algebra generated by two projections [14]. The characterization further shows that the compatibility of A and B can be decided efficiently numerically: the problem

$$\inf\{t\in\mathbb{R} : t\mathbb{1} + G \geq \mathsf{A}(1) + \mathsf{B}(1)\}\,, \tag{10}$$

subject to the positivity constrains of operators (i)-(iii), is a semidefinite program [15].

4. Convexity

The set of all separable states is a convex subset of all bipartite states. This property is a starting point for many theoretical developments. For instance, one can quantify entanglement of a state ω as the least upper bound of all numbers $t \in [0,1]$ such that the mixed state $t\omega + (1-t)\frac{1}{d_A d_B}\mathbb{1}$ is entangled, or alternatively $t\omega + (1-t)\sigma_{\text{sep}}$ is entangled for all separable states σ_{sep}. These and similar ways to quantify entanglement have been developed in great detail [16].

The convexity plays an important role also in investigations of incompatibility. Let us first recall that the set of observables with a fixed outcome set is a convex set. The convex mixture of A and A' with a mixing parameter $0 \leq t \leq 1$ is defined as

$$(t\mathsf{A} + (1-t)\mathsf{A}')(x) = t\mathsf{A}(x) + (1-t)\mathsf{A}'(x)\,. \tag{11}$$

In [10] the following was noted. For two pairs of observables (A,B) and $(\mathsf{A}',\mathsf{B}')$ and a mixing parameter t, define the mixture as

$$t(\mathsf{A},\mathsf{B}) + (1-t)(\mathsf{A}',\mathsf{B}') = (t\mathsf{A} + (1-t)\mathsf{A}', t\mathsf{B} + (1-t)\mathsf{B}')\,. \tag{12}$$

Here A and A' are assumed to have same outcome set and same for B and B', but the two outcome sets can be different. Importantly, if both pairs (A,B) and $(\mathsf{A}',\mathsf{B}')$ are compatible, then also their mixture is compatible. Namely, if G is a joint observable of (A,B) and G' is a joint observable of $(\mathsf{A}',\mathsf{B}')$, then it is straightforward to verify from the definition of joint observable that the mixture $t\mathsf{G} + (1-t)\mathsf{G}'$ is a joint observable of $t(\mathsf{A},\mathsf{B}) + (1-t)(\mathsf{A}',\mathsf{B}')$.

One possible quantification of incompatibility is the least upper bound of all numbers $t \in [0,1]$ such that the mixture $t(\mathsf{A},\mathsf{B}) + (1-t)(\mathsf{I}_m,\mathsf{I}_n)$ is incompatible for all trivial observables I_m and I_n. An m-outcome trivial observable I_m has the form $\mathsf{I}_m(x) = p(x)\mathbb{1}$ for some probability vector p of length m. As with the previously mentioned ways to quantify entanglement, one can vary also this quantification in several ways, depending to what kind of pairs one mixes the quantified pair (A,B) in question. All these kinds of quantifications are based on the convexity of the set of compatible pairs of observables. A recent work [17] reviews several different quantifications while the general convex framework has been clarified in [18].

Any quantification of incompatibility leads naturally to the question of *maximally incompatible* pairs of observables. I remember when I first time started to talk with Paul about this question. It was during my visit in York in 2011 when I told Paul results that I had found with my friends Alessandro Toigo and Claudio Carmeli on incompatibility of two observables related to mutually unbiased bases, later reported in [19]. We then realized that the bound on the incompatibility of such pairs (triangle shaped area (21) in [19, Prop. 6]) has a natural explanation, valid actually in all operational theories [20]. We also had a conjecture that the canonical position and momentum observables are maximally incompatible, but it took some time to prove it [21]. Remarkably, necessary and sufficient conditions for a pair of maximally incompatible dichotomic measurements to exist in a general probabilistic theory have been found [22], and those conditions show how the existence links to the geometry of the state space.

The earlier convexity argument shows at the same time that the set of all joint observables of a compatible pair of observables is convex. It follows that a compatible pair (A,B) has either unique or infinitely many joint observables, the first case urging for further investigation due to its special nature. One can show that if A (or B) is an extreme element in the respective set of m-outcome (n-outcome) observables, then (A,B) has a unique joint observable [23]. Related results have been reported in [24].

5. Witnesses

An *entanglement witness* is a self-adjoint operator W on $\mathcal{H}^A \otimes \mathcal{H}^B$ such that $\text{tr}[\omega W] \geq 0$ for all separable states ω and $\text{tr}[\eta W] < 0$ at least for some entangled state η. A physical interpretation is straightforward; one measures an observable that allows one to calculate the expectation value $\mathrm{tr}\,[\omega W]$ (e.g. the spectral measure of W) and then directly from that number one concludes that ω is entangled if $\mathrm{tr}\,[\omega W] < 0$. A witness measurement cannot verify that ω is separable as each witness takes positive values also for some entangled states. The fact that for every entangled state there is a witness that detects it follows from the convexity of the subset of separable states and the standard separation theorems of convex analysis.

The convexity of the subset of all compatible pairs of observables hints that one should be able to detect incompatibility with *incompatibility witnesses*, analogously to entanglement witnesses. However, as the set of objects is now all pairs of observables with some fixed outcome sets, the physical interpretation is not so straightforward as with bipartite states. To explain the physically relevant form of incompatibility witnesses, let A and B be two observables with m and n outcomes, respectively. Fix labeled sets of test states, $\mathcal{E}^{(1)} = \{\varrho_x^{(1)}\}_{x=1}^m$ and $\mathcal{E}^{(2)} = \{\varrho_y^{(2)}\}_{y=1}^n$. The ability of A to distinguish the states in $\mathcal{E}^{(1)}$ is taken to be the average guessing probability

$$P_{\mathrm{guess}}(\mathcal{E}^{(1)};\mathsf{A}) = \tfrac{1}{m}\sum_{x=1}^{m} \mathrm{tr}\left[\varrho_x^{(1)}\mathsf{A}(x)\right],$$

and $P_{\mathrm{guess}}(\mathcal{E}^{(2)};\mathsf{B})$ is defined similarly. The average of these guessing probabilities defines a real-valued function on pairs of observables,

$$\xi(\mathsf{A},\mathsf{B}) = \tfrac{1}{2}P_{\mathrm{guess}}(\mathcal{E}^{(1)};\mathsf{A}) + \tfrac{1}{2}P_{\mathrm{guess}}(\mathcal{E}^{(2)};\mathsf{B})\,. \tag{13}$$

The essential property of this function is that

$$\xi(t(\mathsf{A},\mathsf{B}) + (1-t)(\mathsf{A}',\mathsf{B}')) = t\xi(\mathsf{A},\mathsf{B}) + (1-t)\xi(\mathsf{A}',\mathsf{B}')\,, \tag{14}$$

which is the basic mathematical requirement for a witness. It can happen that the test states are chosen in a bad way so that ξ does not detect incompatibility, but apart from that unlucky situation there exists a boundary value c_ξ such that $\xi(\mathsf{A},\mathsf{B}) \leq c_\xi$ for all compatible pairs (A,B), while $\xi(\mathsf{A}',\mathsf{B}') > c_\xi$ at least for some incompatible pair $(\mathsf{A}',\mathsf{B}')$. The boundary value has an interpretation as the best guessing probability in the corresponding discrimination task with post-measurement information [25]. Importantly, if we vary the test states and their prior probabilities, then these types of witnesses are enough to detect the incompatibility of any incompatible pair of observables [26]. Related aspects of incompatibility detection have been recently developed and discussed in [27, 28].

As an example, let us consider the qubit case and choose $\mathcal{E}^{(1)} = \{\frac{1}{2}(\mathbb{1} \pm \sigma_1)\}$ and $\mathcal{E}^{(2)} = \{\frac{1}{2}(\mathbb{1} \pm \sigma_2)\}$. We then have

$$\xi(\mathsf{M}^{\alpha,\mathbf{a}},\mathsf{M}^{\beta,\mathbf{b}}) = \tfrac{1}{2}(1 + \tfrac{1}{2}(a_1 + b_2))\,. \tag{15}$$

It can be shown that for these collections of states we have

$$\xi(\mathsf{A},\mathsf{B}) \leq \tfrac{1}{2}(1 + \tfrac{1}{\sqrt{2}}) \tag{16}$$

for any pair of compatible dichotomic qubit observables [25]. We thus see that the defined witness detects some incompatible pairs, but a comparison to (6) and (8) shows that not all are detected. The bound (16) corresponds to a scenario where Alice sends a message in a perfectly distinguishable form (i.e., both sets consist of orthogonal pure states), but Bob needs to perform a measurement before Alice announces which was the used encoding (i.e., σ_1-eigenbases or σ_2-eigenbases). If Bob would know the used encoding before his measurement, then he could use an incompatible pair and we see from (15) that he can then even distinguish the states perfectly, as expected. The values above the bound, i.e., $a_1 + b_2 > \sqrt{2}$, correspond to choices of measurements that are 'too lucky' to be made without knowing the encoding.

Journal of Physics: Conference Series **1638** (2020) 012002 doi:10.1088/1742-6596/1638/1/012002

6. Simple condition

Let ω be a bipartite state. The partial states of ω are defined as $\omega_{[1]} = \text{tr}_2[\omega]$ and $\omega_{[2]} = \text{tr}_1[\omega]$. In general, it is not possible to conclude entanglement or separability of ω from these partial states, but they do however give relevant information that can be used as a quick test of separability. Firstly, the product state $\omega_{[1]} \otimes \omega_{[2]}$ is different to the original state ω unless the latter is already a product state. It is also true that if $\omega_{[1]}$ or $\omega_{[2]}$ is pure, then ω is separable. Further, for a pure state ω this is a necessary and sufficient condition for separability. In conclusion, by calculating the partial states one gets an easy sufficient condition for separability, which is also necessary for a special class of bipartite states.

There exists also a simple condition that implies compatibility and is further necessary for compatibility for a specific type of observables. Namely, suppose that A and B commute, i.e., $[\mathsf{A}(x), \mathsf{B}(y)] = 0$ for all x, y. Then A and B are compatible as the product of two commuting positive operators is positive and hence $\mathsf{G}(x, y) = \mathsf{A}(x)\mathsf{B}(y)$ is a valid joint observable of A and B. Further, if A or B is sharp (i.e., consists of projection operators), then the commutativity is also necessary for A and B being compatible. In this case the product joint observable is their unique joint observable [29].

There is a also more effective sufficient condition for compatibility than commutativity, but which is still easy to check. In [10] I found with Paul that a joint observable for qubit observables $\mathsf{M}^{\mathbf{a}}$ and $\mathsf{M}^{\mathbf{b}}$, assuming they are compatible, can be always chosen as

$$\mathsf{G}(x, y) = \tfrac{1}{2}(\mathsf{M}^{\mathbf{a}}(x)\mathsf{M}^{\mathbf{b}}(y) + \mathsf{M}^{\mathbf{b}}(y)\mathsf{M}^{\mathbf{a}}(x)) \equiv \tfrac{1}{2}\mathsf{M}^{\mathbf{a}}(x) \circ \mathsf{M}^{\mathbf{b}}(y)\,. \tag{17}$$

This motivates the following joint observable ansatz [30]. For any two observables A and B, define

$$J_{\mathsf{A},\mathsf{B}}(x, y) = \tfrac{1}{2}\mathsf{A}(x) \circ \mathsf{B}(y)\,. \tag{18}$$

The map $J_{\mathsf{A},\mathsf{B}}$ satisfies $\sum_y J_{\mathsf{A},\mathsf{B}}(x, y) = \mathsf{A}(x)$ and $\sum_x J_{\mathsf{A},\mathsf{B}}(x, y) = \mathsf{B}(y)$, but the operators $J_{\mathsf{A},\mathsf{B}}(x, y)$ are not necessarily positive. We can take the positivity as a sufficient condition for compatibility: if $J_{\mathsf{A},\mathsf{B}}(x, y) \geq 0$ for all x, y, then A and B are compatible. We call this *Jordan criterion* for compatibility since it uses the Jordan products of operators. Clearly, for a commuting pair the Jordan criterion holds, but it also covers many more cases. Namely, as said before, for a pair of unbiased dichotomic qubit observables the Jordan criterion gives (6) and is thus not only sufficient but also necessary. For biased dichotomic qubit observables the Jordan criterion does not cover all compatible pairs [30]. However, testing several compatibility criteria with random pairs of observables shows that the Jordan criterion performs well [31].

7. Entanglement/incompatibility breaking channels

Both entanglement and incompatibility are properties that are destroyed under the effect of noise. Paul investigated the influence of noise for joint measurements in several works. The concept of *unsharp reality* evolved from these investigations and is clearly explained already in [6]. A comprehensive investigation on the philosophical consequences of the concept is presented in [32].

In the extreme cases entanglement and incompatibility can be erased completely. Let us first recall that a transformation on a quantum system is describe by a quantum channel, which is a trace preserving completely positive map. A quantum channel Λ is called *entanglement breaking* if $(\Lambda \otimes \mathbb{1})(\omega)$ is separable for all bipartite states ω. It is known [33] that this is the case if and only if there exists an observable M and states ϱ_x such that

$$\Lambda(\varrho) = \sum_x \text{tr}\left[\varrho \mathsf{M}(x)\right] \varrho_x\,. \tag{19}$$

Hence, entanglement breaking channels are exactly measure-and-prepare channels.

A channel Λ can be equally well considered in the Heisenberg picture, and we denote this map by Λ^*. For every observable A, the transformed observable $\Lambda^*(\mathsf{A})$ is defined as $\Lambda^*(\mathsf{A})(x) = \Lambda^*(\mathsf{A}(x))$. As in [34], we say that a channel Λ is *incompatibility breaking* if observables $\Lambda^*(\mathsf{A}_1), \ldots, \Lambda^*(\mathsf{A}_r)$ are compatible for any choice of observables $\mathsf{A}_1, \ldots, \mathsf{A}_r$. A channel Λ having the form (19) reads in the Heisenberg picture as

$$\Lambda^*(T) = \sum_x \mathrm{tr}\,[\varrho_x T]\,\mathsf{M}(x)\,, \tag{20}$$

and we thus see that any transformed observable $\Lambda^*(\mathsf{A})$ is a post-processing of M, implying that Λ^* is incompatibility breaking. We conclude that all entanglement breaking channels are incompatibility breaking. Curiously, there are incompatibility breaking channels that are not entanglement breaking [34]. In that sense, incompatibility is more sensitive to noise than entanglement. It appears to be a difficult problem to characterize all incompatibility breaking channels, although in the Gaussian framework they do have a neat form [35].

As a side remark, all incompatibility breaking channels are entanglement breaking if we require them to break not only incompatibility of observables but also incompatibility of channels [36]. An open question is to characterize the subset of devices that determine incompatibility in the sense that their incompatibility is broken only by measure-and-prepare channels.

8. More stringent forms of separability/compatibility

A separable state ω has, by definition, a convex decomposition of the type (1). One can make a finer separation of separable states by looking for more specific forms of these kind of decompositions. In particular, if the states ϱ_i^A (or ϱ_i^B) can be chosen to be orthogonal pure states, then ω is said to have *zero discord*. Separable states with nonzero discord have been shown to have an advantage over zero discord states in some tasks [37]. We have still a smaller subset of separable states if we require that both collections $\{\varrho_i^A\}$ and $\{\varrho_i^B\}$ can be chosen to be orthogonal pure states; these kind of states are sometimes called *fully classical states*.

By the definition, a compatible pair (A, B) has a joint observable. Analogously to separable states, we may thus ask if there is a joint observable of some specific type. One such special type corresponds to the possibility of measuring A without disturbing B, and in this way performing their joint measurement. The *nondisturbance* means that there is a joint observable G of the form

$$\mathsf{G}(x, y) = \mathcal{I}_x(\mathsf{B}(y))\,, \tag{21}$$

where $\mathcal{I}$ is an instrument describing some measurement of A. Mathematically, each $\mathcal{I}_x$ is a completely positive map such that $\mathcal{I}_x(\mathbb{1}) = \mathsf{A}(x)$. The nondisturbance is not a symmetric relation [38]; it can happen that A can be measured without disturbing B but not vice versa. For this reason, we have also a smaller class of *mutually nondisturbing pairs*. For qubit observables both nondisturbance and mutual nondisturbance are equivalent to commutativity [38], but in higher dimensions we hence have these tighter forms of compatibility. A special form of nondisturbance gives also a physical meaning for the commutativity; as proven by Paul and Javed Singh in [39], two observables A and B commute if and only if the Lüders measurement of A (equivalently that of B) does not disturb B.

The nondisturbance condition gives a physically well motivated criteria to divide compatible pairs into finer classes, but this is not all. For instance, one can take *broadcastability* still as a tighter form of compatibility [40], or one can take a different approach and analyze the minimal number of nonzero elements in all joint observables of a given compatible pair [41].

9. More than two systems/observables

Multipartite entanglement refers to entanglement of more than two systems. The main interests in multipartite entanglement are those features that are not present in bipartite entanglement. One such phenomenon is the following. It is possible to have a tripartite entangled state ω such that all of its three bipartite partial states $\omega_{[12]} = \mathrm{tr}_3[\omega]$, $\omega_{[13]} = \mathrm{tr}_2[\omega]$ and $\omega_{[23]} = \mathrm{tr}_1[\omega]$ are separable. In this sense, entanglement is a property of the full state but not seen in any of its parts. A well-known example of this kind of state is the *GHZ-state* of three qubits,

$$\omega = \frac{1}{\sqrt{2}} \left(|000\rangle + |111\rangle \right) ,$$

for which we have

$$\omega_{[12]} = \omega_{[13]} = \omega_{[23]} = \frac{1}{2} \left(|00\rangle\langle 00| + |11\rangle\langle 11| \right) .$$

An analogy of the previous phenomena for incompatibility has been discussed e.g. in [42]. We say that a set $\{\mathsf{A}_1, \ldots, \mathsf{A}_p\}$, $p \geq 3$, is a *p-Specker set* if $\mathsf{A}_1, \ldots, \mathsf{A}_p$ are incompatible but every proper subset is compatible. A subset of a compatible set is compatible, hence it is enough to verify that all subsets containing $p-1$ observables are compatible.

As an example, the triplet of unbiased qubit observables $\mathsf{M}^{t\mathbf{x}}$, $\mathsf{M}^{t\mathbf{y}}$ and $\mathsf{M}^{t\mathbf{z}}$ forms a 3-Specker set for all values $1/\sqrt{3} < t \leq 1\sqrt{2}$ [29]. (As an interesting side remark, it has been shown in [43] that this kind of genuine triplewise incompatibility, i.e., incompatibility that does not reduce to pairwise relations, can be tested in a device-independent way.) By using generators of a Clifford algebra it was shown that a p-Specker set exists for every $p \geq 3$ [44]. The minimal Hilbert space dimension in that construction is $2^{p/2}$ for even p and $2^{(p-1)/2}$ for odd p. The Clifford algebra construction, however, does not give the absolute minimal dimension: a 4-Specker set can be formed by choosing suitable unbiased dichotomic qubit observables [45]. A recent investigation, reported in [46], goes deep into these questions and develops new techniques; it is proven that every p-Specker set exists already for qubit measurements. Specker sets are specific types of *joint measurability structures* and there are still many open questions on this topic; some open questions are listed in [46].

10. Conclusions

In this paper I have described some analogies between quantum incompatibility and entanglement. The plurality and attractiveness of different aspects of quantum incompatibility hopefully convinces one to speak about *theory of quantum incompatibility.* For me, this theory carries the memory of Paul.

Acknowledgments

I am grateful to Claudio Carmeli, Ion Nechita, Marco Piani, Alessandro Toigo and Mario Ziman for many valuable discussions and their useful comments on an earlier version of this paper.

This work was performed as part of the Academy of Finland Centre of Excellence program, Project 312058. Financial support from the Academy of Finland, Project 287750, is also acknowledged.

References

[1] Busch P and Schmidt H-J 2010 Coexistence of qubit effects *Quant. Inf. Proc.* **9** 143–69
[2] Heinosaari T, Miyadera T and Ziman M 2016 An invitation to quantum incompatibility *J. Phys. A: Math. Theor.* **49** 123001
[3] Ali S T, Carmeli C, Heinosaari T and and Toigo A 2009 Commutative POVMs and fuzzy observables *Found. Phys.* **39** 593–612

[4] Peres A 1996 Separability criterion for density matrices *Phys. Rev. Lett.* **77** 1413–15
[5] Horodecki M, Horodecki P and Horodecki R 1996 Separability of mixed states: necessary and sufficient conditions *Phys. Lett.* A **223** 1–8
[6] Busch P 1986 Unsharp reality and joint measurements for spin observables *Phys. Rev.* D **33** 2253–61
[7] Busch P 1987 Some realizable joint measurements of complementary observables *Found. Phys.* **17** 905–37
[8] Busch P and Shilladay C 2006 Complementarity and uncertainty in Mach–Zehnder interferometry and beyond *Phys. Rep.* **435** 1–31
[9] Bullock T and Busch P 2018 Measurement uncertainty relations: characterising optimal error bounds for qubits *J. Phys. A: Math. Theor.* **51** 283001
[10] Busch P and Heinosaari T 2008 Approximate joint measurements of qubit observables *Quant. Inf. Comp.* **8** 797–818
[11] Stano P, Reitzner D, and Heinosaari T 2008 Coexistence of qubit effects *Phys. Rev.* A **78** 012315
[12] Yu A, Liu N-L, Li L and Oh C H 2010 Joint measurement of two unsharp observables of a qubit *Phys. Rev.* A **81** 062116
[13] Busch P 2009 On the sharpness and bias of quantum effects *Found. Phys.* **39** 712
[14] Heinosaari T, Kiukas J and Reitzner D 2014 Coexistence of effects from an algebra of two projections *J. Phys. A: Math. Theor.* **47** 225301
[15] Wolf M M, Perez-Garcia D and Fernandez C 2009 Measurements incompatible in quantum theory cannot be measured jointly in any other no-signaling theory *Phys. Rev. Lett.* **103** 230402
[16] Horodecki R, Horodecki P, Horodecki M, and Horodecki K 2009 Quantum entanglement *Rev. Mod. Phys.* **81** 865–942
[17] Designolle S, Farkas M and Kaniewski J 2019 Incompatibility robustness of quantum measurements: a unified framework *New J. Phys.* **21** 113053
[18] Haapasalo E 2015 Robustness of incompatibility for quantum devices *J. Phys. A: Math. Theor.* **48** 255303
[19] Carmeli C, Heinosaari T and Toigo A 2012 Informationally complete joint measurements on finite quantum systems *Phys. Rev.* A **85** 012109
[20] Busch P, Heinosaari T, Schultz J and Stevens N 2013 Comparing the degrees of incompatibility inherent in probabilistic physical theories *EPL* **103** 10002
[21] Heinosaari T, Schultz J, Toigo A and Ziman M 2014 Maximally incompatible quantum observables *Phys. Lett.* A **378** 1695–9
[22] Jenčová A and Plávala M 2017 Conditions on the existence of maximally incompatible two-outcome measurements in general probabilistic theory *Phys. Rev.* A **96** 022113
[23] Haapasalo E, Heinosaari T and Pellonpää J-P 2014 When do pieces determine the whole? Extreme marginals of a completely positive map *Rev. Math. Phys.* **26** 1450002
[24] Guerini L and Cunha M T 2018 Uniqueness of the joint measurement and the structure of the set of compatible quantum measurements *J. Math. Phys.* **59** 042106
[25] Carmeli C, Heinosaari T and Toigo A 2019 State discrimination with postmeasurement information and incompatibility of quantum measurements *Phys. Rev.* A **98** 012126
[26] Carmeli C, Heinosaari T and Toigo A 2019 Quantum incompatibility witnesses *Phys. Rev. Lett.* **122** 130402
[27] Skrzypczyk P, Šupić I and Cavalcanti D 2019 All sets of incompatible measurements give an advantage in quantum state discrimination *Phys. Rev. Lett.* **122** 130403
[28] Uola R, Kraft T, Shang J, Yu X-D and Gühne O 2019 Quantifying quantum resources with conic programming *Phys. Rev. Lett.* **122** 130404
[29] Heinosaari T, Reitzner D and Stano P 2008 Notes on joint measurability of quantum observables *Found. Phys.* **38** 1133–47
[30] Heinosaari T 2013 A simple sufficient condition for the coexistence of quantum effects *J. Phys. A: Math. Theor.* **46** 152002
[31] Heinosaari T, Jivulescu M.A and Nechita I 2019 Random positive operator valued measures *Preprint* arXiv:1902.04751 [quant-ph]
[32] Busch P and Jaeger G 2010 Unsharp quantum reality *Found. Phys.* **40** 1341–67
[33] Horodecki M, Shor P W and Ruskai M B 2003 Entanglement breaking channels *Rev. Math. Phys.* **15** 629–41
[34] Heinosaari T, Kiukas J, Reitzner D and Schultz J 2015 Incompatibility breaking quantum channels *J. Phys. A: Math. Theor.* **48** 435301
[35] Heinosaari T, Kiukas J, and Schultz J 2015 Breaking Gaussian incompatibility on continuous variable quantum systems *J. Math. Phys.* **56** 082202
[36] Heinosaari T and Miyadera T 2017 Incompatibility of quantum channels *J. Phys. A: Math. Theor.* **50** 135302
[37] Cavalcanti D, Aolita L, Boixo S, Modi K, Piani M and Winter A 2011 Operational interpretations of quantum discord *Phys. Rev.* A **83** 032324

[38] Heinosaari T and Wolf M M 2010 Nondisturbing quantum measurements *J. Math. Phys.* **51** 092201
[39] Busch P and Singh J 1998 Lüders theorem for unsharp quantum measurements *Phys. Lett.* A **249** 10–2
[40] Heinosaari T 2016 Simultaneous measurement of two quantum observables: compatibility, broadcasting, and in-between *Phys. Rev.* A **93** 042118
[41] Skrzypczyk P, Hoban M J, Sainz A B and Linden N 2020 The complexity of compatible measurements *Phys. Rev. Research* **2** 023292
[42] Liang Y C, Spekkens R W and Wiseman H M 2011 Specker's parable of the overprotective seer: a road to contextuality, nonlocality and complementarity *Phys. Rep.* **506** 1–39
[43] Quintino M T, Budroni C, Woodhead E, Cabello A and Cavalcanti D 2019 Device-independent tests of structures of measurement incompatibility *Phys. Rev. Lett.* **123** 180401
[44] Kunjwal R, Heunen C and Fritz T 2014 Quantum realization of arbitrary joint measurability structures *Phys. Rev.* A **89** 052126
[45] Uola R, Luoma K, Moroder T and Heinosaari T 2016 Adaptive strategy for joint measurements *Phys. Rev.* A **94** 022109
[46] Andrejic N and Kunjwal R 2020 Joint measurability structures realizable with qubit measurements: incompatibility via marginal surgery *Preprint* arXiv:2003.00785 [quant-ph]

Journal of Physics: Conference Series **1638** (2020) 012003 doi:10.1088/1742-6596/1638/1/012003

Subspace constraints for joint measurability

Jukka Kiukas

Department of Mathematics, Aberystwyth University, Aberystwyth SY23 3BZ, United Kingdom

E-mail: `jek20@aber.ac.uk`

Abstract. The structure of quantum effects, positive operators of norm at most one, played a central role in the work of Paul Busch on uncertainty, complementarity, and joint measurability in quantum measurement theory. Here we focus on one aspect of this structure, called "strength of an effect along a ray" [Lett. Math. Phys. **47** 329 (1999)], presenting a few observations not explicit in the existing literature. In fact, the strength function turns out to be useful for characterising positivity and complementarity of quantum effects of a suitable block matrix form, and for studying extensions of joint measurements defined on a subspace of codimension one.

1. Introduction

Uncertainty, complementarity, and joint measurability have been central to quantum mechanics since the beginning [1–3], and have acquired renewed interest in recent years in the context of quantum information and correlations. In particular, the existence of *incompatible* observables, i.e., those which cannot be measured using a single device, is a fundamental non-classical feature not only interesting in its own right [4, 5], but also essential for protocols such quantum steering [6–10] and state discrimination [11, 12].

While joint measurability was originally formulated as commutativity of Hermitian operators, the concept only acquired its current richness when applied to generalised observables, POVMs [13–15], which are necessary for practically relevant situations involving noise and open systems. Joint measurability of a set of observables does not require their commutativity; the general definition (e.g. [16]) is stated in terms of classical post-processing from a common observable, which immediately relates the concept to hidden state models for bipartite quantum states [17]. While the general definition is valid for any set of observables, it simplifies if the number of observables is finite - in particular, two observables are jointly measurable if they are marginals of a single common observable, see e.g. [15].

Restriction to two observables is motivated by the study of pairs of maximally incompatible, *complementary*, observables, especially position and momentum, which can be approximated by jointly measurable pairs obtained by introducing noise. In fact, this form of joint measurability appeared most often in the work of Paul Busch; for review see [18]. Similar ideas have appeared in the context of *Mutually Unbiased Bases* (MUB) which represent complementary observables in finite-dimensional systems, see e.g. [19]. Complementarity in the sense of a pair of MUBs formalises the idea that a state with a definite outcome for one of the observables has a completely uniform outcome distribution for the other one. While this works well in the finite-dimensional case, complementarity of continuous variables is more subtle. A considerable part of the work

Journal of Physics: Conference Series **1638** (2020) 012003 doi:10.1088/1742-6596/1638/1/012003

of Paul Busch was aimed at forming and studying an exact definition of complementarity based on these intuitive ideas [18].

Central to Paul's definition of complementarity was the order structure of quantum effects; as described in [18], one aspect of this structure is the notion of "strength of an effect along a ray" in Hilbert space [20]. The purpose of the present paper is to focus on this concept, which turns out to have various uses in the context of complementarity and joint measurability. We first briefly review the definition, making a connection to Schur complements of block matrices, and then provide a characterisation of complementary effects which goes somewhat beyond the existing literature (including the review [18]). Finally, we show how this concept naturally features in the study of joint measurability of pairs of observables having a suitable block structure relative to a subspace of co-dimension one.

2. Strength of an effect along a ray

Let $\mathcal{H}$ be a separable (possibly infinite-dimensional) Hilbert space, and $\mathcal{B}(\mathcal{H})$ the set of bounded operators on $\mathcal{H}$. For each vector $\in \mathcal{H}$ we let $|\psi\rangle\langle\psi| \in \mathcal{B}(\mathcal{H})$ denote the rank-1 operator $\varphi \mapsto \langle\psi|\varphi\rangle\psi$.

An operator $E \in \mathcal{B}(\mathcal{H})$ is called an *effect* if $0 \leq E \leq \mathbb{I}$ where $\mathbb{I}$ is the identity operator. We recall that effects can be operationally associated to detection events, say, of a measurement yielding to a specific outcome. Given a quantum state ρ (a positive operator of unit trace), the probability of the event is given by $\mathrm{tr}[\rho E]$. Each projection P is an effect; in the terminology of Paul Busch, an effect is called *unsharp* [21] if it is not a projection.

Each vector $\psi \in \mathcal{H}$ with $\|\psi\| \leq 1$ determines the rank one effect $|\psi\rangle\langle\psi|$. These are the elementary units from which more complex effects are composed of, hence they are sometimes called *weak atoms* [20]. Accordingly, given an effect E and a fixed vector $\psi \in \mathcal{H}$, one may ask whether E can be decomposed as $E = \lambda|\psi\rangle\langle\psi| + E'$ where $\lambda \geq 0$ and $E' \geq 0$. Operationally, this corresponds to a *fine-graining*: we consider the event associated to E as being split into two mutually exclusive events, assigned to the weak atom $\lambda|\psi\rangle\langle\psi|$, and the remainder E', respectively—indeed, $\mathrm{tr}[\rho(\lambda|\psi\rangle\langle\psi|)] + \mathrm{tr}[\rho E'] = \mathrm{tr}[\rho E]$ for any state ρ. The maximum λ for which such a decomposition holds is called *the strength of E along ψ*; explicitly,

$$\lambda(E, \psi) := \sup\{\lambda \geq 0 \mid \lambda|\psi\rangle\langle\psi| \leq E\}. \tag{1}$$

Note that $\lambda(E, \psi) \in [0, \|\psi\|^{-2}]$, and $\lambda(E, \psi) = \infty$ iff $\psi = 0$. The optimisation problem (1) can be solved explicitly by using the connection between effect order and range inclusion [22]. In fact, there is a $\lambda > 0$ with $\lambda|\psi\rangle\langle\psi| \leq E$, if and only if $\psi \in \mathrm{ran}\, E^{\frac{1}{2}}$. Noting that the restriction of $E^{\frac{1}{2}}$ to $(\ker E)^{\perp}$ is injective, we may define $E^{-\frac{1}{2}}$ as the unbounded operator on $\mathcal{H}$ with domain $\mathrm{ran}\, E^{\frac{1}{2}}$ and range $(\ker E)^{\perp}$. Denoting $\|E^{-\frac{1}{2}}\varphi\| = \infty$ whenever φ is not in the domain, the following result was obtained in [20]:

$$\lambda(E, \psi)^{-1} = \|E^{-\frac{1}{2}}\psi\|^2 \tag{2}$$

for any effect E and vector $\in \mathcal{H}$. We then note that $\mathrm{ran}\, E^{\frac{1}{2}} = \{\psi \in (\ker E)^{\perp} \mid \int_{[0,1]} x^{-1}\langle\psi|\mathsf{E}(dx)\psi\rangle < \infty\}$, where E is the spectral measure of E, and, furthermore,

$$\lambda(E, \psi)^{-1} = \int_{[0,1]} x^{-1}\langle\psi|\mathsf{E}(dx)\psi\rangle, \tag{3}$$

which holds for all $\psi \in \mathcal{H}$ (the left hand side being finite exactly when the right hand side is). In particular, we obtain the following equality from functional calculus.

Lemma 1.

$$\lambda(E, \psi)^{-1} + \lambda(\mathbb{I} - E, \psi)^{-1} = \lambda(E(\mathbb{I} - E), \psi)^{-1}, \tag{4}$$

for each $\psi \in \mathcal{H}$, and each effect $E \in \mathcal{B}(\mathcal{H})$.

3. Positive extensions of positive subspace operators

We now consider the above Hilbert space $\mathcal{H}$ as a subspace of codimension one in the "minimally" enlarged Hilbert space $\tilde{\mathcal{H}} = \mathbb{C} \oplus \mathcal{H}$. As an example, one could think of $\tilde{\mathcal{H}}$ as the Fock space truncated to the first two sectors, the vacuum state $|\mathrm{vac}\rangle$ and the single system space. We write the associated vectors in $\tilde{\mathcal{H}}$, linear functionals $\tilde{\mathcal{H}} \to \mathbb{C}$, and operators in $\mathcal{B}(\tilde{\mathcal{H}})$ in the convenient (slightly nonstandard) matrix notation

$$c|\mathrm{vac}\rangle \oplus \varphi = \begin{pmatrix} c \\ |\varphi\rangle \end{pmatrix} \in \tilde{\mathcal{H}}, \quad \bar{c}\langle\mathrm{vac}| \oplus \langle\varphi| = \begin{pmatrix} \bar{c} & \langle\varphi| \end{pmatrix} : \tilde{\mathcal{H}} \to \mathbb{C}, \quad \tilde{E} = \begin{pmatrix} p & \langle\psi'| \\ |\psi\rangle & E \end{pmatrix} \in \mathcal{B}(\tilde{\mathcal{H}}),$$

where $p, c \in \mathbb{C}$, $\psi, \psi', \varphi \in \mathcal{H}$, and $E \in \mathcal{B}(\mathcal{H})$, and the action of the operator and the functional on a vector is computed by usual matrix multiplication rule and "bra-ket" notation. In the Fock space picture, vectors of the above form represent coherent superpositions of a single system and vacuum. In what follows we sometimes use P_0 to denote the projection on $\tilde{\mathcal{H}}$ with range $\mathcal{H}$.

We then define the *Schur complements* associated to $\tilde{E}$ by

$$\tilde{E}/p := E - p^{-1}|\psi\rangle\langle\psi'|, \qquad \tilde{E}/E := p - \langle E^{-\frac{1}{2}}\psi'|E^{-\frac{1}{2}}\psi\rangle,$$

where the first one is defined when $p \neq 0$, and the second one when $E \geq 0$ and $\psi, \psi' \in \operatorname{ran} E^{\frac{1}{2}}$. In the case where $\dim H < \infty$ these coincide with the usual Schur complements for block matrices [23]. In the general case, they are closely related to the strength function $\lambda(E, \psi)$, which characterises the positivity of $\tilde{E}$ as the following proposition shows.

Proposition 1. *Let*

$$\tilde{E} = \begin{pmatrix} p & \langle\psi| \\ |\psi\rangle & E \end{pmatrix}$$

where $E \geq 0$, $p > 0$, and $\psi \in \mathcal{H}$. Then the following are equivalent:

(i) $\tilde{E} \geq 0$;
(ii) $\tilde{E}/p \geq 0$;
(iii) $p\,\lambda(E, \psi) \geq 1$;
(iv) $\tilde{E}/E$ *is defined and* $\tilde{E}/E \geq 0$.

Proof. From equations (1) and (2), it is clear that $\tilde{E}/p \geq 0$ iff $p^{-1} \leq \lambda(E, \psi)$ iff $p - \|E^{-\frac{1}{2}}\psi\|^2 \geq 0$ iff $\tilde{E}/E \geq 0$. The equivalence between (i) and (ii) can be found in [24, p.27]. □

We note that this result is well-known in the finite-dimensional case [23]. In general, the following corollary is immediate.

Corollary 1. *In the notation of the preceding proposition, if $\tilde{E} \geq 0$, then $\|\psi\|^2 \leq p\|E\|$.*

Proof.

$$\|\psi\|^2 = \|E^{\frac{1}{2}}E^{-\frac{1}{2}}\psi\|^2 \leq \|E^{\frac{1}{2}}\|^2\lambda(E, \psi)^{-1} \leq \|E^{\frac{1}{2}}\|^2 p = \|E\|p\,.$$

□

Journal of Physics: Conference Series **1638** (2020) 012003 doi:10.1088/1742-6596/1638/1/012003

4. Coherent extensions of effects and connection to complementarity

Any effect $E \in \mathcal{B}(\mathcal{H})$ trivially defines an effect on $\tilde{\mathcal{H}}$ as $\mathcal{H}$ is a subspace. However, if we require that the resulting effect be used to *detect (vacuum) coherence* in states, that is, distinguish coherent superposition states between the vacuum and $\mathcal{H}$ from classical mixtures, the extension is not entirely straightforward. We first define, for each $p > 0$, $\psi \in \mathcal{H}$, a map $I_{p,\psi} : \mathcal{B}(\mathcal{H}) \to \mathcal{B}(\tilde{\mathcal{H}})$ via

$$I_{p,\psi}(E) = \begin{pmatrix} p & \langle\psi| \\ |\psi\rangle & E \end{pmatrix}.$$

This extension map is called *coherent* if $\psi \neq 0$, and we call ψ the associated *coherence*. Clearly, a coherent extension is needed for detecting coherence in states, as $\mathrm{tr}[\rho I_{p,0}(E)] = \mathrm{tr}[\rho' I_{p,0}(E)]$ for the density operators of the form

$$\rho = \begin{pmatrix} q & \langle\xi| \\ |\xi\rangle & \rho_0 \end{pmatrix}, \qquad \rho' = \begin{pmatrix} q & 0 \\ 0 & \rho_0 \end{pmatrix}.$$

By Proposition 1, $I_{p,\psi}(E) \geq 0$ if and only if $p\lambda(E,\psi) \geq 1$. We now easily obtain the following.

Proposition 2. *Let $E \in \mathcal{B}(\mathcal{H})$ be an effect, and $\psi \in \mathcal{H} \setminus \{0\}$. The following are equivalent:*

(i) $I_{p,\psi}(E)$ is an effect for some $p \in [0,1]$;

(ii) $\lambda(E(\mathbb{I}-E),\psi) \geq 1$.

In this case, $I_{p,\psi}(E)$ is an effect if and only if $\lambda(E,\psi)^{-1} \leq p \leq 1-\lambda(\mathbb{I}-E,\psi)^{-1}$.

Proof. Denote $a := \lambda(E,\psi)^{-1}$ and $b := 1-\lambda(\mathbb{I}-E,\psi)^{-1}$, and note that $a > 0$ and $b < 1$ because $\neq 0$. Now, for any $p \in [0,1]$ we have $\mathbb{I} - I_{p,\psi}(E) = I_{1-p,-\psi}(\mathbb{I}-E)$. Hence $I_{p,\psi}(E)$ is an effect iff $p \in (0,1)$ and $p^{-1} \leq \lambda(E,\psi)$ and $(1-p)^{-1} \leq \lambda(\mathbb{I}-E,\psi)$. But this happens exactly when $a \leq p \leq b$. Hence, (i) is equivalent to $a \leq b$, which by Lemma 1, is equivalent to (ii). In this case $I_{p,\psi}(E)$ is an effect exactly when $p \in [a,b]$. □

Corollary 2. *If P is a projection, it has no coherent effect extension $I_{p,\psi}(P)$. In other words, only unsharp effects can be coherently extended into effects.*

Proof. Now $P(\mathbb{I}-P) = 0$ so $\lambda(E(\mathbb{I}-E),\psi) \geq 1$ only when $\psi = 0$, and hence the claim follows from Proposition 2. Alternatively (and more explicitly), we note that $I_{p,\psi}(P) \geq 0$ and $\mathbb{I} - I_{p,\psi}(P) = I_{1-p,-\psi}(\mathbb{I}-P) \geq 0$ implies $\psi \in \mathrm{ran}\,P \cap \mathrm{ran}\,(\mathbb{I}-P) = \{0\}$. □

Complementarity of quantum effects was one of the key notions in the work of Paul Busch. Here we only quote the basics from the review [18], before proceeding to show how complementarity is related to the extension of subspace effects.

Two effects $E, F \in \mathcal{B}(\mathcal{H})$ are called *complementary*, if there is no nonzero effect $A \in \mathcal{B}(\mathcal{H})$ such that $A \leq E$ and $A \leq F$. The following characterisation can be found in [18] (see references therein for the original papers).

Proposition 3. *For two effects $E, F \in \mathcal{B}(\mathcal{H})$, the following are equivalent:*

(i) E and F are complementary.

(ii) $\mathrm{ran}\,E^{\frac{1}{2}} \cap \mathrm{ran}\,F^{\frac{1}{2}} = \{0\}$*;*

(iii) There is no $\psi \in \mathcal{H} \setminus \{0\}$ such that $|\psi\rangle\langle\psi| \leq E$ and $|\psi\rangle\langle\psi| \leq F$;

We can now state the following result, which reveals another aspect of complementarity.

Proposition 4. *For two effects $E, F \in \mathcal{B}(\mathcal{H})$, the following are equivalent:*

(i) E and F are complementary;

Journal of Physics: Conference Series **1638** (2020) 012003 doi:10.1088/1742-6596/1638/1/012003

(ii) $\lambda(E,\psi)=0$ *or* $\lambda(F,\psi)=0$ *for each* $\psi\in\mathcal{H}\setminus\{0\}$*;*

(iii) The coherent extensions $I_{p,\psi}(E)$ *and* $I_{p,\psi}(F)$ *are not both positive for any choice of* $\in\mathcal{H}\setminus\{0\}$, $p\in[0,1]$.

Proof. Proposition 3 shows that (i) is equivalent to (ii), and the equivalence of (ii) and (iii) is clear from Proposition 1. □

Proposition 5. *For two effects* $E,F\in\mathcal{B}(\mathcal{H})$*, the following are equivalent:*

(i) The effects $E(\mathbb{I}-E)$ *and* $F(\mathbb{I}-F)$ *are complementary;*

(ii) The coherent extensions $I_{p,\psi}(E)$ *and* $I_{p',\psi}(F)$ *are not both effects for any choice of* $\in\mathcal{H}\setminus\{0\}$, $p,p'\in[0,1]$.

Proof. Follows immediately from Proposition 2. □

5. Coherent extensions of operations, observables, and instruments

In this section we consider constraints for extending the standard quantum objects related to measurement processes. We also give an operational interpretation to such extensions within the above mentioned Fock space context, which allows us to describe the idea that a measurement may leave the system in a coherent superposition with the vacuum after the measurement.

5.1. Operations

The basic definitions can be found in [15]. An *operation* is a conditional transformation between the state spaces of two quantum systems. Denoting by $\mathcal{H}$ and $\mathcal{H}'$ the corresponding Hilbert spaces, an operation is any completely positive normal linear map $\Phi:\mathcal{B}(\mathcal{H})\mapsto\mathcal{B}(\mathcal{H}')$ with $\Phi(\mathbb{I})\leq\mathbb{I}$. Here the effect $\Phi(\mathbb{I})$ (called the effect of Φ) has the following operational meaning: for any state ρ on $\mathcal{H}'$, the number $\mathrm{tr}[\rho\Phi(\mathbb{I})]$ is the probability of the event which triggers the transformation (typically an occurrence of a specific outcome in a measurement). Every operation can be decomposed in terms of its *Kraus operators* $K_k:\mathcal{H}\to\mathcal{H}'$ as $\Phi(A)=\sum_k K_k^*AK_k$, where the sum may be (countably) infinite, converging in the weak operator topology. An operation Φ is called *channel* if $\Phi(\mathbb{I})=\mathbb{I}$. Clearly, if E is the effect of an operation Φ, it is also the effect of $\Phi\circ\Lambda$ where Λ is any channel.

An operation is called *pure* if it can be decomposed using only one Kraus operator K. In this case the transformation preserves pure states, acting as $\psi\mapsto K\psi$ in the Hilbert space level. We note that *any* effect $E\in\mathcal{B}(\mathcal{H})$ can be decomposed in various ways as $E=K^*K$ with $K:\mathcal{H}\to\mathcal{H}'$ where $\mathcal{H}'$ can be any Hilbert space of dimension at least the rank of E. Operationally, these correspond to different pure operations whose triggering event is described by E. The case where $E=|\psi\rangle\langle\psi|$ for some $\psi\in\mathcal{H}$ is special in that the final normalised state after the operation is just the weak atom $\|\psi\|^{-2}|\psi\rangle\langle\psi|$, i.e., all information on the initial state has been lost in the transformation.

Consider now a fine-graining of E discussed in section 2, that is, $E=|\eta\rangle\langle\eta|+E'$, where E' is an effect. Hence we must have $\lambda(E,\eta)\geq 1$, and in particular, $\eta\in\operatorname{ran}E^{\frac{1}{2}}$. Take any Kraus operator $K:\mathcal{H}\to\mathcal{H}'$ such that $E=K^*K$. It follows (see [18, Lemma 1]) that there is a unit vector $\varphi\in\mathcal{H}'$ such that

$$K^*\varphi=\sqrt{\lambda(E,\eta)}\eta.$$

Denoting $\mu=\lambda(E,\eta)^{-1}$ we define $K_1:=\sqrt{\mu}|\varphi\rangle\langle\varphi|$ and $K_0:=\sqrt{1-\mu}|\varphi\rangle\langle\varphi|+\mathbb{I}-|\varphi\rangle\langle\varphi|$; then $K_1^*K_1=\mu|\varphi\rangle\langle\varphi|$, $K_0^*K_0=\mathbb{I}-\mu|\varphi\rangle\langle\varphi|$, so we may define a channel $\Lambda(A)=K_0^*AK_0+K_1^*AK_1$ with two Kraus operators, corresponding to the above decomposition: $E'=K^*K_0^*K_0K$ and $|\eta\rangle\langle\eta|=K^*K_1^*K_1K$. Hence the effect E can be associated to a sequential detection where exactly one of the two pure operations corresponding to K_0 and K_1 are performed after K,

Journal of Physics: Conference Series **1638** (2020) 012003 doi:10.1088/1742-6596/1638/1/012003

with respective probabilities μ and $1-\mu$. In particular, the choice between the two is entirely classical, i.e., *incoherent*, as described by the operation $\Phi(A) = K^*\Lambda(A)K$.

The detection can be modified into a coherent one using the Fock space extension considered above: we first look at the pure operation acting on a vector $\phi \in \mathcal{H}$ in two stages as follows:

$$\begin{aligned}\phi &\mapsto K\phi = \langle\varphi|K\phi\rangle\varphi + (\mathbb{I} - |\varphi\rangle\langle\varphi|)K\phi \\ &\mapsto \sqrt{\mu}\langle\varphi|K\phi\rangle|\text{vac}\rangle \oplus (\sqrt{1-\mu}\langle\varphi|K\phi\rangle\varphi + (\mathbb{I} - |\varphi\rangle\langle\varphi|)K\phi) = \langle\eta|\phi\rangle|\text{vac}\rangle \oplus K_0K\phi.\end{aligned}$$

Here, following the application of K, the component of $K\phi$ along φ transforms into a superposition of the vacuum and φ (with amplitudes squaring to the probabilities μ and $1-\mu$ appearing above), while the rest remains unaffected. The case $\mu = 1$, (i.e., $\lambda(E,\eta) = 1$) corresponds to the situation where the entire component along φ gets absorbed. Assuming also $|\text{vac}\rangle \mapsto 0$, this operator reads

$$\tilde{K} = \begin{pmatrix} 0 & \langle\eta| \\ 0 & K_0K \end{pmatrix},$$

and one can readily check that $\tilde{K}^*\tilde{K} = 0 \oplus E = E$, that is, $\tilde{K}$ defines a pure operation acting on the extended Hilbert space but whose effect is again E. Hence, instead of the incoherent operation Φ with two Kraus operators acting on the single system subspace, this operation combines the two processes coherently.

The detection corresponding to the new operation can *create* coherence, but cannot *detect* existing coherence, as its *effect* E is still incoherent. This is because the operation discards any existing vacuum contribution as $\tilde{K}|\text{vac}\rangle = 0$. In order to have a coherent effect, we only need to replace this by $\tilde{K}|\text{vac}\rangle = u|\text{vac}\rangle$ where $u \in \mathbb{C}$ is nonzero. This gives

$$\tilde{K} = \begin{pmatrix} u & \langle\eta| \\ 0 & K_0K \end{pmatrix}, \tag{5}$$

leading to

$$\tilde{K}^*\tilde{K} = \begin{pmatrix} \overline{u} & 0 \\ |\eta\rangle & K^*K_0^* \end{pmatrix}\begin{pmatrix} u & \langle\eta| \\ 0 & K_0K \end{pmatrix} = \begin{pmatrix} |u|^2 & \overline{u}\langle\eta| \\ u|\eta\rangle & |\eta\rangle\langle\eta| + E' \end{pmatrix} = \begin{pmatrix} |u|^2 & \overline{u}\langle\eta| \\ u|\eta\rangle & E \end{pmatrix} = I_{|u|^2,u\eta}(E),$$

and hence showing that we have indeed obtained a coherent positive extension of E. (While positivity is clear by construction, we also see it from the constraint on the strength function, as $\lambda(E,\eta) = |u|^2\,\lambda(E,u\eta) \geq 1$.) However, an extra constraint now emerges, as the coherent extension $I_{|u|^2,u\eta}(E)$ is not necessarily an effect (see Proposition 2); we also need $(1-|u|^2)\,\lambda(\mathbb{I}-E,u\eta) \geq 1$, that is

$$\frac{|u|^2}{1-|u|^2} \leq \lambda(\mathbb{I}-E,\eta).$$

When this condition holds, the operator (5) describes a pure operation whose effect is $I_{|u|^2,u\eta}(E)$, a coherent effect extension of E, and the complementary effect $I_{1-|u|^2,-u\eta}(\mathbb{I}-E)$ can be associated with a pure operation of the same kind.

In conclusion, we have shown that any coherent effect extension can be associated to a detection event where the system, when detected, proceeds into a coherent superposition of vacuum state and "no-absorption" relative to a single vector state. We also remark that *all* coherent effect extensions $I_{p,\psi}(E)$ of E can be associated in this way to coherent pure operations of the form (5), with $\psi = u\eta$ and $p = |u|^2$.

5.2. Observables

In the preceding section we considered coherent extensions of single effects and operations. In general, *quantum measurements* are described by collections of these, so we may use the above procedure to extend them as well. Beginning at the general level, we fix Ω be the set of outcomes for the measurements, with $\mathcal{A}$ a σ-algebra of subsets of Ω.

An *observable* on the Hilbert space $\mathcal{H}$ is a normalised positive operator valued measure (POVM), that is, a weakly (or, equivalently, strongly) σ-additive map $\mathsf{E} : \mathcal{A} \to \mathcal{B}(\mathcal{H})$, such that $\mathsf{E}(X) \geq 0$ for each $X \in \mathcal{A}$, and $\mathsf{E}(\Omega) = \mathbb{I}$. The set Ω represents possible outcomes of the observable, in the sense that the event of detecting an outcome in a set $X \in \mathcal{A}$ is associated to the effect $\mathsf{E}(X)$.

Similarly, we can of course define observables on the extended Hilbert space $\tilde{\mathcal{H}}$. Given such an observable $\tilde{\mathsf{E}}$, we can clearly compress it to an observable E on the subspace $\mathcal{H}$ by setting $\mathsf{E}(X) = P_0\tilde{\mathsf{E}}(X)P_0$. We call this the *subspace observable* of $\tilde{\mathsf{E}}$.

Conversely, given an observable in $\mathcal{H}$, it is natural to ask whether it can be extended to an observable on the Hilbert space $\tilde{\mathcal{H}}$ by applying the procedure discussed for effects above. Clearly, we need a probability measure $p : \mathcal{A} \to [0, 1]$, and a (norm) σ-additive vector measure $\Psi : \mathcal{A} \to \mathcal{H}$ such that $\Psi(\Omega) = 0$ and $p(X)\,\lambda(\mathsf{E}(X), \Psi(X)) \geq 1$ (see Proposition 1) for each $X \in \mathcal{A}$, as then

$$\tilde{\mathsf{E}}(X) := \begin{pmatrix} p(X) & \langle\Psi(X)| \\ |\Psi(X)\rangle & \mathsf{E}(X) \end{pmatrix}$$

defines an observable $\tilde{\mathsf{E}} : \mathcal{A} \to \mathcal{B}(\tilde{\mathcal{H}})$. Note that now a single positivity constraint for each $\tilde{\mathsf{E}}(X)$ is sufficient, as the overall normalisation condition $\Psi(\Omega) = 0$ (together with $\mathsf{E}(\Omega) = \mathbb{I}$) then forces it to be an effect. From Corollary 1 we get a necessary condition

$$\|\Psi(X)\|^2 \leq \|\mathsf{E}(X)\|p(X) \leq p(X),$$

which shows that the vector measure Ψ is p-continuous and of bounded variation. Since Hilbert spaces have the Radon-Nikodym property, it follows [25] that Ψ has a density with respect to p. However, E does not need to have an operator valued density relative to p, and the extension problem appears somewhat intractable in full generality. The problem simplifies if both E and p have a densities relative to fixed (not necessarily finite) positive scalar measure μ; suppose therefore that $p(X) = \int_X p_x d\mu(x)$, $\mathsf{E}(X) = \int_X E_x d\mu(x)$ (weakly), where $x \mapsto p_x$ is a positive measurable function, and $x \mapsto E_x$ is a (weakly) measurable $\mathcal{B}(\mathcal{H})$-valued function such that $E_x \geq 0$ for each x. Since Ψ has a density relative to p, it follows that $\Psi(X) = \int_X \psi_x d\mu(x)$ for some measurable $\mathcal{H}$-valued function $x \mapsto \psi_x$. The above extension then takes the form

$$\tilde{\mathsf{E}}(X) := \int_X \begin{pmatrix} p_x & \langle\psi_x| \\ |\psi_x\rangle & E_x \end{pmatrix} d\mu(x),$$

where the integrand is a weakly measurable $\mathcal{B}(\tilde{\mathcal{H}})$-valued function, the integral is understood in the weak sense, and the constraints for normalisation and positivity take the form

$$\int_\Omega \psi_x d\mu(x) = 0, \qquad p_x\,\lambda(E_x, \psi_x) \geq 1 \quad \text{for all } x \in \Omega. \tag{6}$$

One could proceed further with these constraints alone. However, we continue with quantum instruments, which provide a more concrete view to the extension problem.

5.3. Instruments

An *instrument* describes the measurement process, including the state change conditional to the outcome being observed. Mathematically, an instrument $\mathcal{I}$ associates to each set $X \in \mathcal{A}$ an operation $\mathcal{I}(X)$ on $\mathcal{B}(\mathcal{H})$, such that $X \mapsto \mathcal{I}(X)$ is weakly σ-additive, and $\mathcal{I}(\Omega)$ is a channel. This channel is interpreted as the transformation made on the system as a result of the measurement, without conditioning on a particular outcome. For an instrument $\mathcal{I}$, the map $X \mapsto \mathcal{I}(X)(\mathbb{I})$ is an observable, interpreted as the observable measured by the instrument.

An observable of the form $\mathsf{E}(X) = \int_X E_x d\mu(x)$ considered above naturally arises from the *Lüders* instrument

$$\mathcal{I}(X)(A) = \int_X K_x^* A K_x \, d\mu(x), \tag{7}$$

where $x \mapsto K_x$ is measurable, and satisfies $E_x = K_x^* K_x$ for each $x \in \Omega$ [1]. We can now apply the construction of the preceding section pointwise for each x, so as to extend this into a Lüders instrument on $\mathcal{B}(\tilde{\mathcal{H}})$:

$$\tilde{\mathcal{I}}(X)(\tilde{A}) := \int_X \begin{pmatrix} \overline{u}_x & 0 \\ |\eta_x\rangle & K_x^* K_{0,x}^* \end{pmatrix} \tilde{A} \begin{pmatrix} u_x & \langle \eta_x | \\ 0 & K_{0,x} K_x \end{pmatrix} d\mu(x). \tag{8}$$

Here we require that $x \mapsto u_x$ is measurable, with $\int |u_x|^2 d\mu(x) = 1$, $\int u_x \eta_x d\mu(x) = 0$, and $\lambda(\eta_x, E_x) \geq 1$ for each x. The map $K_{0,x}$ is defined as in the preceding section, pointwise using K_x and η_x, and we assume this is done in a measurable way. The interpretation is that conditional on the outcome of the measurement being x, the system has the possibility of subsequently absorbed against the vector state φ_x defined by $K_x^* \varphi_x = \sqrt{\lambda(E_x, \psi_x)}\eta_x$ as described above.

6. Joint measurability and subspace compatibility

A pair of observables E and F on $\mathcal{H}$ with outcome sets Ω_1 and Ω_2, respectively, are said to be *jointly measurable* or *compatible*, if there exists an observable G with outcome set $\Omega_1 \times \Omega_2$, such that $\mathsf{G}(X \times \Omega_2) = \mathsf{E}(X)$ for all X, and $\mathsf{G}(\Omega_1 \times Y) = \mathsf{F}(Y)$ for all Y. The definition can be modified to include arbitrary collections of observables, but pairs are sufficient for our purpose. A concrete way of constructing joint measurements is through sequential combination: if $\mathcal{I}$ is an instrument with observable E, and F is an observable, then (under fairly general conditions, see [15]) the sequential measurement defines a joint observable $\mathsf{G}(X \times Y) := \mathcal{I}(X)(\mathsf{F}(Y))$ with $\mathsf{G}(X \times \Omega_2) = \mathsf{E}(X)$. There other marginal $\mathsf{G}(\Omega_1 \times Y)$ is by construction jointly measurable with E. Of course, the latter does not usually equal F, due to the disturbance caused by the first measurement.

Obviously, the same considerations apply to observables on the extended Hilbert space $\tilde{\mathcal{H}}$. Now, we introduce the following concept: two observables on the extended Hilbert space $\tilde{\mathcal{H}}$ are called *subspace compatible* if their subspace observables (see the preceding subsection) are compatible. The following observation is immediate:

Proposition 6. *If $\tilde{\mathsf{E}}$ and $\tilde{\mathsf{F}}$ are compatible observables on $\tilde{\mathcal{H}}$, then they are also subspace compatible.*

Proof. If $\tilde{\mathsf{G}}$ is a joint observable for $\tilde{\mathsf{E}}$ and $\tilde{\mathsf{F}}$, then the compression $\mathsf{G}(\cdot) = P_0 \tilde{\mathsf{G}}(\cdot) P_0$ is a joint observable for the subspace observables $X \mapsto P_0 \tilde{\mathsf{E}}(X) P_0$ and $Y \mapsto P_0 \tilde{\mathsf{F}}(Y) P_0$. □

The converse problem—whether given subspace compatible observables are also compatible—is interesting, and clearly nontrivial in general. Here we only show how observables jointly measurable with a given coherently extended subspace observable may arise, and then proceed with an example in the next section.

Fix an observable E in $\mathcal{H}$, and let $\tilde{\mathsf{E}}$ be one of its coherent extensions as described in the preceding section. A very general way of constructing observables jointly measurable with $\tilde{\mathsf{E}}$ is

[1] We do not consider here the conditions which ensure measurability in case E_x is given and K_x is obtained by pointwise decomposition; in most relevant cases, K_x is given explicitly.

by dilation (see [26]). In our case, this can be achieved by using the decomposition appearing in (8). We define

$$\tilde{\mathsf{G}}(X\times Y)=\int_X\begin{pmatrix}\overline{u}_x & 0\\ |\eta_x\rangle & K_x^*K_{0,x}^*\end{pmatrix}\tilde{\mathsf{H}}_x(Y)\begin{pmatrix}u_x & \langle\eta_x|\\ 0 & K_{0,x}K_x\end{pmatrix}d\mu(x),$$

where $\tilde{\mathsf{H}}_x(Y)$ is an observable with a fixed outcome set Ω', for each $x\in\Omega$ (and we again assume that the dependence on x is sufficiently measurable). Then $\tilde{\mathsf{G}}(X\times\Omega')=\tilde{\mathsf{E}}(X)$ for each X by construction, and hence by computing the other marginal we find an observable $\tilde{\mathsf{F}}(Y):=\tilde{\mathsf{G}}(\Omega\times Y)$ jointly measurable with $\tilde{\mathsf{E}}$. In fact, under certain conditions *all* observables jointly measurable with $\tilde{\mathsf{E}}$ can be constructed in this way [26].

A particular but practically relevant special case is obtained by choosing $\tilde{\mathsf{H}}_x(Y)$ independent of x, in which case $\tilde{\mathsf{G}}(X\times Y)=\tilde{\mathcal{I}}(X)(\tilde{\mathsf{F}}(Y))$, that is, the joint observable can be implemented as a sequential measurement consisting of the above absorption implementation of $\tilde{\mathsf{E}}$, followed by a measurement of $\tilde{\mathsf{F}}$. Explicitly, let

$$\tilde{\mathsf{H}}(Y)=\begin{pmatrix}q(Y) & \langle\Phi(Y)|\\ |\Phi(Y)\rangle & \mathsf{H}(Y)\end{pmatrix};$$

then the joint measurement is

$$\begin{aligned}\tilde{\mathsf{G}}(X\times Y)&=\int_X\begin{pmatrix}\overline{u}_x & 0\\ |\eta_x\rangle & K_x^*K_{0,x}^*\end{pmatrix}\begin{pmatrix}q(Y) & \langle\Phi(Y)|\\ |\Phi(Y)\rangle & \mathsf{H}(Y)\end{pmatrix}\begin{pmatrix}u_x & \langle\eta_x|\\ 0 & K_{0,x}K_x\end{pmatrix}d\mu(x)\\ &=\int_X\begin{pmatrix}\overline{u}_x & 0\\ |\eta_x\rangle & K_x^*K_{0,x}^*\end{pmatrix}\begin{pmatrix}u_x\,q(Y) & q(Y)\langle\eta_x|+\langle\Phi(Y)|K_{0,x}K_x\\ |u_x\Phi(Y)\rangle & |\Phi(Y)\rangle\langle\eta_x|+\mathsf{H}(Y)K_{0,x}K_x\end{pmatrix}d\mu(x)\\ &=\int_X\begin{pmatrix}|u_x|^2\,q(Y) & \langle q(Y)\psi_x+u_xK_x^*K_{0,x}^*\Phi(Y)|\\ |q(Y)\psi_x+u_xK_x^*K_{0,x}^*\Phi(Y)\rangle & M_x(Y)\end{pmatrix}d\mu(x)\\ &=\begin{pmatrix}p(X)q(Y) & \langle q(Y)\Psi(X)+K(X)\Phi(Y)|\\ |q(Y)\Psi(X)+K(X)\Phi(Y)\rangle & \mathsf{G}(X\times Y)\end{pmatrix},\end{aligned}$$

where $K(X):=\int_X u_xK_x^*K_{0,x}^*d\mu(x)$, and the subspace observable is given by

$$\mathsf{G}(X\times Y):=\int_X M_x(Y)d\mu(x)$$

with

$$\begin{aligned}M_x(Y)&=q(Y)|\eta_x\rangle\langle\eta_x|+I_x(Y)+K_x^*K_{0,x}^*H(Y)K_{0,x}K_x\\ &=q(Y)\mu_xK_x^*|\varphi_x\rangle\langle\varphi_x|K_x+K_x^*K_{0,x}^*H(Y)K_{0,x}K_x+I_x(Y)\\ &=K_x^*\left(q(Y)\mu_x|\varphi_x\rangle\langle\varphi_x|+K_{0,x}^*H(Y)K_{0,x}+I_x(Y)\right)K_x\end{aligned}$$

and $I_x(Y)=2\mathrm{Re}(|K_{0,x}^*\Phi(Y)\rangle\langle\sqrt{\mu}\varphi_x|)$. From this we can understand the structure of the joint observable, the central feature of which is that the particle is only subjected to the subspace part of the second measurement if it does not get absorbed into the vacuum by the first measurement.

First of all, the coherence $q(Y)\Psi(X)+K(X)\Phi(Y)$ is a superposition of the coherence of $\Psi(X)$ the first measurement, and the coherence $\Phi(Y)$ of the second measurement transformed by $K(X)$ conditional to the first measurement taking place without absorption. Note that $K(X)$ is indeed the coherent combination of all the conditional transformations associated to the no-absorption.

Secondly, the subspace observable $\mathsf{G}(X\times Y)$ has the following structure: the term $q(Y)\mu_x|\varphi_x\rangle\langle\varphi_x|$ corresponds to the event of the system being absorbed into the vacuum in

Journal of Physics: Conference Series **1638** (2020) 012003 doi:10.1088/1742-6596/1638/1/012003

the first measurement, with the subsequent measurement $q(Y)$ of the vacuum component in the second measurement. The second term $K_{0,x}^* H(Y) K_{0,x}$ corresponds to the event of no-absorption, consisting of the associated conditional transformation $K_{0,x}$ followed by the subspace component $\mathsf{H}(Y)$ of the second measurement. Finally, $I_x(Y)$ is the interference term for the coherences arising from the two choices, the coherence of the second measurement modified by the no-absorption transform of the first measurement, and absorption in the first measurement. In particular, the interference term only appears if the second measurement is coherent.

As a by-product, we now obtain a new family of observables in $\mathcal{H}$ jointly measurable with the original observable E we started with. Indeed, by Proposition 6, the subspace observable of $\tilde{\mathsf{F}}$ must be jointly measurable with E since the latter is the subspace observable of the first marginal $\tilde{\mathsf{E}}$. Explicitly, the observable

$$\mathsf{F}'(Y) := \int_\Omega K_x^* \left(q(Y)\mu_x|\varphi_x\rangle\langle\varphi_x| + K_{0,x}^* \mathsf{H}(Y) K_{0,x} + I_x(Y) \right) K_x d\mu(x)$$

is jointly measurable with E, and the joint measurement is the G constructed above. This describes the adjustment in the usual sequential Lüders measurement due to the absorption process as just described.

7. Example: unsharp position and momentum

In this final section we illustrate the use of the above absorption-supplemented measurement model, in the case of a standard model position measurement followed by the sharp momentum measurement.

7.1. Supplementing the standard model of position measurement with rank one absorption

Let $\mathcal{H} = L^2(\mathbb{R})$, and Q the usual position operator, with spectral measure E^Q. Since Q is projection valued, it cannot be extended coherently into $\tilde{\mathcal{H}}$ (see Corollary 2), that is, the only way possible extensions are of the form

$$\tilde{\mathsf{E}}^Q(X) = \begin{pmatrix} p(X) & 0 \\ 0 & \mathsf{E}^Q(X) \end{pmatrix},$$

where p is some probability measure. However, realistic position measurements are unsharp, i.e., not projection valued, which makes it possible to extend them using the above procedure.

Fix a function $f \in L^2(\mathbb{R})$, and assume that f is (essentially) bounded. We consider a position measurement of the form (7) where μ is the Lebesgue measure on the real line $\Omega = \mathbb{R}$, and $K_x = f(Q-x)$ is the operator of multiplication by the bounded function $y \mapsto f(y-x)$, that is,

$$\mathcal{I}(X)(A) = \int_X f(Q-x)^* A f(Q-x) dx. \tag{9}$$

Such an instrument is obtained, for instance, from the *standard model* of position measurement (again we refer to [15] and the references therein). This measurement approximately localizes the particle in the sense that the post-measurement state conditional on the outcome x has the wave function proportional to $y \mapsto f(y-x)\psi(y)$ (which is in $L^2(\mathbb{R})$ for almost all x). In particular, if f peaks at zero, the conditional state after the measurement yielding value x is concentrated around x in the position space, indicating that the outcome is likely to occur there. The measured observable

$$\mathsf{E}(X) = \int_X f(Q-x)^* f(Q-x) dx \tag{10}$$

is called an unsharp position observable.

We can now supplement this measurement with rank one absorption. We only consider the simple case in which the normalisation is ensured by taking the coherence of the approximating position observable to be proportional to a single state χ. This requires the condition $\chi \in \operatorname{ran} \sqrt{K_x^* K_x}$ for all $x \in \mathbb{R}$, which reads as

$$\lambda(E_x, \chi)^{-1} = \|E_x^{-\frac{1}{2}} \chi\|^2 = \int_{-\infty}^{\infty} \frac{|\chi(y)|^2}{|f(y-x)|^2} dy < \infty.$$

In order to illustrate this, we consider the Gaussian case, with $\chi(x) = \left(\frac{a}{\pi}\right)^{\frac{1}{4}} e^{-ax^2/2}$, and $f(x) = \left(\frac{\lambda}{\pi}\right)^{\frac{1}{4}} e^{-\lambda x^2/2}$, where $a > 0$ and $\lambda > 0$ can be adjusted to change the spreads of χ and the localization operators. We recall that in the standard model of position measurement, λ can be associated to the scaling of the pointer function of the measurement device [15]. In our case the supports are full, and the possibility of absorption is determined by the relative spread

$$\lambda(E_x, \chi)^{-1} = \sqrt{\frac{a}{\lambda}} \int_{-\infty}^{\infty} e^{-ay^2 + \lambda(y-x)^2} dy = e^{\lambda x^2} \sqrt{\frac{a}{\lambda}} \int_{-\infty}^{\infty} e^{-(a-\lambda)y^2 - 2\lambda xy} dy = \frac{\sqrt{\pi r}}{\lambda} e^{rx^2},$$

where $r := \lambda a (a-\lambda)^{-1}$. Clearly, $\lambda(E_x, \chi)^{-1} < \infty$ exactly when $a > \lambda$. Hence, the spread of χ must be strictly smaller than the spread of the localisation function f of the measurement, to ensure that χ is in the range of each $f(Q-x)$. Next, we need to modulate χ with an x-dependent factor to fulfil the condition for positivity: we define $\eta_x = g(x)\chi$, and choose the coherence function $g(x)$ so that $\lambda(E_x, \eta_x) \geq 1$, that is, $|g(x)|^2 \leq \lambda(E_x, \chi)$, which reads

$$|g(x)|^2 \leq \frac{\lambda}{\sqrt{\pi r}} e^{-rx^2}, \qquad \text{for all } x \in \mathbb{R}.$$

In order to achieve maximal coherence (and also to simplify the result), we choose $g(x) = \lambda^{\frac{1}{2}} (\pi r)^{-\frac{1}{4}} e^{-\frac{1}{2} r x^2}$, so that $\lambda(E_x, \eta_x) \geq 1$ for each x. Finally, we need to ensure normalisation, choosing u_x so that $\int \psi_x dx = 0$; this condition now reads $\int u_x g(x) dx = 0$. Since g is even, any odd function $x \mapsto u_x$ with $\int |u_x|^2 dx = 1$ works here. Perhaps the simplest choice which combines easily with g is $u_x = (\pi/r)^{-\frac{1}{4}} e^{-rx^2/2} \operatorname{sign}(x)$, leading to

$$\psi_x = u_x \eta_x = \sqrt{\lambda/\pi} e^{-rx^2} \operatorname{sign}(x) \chi.$$

Note that due to maximal coherence, $K_{0,x} = \mathbb{I} - |\varphi_x\rangle\langle\varphi_x|$ is a projection. We can now find the absorption "sink" unit vector φ_x given by the condition $K_x^* \varphi_x = \eta_x$; explicitly,

$$\varphi_x(y) = (a\lambda/(r\pi))^{\frac{1}{4}} e^{-\frac{1}{2}\frac{\lambda a}{r}\left(y + \frac{r}{a}x\right)^2} = \varphi_0\left(y + \frac{r}{a}x\right),$$

that is, $\varphi_x = U_x \varphi_0$ where U_x is the scaled translation semigroup. Hence, the sink associated to outcome x is obtained from the sink at $x = 0$ by translation. Next, we define $B_x := U_x^* f(Q-x)$, which is a combination of localisation around x and translation by x. This allows us to write

$$g(x)\chi = \eta_x = B_x^* \varphi_0, \qquad \text{for all } x. \tag{11}$$

Note also that $B_x^* B_x = f(Q-x)^2$. Next, letting $R := \mathbb{I} - |\varphi_0\rangle\langle\varphi_0|$ we have $K_{0,x} K_x = U_x R B_x$, and we can write the instrument as

$$\tilde{\mathcal{I}}(X)(\tilde{A}) = \int_X \begin{pmatrix} u_x & 0 \\ |g(x)\chi\rangle & B_x^* R U_x^* \end{pmatrix} \tilde{A} \begin{pmatrix} u_x & \langle g(x)\chi| \\ 0 & U_x R B_x \end{pmatrix} dx, \tag{12}$$

and the associated coherent extension of the position observable (10) is given explicitly by

$$\tilde{\mathsf{E}}(X) = \int_X \begin{pmatrix} |u_x|^2 & u_x g(x)\langle\chi| \\ u_x g(x)|\chi\rangle & f(Q-x)^2 \end{pmatrix} dx = \left(\frac{\lambda}{\pi}\right)^{\frac{1}{2}} \int_X \begin{pmatrix} (r/\lambda)^{\frac{1}{2}} & \mathrm{sign}(x)\langle\chi| \\ \mathrm{sign}(x)|\chi\rangle & e^{-\lambda(Q-x)^2} e^{rx^2} \end{pmatrix} e^{-rx^2}\, dx. \tag{13}$$

One can easily check directly that this is indeed an observable whose compression to $\mathcal{H}$ (i.e., the lower right block of the matrix) is (10).

7.2. Approximate joint measurements of extended position and momentum

A straightforward way of implementing approximate joint measurements of position and momentum is their sequential combination. In particular, measuring first an unsharp position using the instrument (9), and then subsequently a sharp momentum observable $\mathsf{E}^P(Y)$, yields the joint observable

$$\mathsf{G}(X\times Y) = \int_X f(Q-x)\mathsf{E}^P(Y)f(Q-x)dx$$

whose marginal observables (10) and

$$\mathsf{F}(Y) := \int f(Q-x)\mathsf{E}^P(Y)f(Q-x)dx \tag{14}$$

are unsharp versions of position and momentum [15]. From the point of view of the extended Hilbert space, any extensions of these observables are therefore (by definition) subspace compatible. We now proceed to find a "distorted" momentum observable, whose incoherent extension is not only subspace compatible with (10), but also compatible with the above constructed coherent extension (13).

A natural way to do this is to apply the above instrument to an extension of the sharp momentum observable E^P. Due to being sharp, the latter can only be extended incoherently, and if we wish to keep the extension sharp, there is only one option:

$$\tilde{\mathsf{E}}^P(Y) = \begin{pmatrix} 1 & 0 \\ 0 & \mathsf{E}^P(Y) \end{pmatrix}.$$

Now, measuring first $\tilde{\mathsf{E}}$ using the above constructed instrument (12), and then measuring the sharp observable $\tilde{\mathsf{E}}^P$ gives the joint observable $\tilde{\mathsf{G}}(X\times Y) = \tilde{\mathcal{I}}(X)(\tilde{\mathsf{E}}^P(Y))$, whose first marginal is $\tilde{\mathsf{E}}$. It is interesting to find the second marginal $\tilde{\mathsf{F}}$, which is then by construction jointly measurable with $\tilde{\mathsf{E}}$. First, we compute the joint observable

$$\begin{aligned}
\tilde{\mathsf{G}}(X\times Y) &:= \int_X \begin{pmatrix} u_x & 0 \\ |g(x)\chi\rangle & B_x^* R U_x^* \end{pmatrix} \begin{pmatrix} 1 & 0 \\ 0 & \mathsf{E}^P(Y) \end{pmatrix} \begin{pmatrix} u_x & \langle g(x)\chi| \\ 0 & U_x R B_x \end{pmatrix} dx \\
&= \int_X \begin{pmatrix} u_x & 0 \\ |g(x)\chi\rangle & B_x^* R U_x^* \end{pmatrix} \begin{pmatrix} u_x & \langle g(x)\chi| \\ 0 & \mathsf{E}^P(Y) U_x R B_x \end{pmatrix} dx \\
&= \int_X \begin{pmatrix} u_x^2 & \langle u_x g(x)\chi| \\ |u_x g(x)\chi\rangle & g(x)^2|\chi\rangle\langle\chi| + B_x^* R U_x^* E^P(Y) U_x R B_x \end{pmatrix} dx \\
&= \left(\frac{\lambda}{\pi}\right)^{\frac{1}{2}} \int_X \begin{pmatrix} (r/\lambda)^{\frac{1}{2}} & \mathrm{sign}(x)\langle\chi| \\ \mathrm{sign}(x)|\chi\rangle & M_x(Y) \end{pmatrix} e^{-rx^2}\, dx,
\end{aligned}$$

where

$$M_x(Y) = \sqrt{\frac{\lambda}{r}}|\chi\rangle\langle\chi| + e^{rx^2}\sqrt{\frac{\pi}{\lambda}} B_x^* R \mathsf{E}^P(Y) R B_x,$$

Journal of Physics: Conference Series **1638** (2020) 012003 doi:10.1088/1742-6596/1638/1/012003

and we have used the fact that U_x commutes with $\mathsf{E}^P(Y)$. In particular, the joint observable is coherent, with the same coherences as $\tilde{\mathsf{E}}$, and its subspace observable is

$$\mathsf{G}(X \times Y) := \left(\frac{\lambda}{\pi}\right)^{\frac{1}{2}} \int_X M_x(Y) e^{-rx^2} dx = \int_X B_x^* (|\varphi_0\rangle\langle\varphi_0| + R\mathsf{E}^P(Y)R) B_x dx.$$

Now $\mathsf{G}(X \times \mathbb{R}) = \mathsf{E}(X)$, the standard position observable, but the other marginal is different from (14), and given instead by

$$\mathsf{F}'(Y) = \int_{\mathbb{R}} B_x^* \left(|\varphi_0\rangle\langle\varphi_0| + R\mathsf{E}^P(Y)R\right) B_x dx.$$

We now observe how the structure of the joint observable reflects the detection scheme—the momentum measurement is in fact performed only conditional on the no-absorption event (represented by the projection R), in accordance with the obvious intuition: if the particle gets absorbed during the position measurement, it cannot experience the following momentum measurement. Finally, we note that the marginal of the extended joint observable is

$$\tilde{\mathsf{F}}(Y) := \tilde{\mathsf{G}}(\mathbb{R} \times Y) = \begin{pmatrix} 1 & 0 \\ 0 & \mathsf{F}'(Y) \end{pmatrix},$$

that is, the momentum observable is distorted, but remains incoherent.

References

[1] Heisenberg W 1927 *Z. Phys.* **43** 172–198
[2] Bohr N 1935 *Phys. Rev.* **48** 696–702
[3] Ludwig G 1964 *Z. Phys.* **181** 233–260
[4] Busch P, Heinosaari T, Schultz J and Stevens N 2013 *EPL* **103** 10002
[5] Wolf M, Perez-Garcia D and Fernandez C 2009 *Phys. Rev. Lett.* **103** 230402
[6] Uola R, Moroder T and Gühne O 2014 *Phys. Rev. Lett.* **113** 160403
[7] Quintino M, Vertesi T and Brunner N 2014 *Phys. Rev. Lett.* **113** 160402
[8] Uola R, Budroni C, Gühne O and Pellonpää J-P 2015 *Phys. Rev. Lett.* **115** 230402
[9] Heinosaari T, Kiukas J, Reitzner D and Schultz J 2015 *J. Phys. A: Math. Theor.* **48** 435301
[10] Kiukas J, Budroni C, Uola R and Pellonpää J-P 2017 *Phys. Rev.* A **96** 042331
[11] Carmeli C, Heinosaari T and Toigo A 2019 *Phys. Rev. Lett.* **122** 130402
[12] Skrzypczyk P, Supic I and Cavalcanti D 2019 *Phys. Rev. Lett.* **122** 130403
[13] Ludwig G 1985 *An Axiomatic Basis for Quantum Mechanics I* (Berlin: Springer)
[14] Werner R 1986 *J. Math. Phys.* **27** 793–803
[15] Busch P, Lahti P, Pellonpää J-P and Ylinen K 2016 *Quantum Measurement* (Berlin: Springer)
[16] Twareque Ali S, Carmeli C, Heinosaari T and Toigo A 2009 *Found. Phys.* **39** 593–612
[17] Werner R F 1989 *Phys. Rev.* A **40** 4277–81
[18] Kiukas J, Lahti P, Pellonpää J-P and Ylinen K 2019 *Found. Phys.* **49** 506–31
[19] Durt T, Englert B-G, Bengtsson I and Życzkowski K 2010 *Int. J. Quantum Inf.* **8** 535–640
[20] Busch P and Gudder S 1999 *Lett. Math. Phys.* **47** 329–37
[21] Busch P 1985 Can quantum theoretical reality be considered sharp? *Recent Developments in Quantum Logic* ed P Mittelstaedt and E-W Stachow (Mannheim: Bibliographisches Institut) pp 81–101
[22] Douglas R 1966 *Proc. Am. Math. Soc.* **17** 413–6
[23] Boyd S and Vandenberghe L 2004 *Convex Optimization* (Cambridge: Cambridge University Press)
[24] Paulsen V 2002 *Completely Bounded Maps and Operator Algebras* (Cambridge: Cambridge University Press)
[25] Diestel J and Uhl Jr J J 1977 *Vector Measures* (Mathematical Surveys vol 15) (Providence, RI: American Mathematical Society)
[26] Haapasalo E and Pellonpää J-P 2017 *J. Math. Phys.* **58** 122104

Mathematical Foundations of Quantum Mechanics - *in memoriam* Paul Busch IOP Publishing
Journal of Physics: Conference Series **1638** (2020) 012004 doi:10.1088/1742-6596/1638/1/012004

Pairwise coexistence of effects versus coexistence

Werner Stulpe
Aachen University of Applied Sciences, Jülich Campus, D-52428 Jülich, Germany
E-mail: stulpe@fh-aachen.de

Abstract. The concept of coexistence of quantum mechanical effects is reviewed. We distinguish between coexistence and pairwise coexistence and give an example showing the non-equivalence the two concepts. A theorem on some closures of sets of pairwise coexistent effects is proved. Some proofs of the well-known fact that any set of pairwise coexistent (mutually compatible) sharp effects is coexistent are considered. In particular, a proof is presented that is based on the statement that every countably complete Boolean lattice of sharp effects is closed in some topology.

1. Introduction

One of the most crucial facts of quantum mechanics is the existence of observables that cannot be measured jointly. This issue has, from the early days of the theory up to the present, been investigated extensively (see, e.g., [1–3]). As is well known, the notion of an observable can be traced back to that of simple observables with only two outcomes, the effects. We review the concepts for the joint measurability of simple observables and present some results the author did not find elsewhere.

In quantum mechanics, an *effect* is described by and identified with a positive self-adjoint operator A acting in a separable complex Hilbert space $\mathcal{H}$ satisfying $\|A\| \leq 1$, i.e., $0 \leq A \leq I$, I being the unit operator. The set $\mathcal{E}(\mathcal{H})$ of all effects is a closed convex subset of the real Banach space $\mathcal{B}_s(\mathcal{H})$ of all bounded self-adjoint operators. It is well known that the extreme points of $\mathcal{E}(\mathcal{H})$ are the orthogonal projections P; an effect $P \in \partial_e\mathcal{E}(\mathcal{H})$, $\partial_e\mathcal{E}(\mathcal{H})$ denoting the extreme boundary of $\mathcal{E}(\mathcal{H})$, is called *sharp* whereas the other effects are called *unsharp*.[1]

The *states* are given by the density operators W; W is a positive self-adjoint trace-class operator of trace 1. The real vector space $\mathcal{T}_s(\mathcal{H})$ of all self-adjoint trace-class operators is, equipped with the trace norm, also a real Banach space. The set $\mathcal{S}(\mathcal{H})$ of all density operators is a closed convex subset of $\mathcal{T}_s(\mathcal{H})$ whose extreme points are the one-dimensional orthogonal projections $P_\psi := |\psi\rangle\langle\psi|$; a state $P_\psi \in \mathcal{S}(\mathcal{H})$ is called *pure*. The *probability* for the occurrence of an effect $A \in \mathcal{E}(\mathcal{H})$ (respectively, for 1 where the two outcomes of A are 0 and 1) in a state $W \in \mathcal{S}(\mathcal{H})$ is given by $\operatorname{tr} WA$.

According to the bilinear functional $(V, A) \mapsto \operatorname{tr} VA$ where $V \in \mathcal{T}_s(\mathcal{H})$ and $A \in \mathcal{B}_s(\mathcal{H})$ are arbitrary, $\mathcal{B}_s(\mathcal{H})$ is isometrically isomorphic to the dual space $\mathcal{T}_s(\mathcal{H})'$. Thus, the weak-* topology $\sigma(\mathcal{B}_s(\mathcal{H}), \mathcal{T}_s(\mathcal{H}))$ is defined. We call this topology briefly the *σ-topology*; it is weakest topology such that the elements of $\mathcal{T}_s(\mathcal{H})$, considered as linear functionals on $\mathcal{B}_s(\mathcal{H})$, are continuous. Since

[1] The terms "effect" and "coexistence" were introduced by Ludwig [4–7], whereas the terms "sharp" and "unsharp" for effects and observables were introduced by Busch. Ludwig calls a sharp effect a *decision effect.*

Journal of Physics: Conference Series **1638** (2020) 012004 doi:10.1088/1742-6596/1638/1/012004

$\sigma(\mathcal{B}_s(\mathcal{H}), \mathcal{T}_s(\mathcal{H})) = \sigma(\mathcal{B}_s(\mathcal{H}), \mathcal{S}(\mathcal{H}))$, a neighborhood base of $A \in \mathcal{B}_s(\mathcal{H})$ is given by the open sets

$$U(A; W_1, \ldots, W_n; \varepsilon) := \left\{ \widetilde{A} \in \mathcal{B}_s(\mathcal{H}) \,\middle|\, |\mathrm{tr}\, W_j \widetilde{A} - \mathrm{tr}\, W_j A| < \varepsilon \text{ for } j = 1, \ldots, n \right\} \tag{1}$$

where $W_j \in \mathcal{S}(\mathcal{H})$ and $\varepsilon > 0$. Note that the σ-topology can be interpreted as the topology of the physical approximation of effects. Namely, an effect $A \in \mathcal{E}(\mathcal{H})$ is physically approximated by $\widetilde{A} \in \mathcal{E}(\mathcal{H})$ if in many (but finitely many) states W_j the probabilities $\mathrm{tr}\, W_j \widetilde{A}$ differ from $\mathrm{tr}\, W_j A$ by an amount less than a small $\varepsilon > 0$. This statement can be tested experimentally and can be characterized mathematically by $\widetilde{A} \in U(A; W_1, \ldots, W_n; \varepsilon)$.

The σ-topology is the topology on $\mathcal{B}_s(\mathcal{H})$ that is induced by the ultraweak operator topology on $\mathcal{B}_s(\mathcal{H})$, i.e., $\sigma(\mathcal{B}_s(\mathcal{H}), \mathcal{T}_s(\mathcal{H})) = \sigma(\mathcal{B}(\mathcal{H}), \mathcal{T}(\mathcal{H})) \cap \mathcal{B}_s(\mathcal{H})$, $\mathcal{B}(\mathcal{H})$ and $\mathcal{T}(\mathcal{H})$ denoting the complex Banach spaces of all bounded, resp., trace-class operators. The weak operator topology on $\mathcal{B}_s(\mathcal{H})$ is just the topology $\sigma(\mathcal{B}_s(\mathcal{H}), \partial_e \mathcal{E}(\mathcal{H})))$. On bounded sets the σ-topology and the weak operator topology coincide. This can be concluded by means of the neighborhoods (1) and the corresponding neighborhood base of the weak operator topology or, alternatively, by means of the following compactness argument. According to the Banach-Alaoglu theorem, the closed unit ball of $\mathcal{B}_s(\mathcal{H})$, $B_{\mathcal{B}_s(\mathcal{H})} := \{A \in \mathcal{B}_s(\mathcal{H}) \mid \|A\| \leq 1\}$, is σ-compact. The weak operator topology is Hausdorff and weaker than the σ-topology, hence, on $B_{\mathcal{B}_s(\mathcal{H})}$ both topologies must be equal.

The contemporary conception of an *observable* is that of a normalized effect-valued measure $\boldsymbol{F}$ on some measurable space (M, Ξ) where M is interpreted to be the *value space* of $\boldsymbol{F}$, i.e., a space encompassing the possible measuring values. In particular, in quantum mechanics $\boldsymbol{F}$ is a normalized positive-operator-valued measure on (M, Ξ), i.e., a map $\boldsymbol{F}\colon \Xi \to \mathcal{E}(\mathcal{H})$, $B \mapsto \boldsymbol{F}(B)$, satisfying $\boldsymbol{F}(\emptyset) = 0$, $\boldsymbol{F}(M) = I$, and $\boldsymbol{F}(\bigcup_{j=1}^{\infty} B_j) = \sum_{j=1}^{\infty} B_j$ where the measurable sets B_j are mutually disjoint and the sum converges in the σ-topology, for instance. The set $R(\boldsymbol{F}) := \boldsymbol{F}(\Xi) = \{\boldsymbol{F}(B) \mid B \in \Xi\}$ is the range of $\boldsymbol{F}$. An observable $\boldsymbol{E}\colon \Xi \to \mathcal{E}(\mathcal{H})$ is called *sharp* if $R(\boldsymbol{E}) \subseteq \partial_e \mathcal{E}(\mathcal{H})$ and can be considered as a map $\boldsymbol{E}\colon \Xi \to \partial_e \mathcal{E}(\mathcal{H})$, otherwise an observable is called *unsharp*.

In Section 2, we recall the concept of the coexistence of a set of effects and define the pairwise coexistence of a set of effects. We show that a set of pairwise coexistent effects is not necessarily coexistent and prove some closure properties of pairwise coexistence. In Section 3 we give a proof of the well-known fact that two sharp effects are coexistent if and only if they commute; two coexistent sharp effects are also called compatible. Furthermore, following Gudder ([1], [8, pp. 53–129]), we use the concept of the compatant of a set of sharp effects to show that every set of mutually compatible sharp effects is contained in some complete Boolean sublattice of $\partial_e \mathcal{E}(\mathcal{H})$.

It is well known that any set of mutually compatible sharp effects is contained in the range of a (sharp) observable. That is, according to the definition in Section 2, any set of pairwise coexistent sharp effects is coexistent, in contrast to the case of arbitrary effects. The statement of the penultimate sentence can quickly be proved as follows. Let S be a set of mutually compatible sharp effects. Then S is a set of commuting orthogonal projections of the separable Hilbert space $\mathcal{H}$ and, according to a theorem of von Neumann [9, 10], there exists a (bounded) self-adjoint operator A acting in $\mathcal{H}$ such that all $P \in S$ are functions of A, i.e.,

$$P = f_P(A) = \int f_P \, \mathrm{d}\boldsymbol{E} = \int f_P(\lambda) \, \boldsymbol{E}(\mathrm{d}\lambda) = \int \lambda \, (\boldsymbol{E} \circ f_P^{-1})(\mathrm{d}\lambda) \tag{2}$$

where $f_P\colon \mathbb{R} \to \mathbb{R}$ is a bounded Borel-measurable function, $\boldsymbol{E}$ the spectral measure of A, and $(\boldsymbol{E} \circ f_P^{-1})(B) = \boldsymbol{E}(f_P^{-1}(B))$, $B \subseteq \mathbb{R}$ being a Borel set. From the representation of P by the last integral of (2) it follows that $\boldsymbol{E} \circ f_P^{-1}$ is the spectral measure of the projection P and therefore $P = \boldsymbol{E}(f_P^{-1}(\{1\})) \in R(\boldsymbol{E})$. Hence, every $P \in S$ is in the range of the sharp observable $\boldsymbol{E}$ and

$S \subseteq R(\boldsymbol{E})$. Alternatively, this can be concluded from $\int f_P \, \mathrm{d}\boldsymbol{E} = P = P^2 = \int f_P^2 \, \mathrm{d}\boldsymbol{E}$ which implies that $f_P = f_P^2$ $\boldsymbol{E}$-almost everywhere. Thus, f_P can be chosen to be the characteristic function χ_B of some Borel set B which entails that $P = \int \chi_B \, \mathrm{d}\boldsymbol{E} = \boldsymbol{E}(B) \in R(\boldsymbol{E})$.

Another proof of $S \subseteq R(\boldsymbol{E})$, S as in the preceding paragraph, is due to Varadarajan given in the context of quantum logic (see [11]). This proof is based on a representation theorem of Loomis [12] which states that, given a countably complete Boolean lattice $\mathcal{L}$ (Boolean σ-algebra), there exists a measurable space (M, Ξ) and a σ-homomorphism of Ξ onto $\mathcal{L}$.

Since both the mentioned theorem of von Neumann, as well as the theorem of Loomis, are considerably deep and, moreover, the latter is beyond the scope of Hilbert space, it is worth presenting a proof of $S \subseteq R(\boldsymbol{E})$, S as before, that uses only the theory of Hilbert space and bounded linear operators acting in Hilbert space and does not presuppose von Neumann's theorem. This is performed in Sections 4 and 5. It turns out that even such a proof is quite involved so that, against the author's original plan, further topics concerning the pairwise coexistence of effects cannot be included in this paper.

In Section 4 we prove several statements on countably complete Boolean sublattices of $\partial_e\mathcal{E}(\mathcal{H})$ which prepare the proof of $S \subseteq R(\boldsymbol{E})$, S as before. The central point is the corollary of Theorem 4 which establishes that every countably complete Boolean sublattice $\mathcal{L}$ of $\partial_e\mathcal{E}(\mathcal{H})$ is closed in the topology on $\partial_e\mathcal{E}(\mathcal{H})$ that is induced by the σ-topology. This property of $\mathcal{L}$ and its σ-separability entail that $\mathcal{L}$ is countably generated, and this result enables the envisaged proof which is given in Section 5. Much of the subject matter of Sections 4 and 5 is a review.

2. Coexistence

According to Ludwig [2, 7], we call two effects $F, G \in \mathcal{E}(\mathcal{H})$ *coexistent* if there exist effects $F', G', H \in \mathcal{E}(\mathcal{H})$ such that

$$\begin{aligned} F &= F' + H \\ G &= G' + H \end{aligned} \tag{3}$$

and $F' + G' + H \leq I$. The decomposition (3) is in general not unique. It is easy to see that two effects are coexistent if and only if they are contained in the range of some observable $\boldsymbol{F}$, i.e., $F, G \in R(\boldsymbol{F}) = \boldsymbol{F}(\Xi)$. So two coexistent effects can be measured jointly by one measuring device. We notice the following simple statements:

(a) The trivial effects 0 and I are coexistent with every $F \in \mathcal{E}(\mathcal{H})$.
(b) The pair (F, F), $F \in \mathcal{E}(\mathcal{H})$, is coexistent.
(c) Any $F, G \in \mathcal{E}(\mathcal{H})$ satisfying $F \leq G$ are coexistent.
(d) The effects F and G are coexistent if $F + G \in \mathcal{E}(\mathcal{H})$.
(e) Any $F, G \in \frac{1}{2}\mathcal{E}(\mathcal{H}) = \frac{1}{2}[0, I] = \{A \in \mathcal{E}(\mathcal{H}) \,|\, 0 \leq A \leq \frac{1}{2}I\}$ are coexistent.
(f) If $F, G \in \mathcal{E}(\mathcal{H})$ are coexistent, then F and $I - G$ also.

Corresponding to the interpretation of the coexistence of effects as their joint measurability by one measuring device, a set S is called *coexistent* if S is contained in the range of some observable $\boldsymbol{F}$, i.e., $S \subseteq R(\boldsymbol{F})$. We call $S \subseteq \mathcal{E}(\mathcal{H})$ *pairwise coexistent* if any two effects $F, G \in S$ are coexistent. Clearly, coexistence of set of effects implies pairwise coexistence, but the converse does in general not hold, as an example of a set of three effects shows. The following simple lemma prepares this example and is also useful for later considerations.

Lemma 1. *If $A \in \mathcal{E}(\mathcal{H})$ and $P \in \partial_e\mathcal{E}(\mathcal{H})$ are related by $A \leq P$, then $A = PA = AP$.*

Proof. For $\chi \in R(P)^\perp$ where $R(P)$ denotes the range $P\mathcal{H}$ of P, we have that

$$\|A^{\frac{1}{2}}\chi\|^2 = \langle\chi|A\chi\rangle \leq \langle\chi|P\chi\rangle = 0$$

and so $A\chi = 0$. For $\varphi \in R(P)$ it then follows, for all $\chi \in R(P)^\perp$, that

$$\langle \chi | A\varphi \rangle = \langle A\chi | \varphi \rangle = 0,$$

i.e., $A\varphi \in R(P)$. Hence, for any $\psi \in \mathcal{H}$, $\psi = \varphi + \chi$ with $\varphi \in R(P)$ and $\chi \in R(P)^\perp$, we obtain

$$A\psi = A\varphi = PA\varphi = PA\varphi + PA\chi = PA\psi.$$

Thus, $A = PA$ and $A = A^* = AP$. □

Now let $\varphi_1, \varphi_2 \in \mathcal{H}$ be two orthogonal unit vectors. Define the vectors

$$\psi_1 := \varphi_1, \qquad \psi_2 := \frac{1}{2}\sqrt{3}\varphi_1 + \frac{1}{2}\varphi_2, \qquad \psi_3 := \frac{1}{2}\varphi_1 + \frac{1}{2}\sqrt{3}\varphi_2$$

and the effects

$$G_j := \frac{1}{2}P_{\psi_j} = \frac{1}{2}|\psi_j\rangle\langle\psi_j|,$$

$j = 1, 2, 3$. According to the above statement (e), the set $S := \{G_1, G_2, G_3\}$ is pairwise coexistent. We show that, however, S is not coexistent. An easy calculation yields

$$\langle \psi_2 | (G_1 + G_2 + G_3)\psi_2 \rangle = \frac{5}{4} > 1,$$

consequently, $G_1 + G_2 + G_3 \not\leq I$. Assume S were coexistent, i.e., there would exist an observable an observable $\boldsymbol{F}$ and sets $B_j \in \Xi$ such that $G_j = \boldsymbol{F}(B_j)$. We obtain that, for $j \neq k$,

$$\boldsymbol{F}(B_j \cap B_k) \leq \boldsymbol{F}(B_j) = \frac{1}{2}P_{\psi_j}$$
$$\boldsymbol{F}(B_j \cap B_k) \leq \boldsymbol{F}(B_k) = \frac{1}{2}P_{\psi_k}.$$

So $2\boldsymbol{F}(B_j \cap B_k) \leq P_{\psi_j}$, and by means of the lemma it follows that $\boldsymbol{F}(B_j \cap B_k) = P_{\psi_j}\boldsymbol{F}(B_j \cap B_k) = P_{\psi_j}\boldsymbol{F}(B_j \cap B_k)P_{\psi_j} \leq \frac{1}{2}P_{\psi_j}$. Thus, $\boldsymbol{F}(B_j \cap B_k) = \lambda_{jk}P_{\psi_j}$ with some $\lambda_{jk} \leq \frac{1}{2}$; analogously, $\boldsymbol{F}(B_j \cap B_k) = \mu_{jk}P_{\psi_k}$ with some $\mu_{jk} \leq \frac{1}{2}$. In particular, we have that $\lambda_{jk}P_{\psi_j} = \mu_{jk}P_{\psi_k}$ which entails $\lambda_{jk} = \mu_{jk} = 0$ since ψ_j and ψ_k are linearly independent. Hence, $\boldsymbol{F}(B_j \cap B_k) = 0$ for all $j \neq k$, in consequence, $\boldsymbol{F}(B_1 \cap B_2 \cap B_3) = 0$ also. Therefore,

$$\begin{aligned} \boldsymbol{F}(B_1 \cup B_2 \cup B_3) &= \boldsymbol{F}(B_1) + \boldsymbol{F}(B_2) + \boldsymbol{F}(B_3) - \boldsymbol{F}(B_1 \cap B_2) - \boldsymbol{F}(B_2 \cap B_3) - \boldsymbol{F}(B_3 \cap B_1) \\ &\qquad + \boldsymbol{F}(B_1 \cap B_2 \cap B_3) \\ &= G_1 + G_2 + G_3 \\ &\not\leq I \end{aligned}$$

which contradicts $\boldsymbol{F}(B_1 \cup B_2 \cup B_3) \leq I$. We have shown that pairwise coexistence does not imply coexistence.[2]

From $\|A\| = \sup_{\|\varphi \leq 1\|} |\langle \varphi | A\varphi \rangle|$ it follows that the unit ball $B_{\mathcal{B}_s(\mathcal{H})} = \{A \in \mathcal{B}_s(\mathcal{H}) \mid \|A\| \leq 1\}$ is equal to the order-unit interval $[-I, I]$. Therefore, $\mathcal{E}(\mathcal{H}) = [0, I] = \frac{1}{2}([-I, I] + I) = \frac{1}{2}(B_{\mathcal{B}_s(\mathcal{H})} + I)$, and the σ-compactness of $B_{\mathcal{B}_s(\mathcal{H})}$ entails that of the convex set $\mathcal{E}(\mathcal{H})$. The latter implies that, according to the Krein-Milman theorem, $\mathcal{E}(\mathcal{H}) = \overline{\operatorname{conv} \partial_e \mathcal{E}(\mathcal{H})}^\sigma$ where the extreme boundary consists of the sharp effects. Moreover, even the stronger statement $\mathcal{E}(\mathcal{H}) = \overline{\partial_e \mathcal{E}(\mathcal{H})}^\sigma$ is true,

[2] This example was, without the details of the proof, already published in the author's paper [13, p. 12]. A similar example is given in [14, Theorem 1].

provided that $\dim \mathcal{H} = \infty$ [15, pp. 20–21]. From this we can further conclude that the norm-topological boundary $\partial\mathcal{E}(\mathcal{H}) = \frac{1}{2}(S_{\mathcal{B}_s(\mathcal{H})} + I)$, $S_{\mathcal{B}_s(\mathcal{H})}$ being the unit sphere $S_{\mathcal{B}_s(\mathcal{H})} := \{A \in \mathcal{B}_s(\mathcal{H}) \mid \|A\| = 1\}$, is σ-dense in $\mathcal{E}(\mathcal{H})$; alternatively, $\overline{\partial\mathcal{E}(\mathcal{H})}^{\sigma} = \mathcal{E}(\mathcal{H})$ follows from the fact that $S_{\mathcal{B}_s(\mathcal{H})}$ is σ-dense in $B_{\mathcal{B}_s(\mathcal{H})}$.

It is well known that, for a separable Banach space $\mathcal{V}$, the weak-$*$ topology on the closed unit ball $B_{\mathcal{V}'}$ of the dual space $\mathcal{V}'$, $\sigma(\mathcal{V}', \mathcal{V}) \cap B_{\mathcal{V}'}$, is second countable (a straightforward proof is given in [13]). So w.r.t. $\sigma(\mathcal{V}', \mathcal{V}) \cap B_{\mathcal{V}'}$, the topological concepts can be characterized using sequences. In particular, the weak-$*$ compactness of $B_{\mathcal{V}'}$ is equivalent to the weak-$*$ sequential compactness. Since our Hilbert space is separable, the space $\mathcal{T}_s(\mathcal{H})$ of the self-adjoint operators is separable, and the above remarks apply to the topology on the unit ball $B_{\mathcal{B}_s(\mathcal{H})}$ induced by our σ-topology. In consequence, the topology $\sigma \cap \mathcal{E}(\mathcal{H})$ is second countable as well, and $\mathcal{E}(\mathcal{H})$ is sequentially $\sigma \cap \mathcal{E}(\mathcal{H})$-compact and therefore also sequentially σ-compact.

The question arises what one can say about the convex hull and the σ-closure of a set S of pairwise coexistent effects. In particular, two effects $F, G \in \overline{S}^{\sigma} \setminus S$ can, in the σ-topology, be approximated by two coexistent effects of S. Since the σ-topology describes the physical approximation of effects by other ones, F and G should be coexistent. The following theorem confirms this expectation.

Theorem 1. *The convex hull and the σ-closure of a set S of pairwise coexistent effects are pairwise coexistent. In particular, the smallest σ-closed convex set containing S, $\overline{\operatorname{conv} S}^{\sigma}$, is pairwise coexistent.*

Proof. The convex hull of S consists of the finite convex linear combinations of elements of S. We have to show that any two effects $F = \sum_{j=1}^{m} \alpha_j F_j$ and $G = \sum_{k=1}^{n} \beta_k F_k$ where $F_j, G_k \in S$, $\alpha_j, \beta_k \geq 0$, and $\sum_{j=1}^{m} \alpha_j = \sum_{k=1}^{n} \beta_k = 1$, are coexistent. The effects F and G can be written in the form $F = \sum_{l=1}^{N} \delta_l \widetilde{F}_l$ and $G = \sum_{l=1}^{N} \delta_l \widetilde{G}_l$ where $\delta_l \geq 0$, $\sum_{l=1}^{N} \delta_l = 1$, each $\widetilde{F}_l$ equals some F_j, and each $\widetilde{G}_l$ equals some G_k (for more details of this part of the proof, see [13]). Now, from the coexistence of $\widetilde{F}_l$ and $\widetilde{G}_l$,

$$\widetilde{F}_l = A_{l1} + A_{l0}$$
$$\widetilde{G}_l = A_{l2} + A_{l0},$$

$A_{l1}, A_{l2}, A_{l0} \in \mathcal{E}(\mathcal{H})$, $A_{l1} + A_{l2} + A_{l0} \leq I$, we obtain that

$$F = \sum_{l=1}^{N} \delta_l A_{l1} + \sum_{l=1}^{N} \delta_l A_{l0} =: A_1 + A_0$$
$$G = \sum_{l=1}^{N} \delta_l A_{l2} + \sum_{l=1}^{N} \delta_l A_{l0} =: A_2 + A_0$$

where $A_1, A_2, A_0 \in \mathcal{E}(\mathcal{H})$ and $A_1 + A_2 + A_0 = \sum_{l=1}^{N} \delta_l (A_{l1} + A_{l2} + A_{l0}) \leq I$. Hence, F and G are coexistent.

Let $F, G \in \overline{S}^{\sigma}$. Then there exist sequences $\{F_n\}_{n\in\mathbb{N}}$, $\{G_n\}_{n\in\mathbb{N}}$ in S and sequences $\{F'_n\}_{n\in\mathbb{N}}$, $\{G'_n\}_{n\in\mathbb{N}}$, and $\{H_n\}_{n\in\mathbb{N}}$ in $\mathcal{E}(\mathcal{H})$ such that

$$\begin{aligned} F &= \sigma\text{-}\lim_{n\to\infty} F_n \\ G &= \sigma\text{-}\lim_{n\to\infty} G_n \end{aligned} \tag{4}$$

and

$$\begin{aligned} F_n &= F'_n + H_n \\ G_n &= G'_n + H_n \end{aligned} \tag{5}$$

where $F'_n + G'_n + H_n \leq I$. Due to the sequential σ-compactness of $\mathcal{E}(\mathcal{H})$, there exists a subsequence $\{F'_{n_j}\}_{j\in\mathbb{N}}$ of $\{F'_n\}_{n\in\mathbb{N}}$ converging to some $F' \in \mathcal{E}(\mathcal{H})$ in the σ-topology. We write $F'_{n(j)} := F'_{n_j}$. Further, there is a subsequence $\{H_{n(j(k))}\}_{k\in\mathbb{N}}$ of $\{H_{n(j)}\}_{j\in\mathbb{N}}$ such that $H_{n(j(k))} \xrightarrow{\sigma} H$ as $k \to \infty$; finally, there is a subsequence $\{G'_{n(j(k(l)))}\}_{l\in\mathbb{N}}$ of $\{G'_{n(j(k))}\}_{k\in\mathbb{N}}$ with $G'_{n(j(k(l)))} \xrightarrow{\sigma} G'$ as $l \to \infty$. Abbreviating $n(j(k(l)))$ simply by n_l, Eqs. (5) imply that

$$\begin{aligned} F_{n_l} &= F'_{n_l} + H_{n_l} \\ G_{n_l} &= G'_{n_l} + H_{n_l} \end{aligned} \tag{6}$$

where $F'_{n_l} + G'_{n_l} + H_{n_l} \leq I$. By (4) and the limit behaviour of the considered subsequences as $l \to \infty$, it follows from (6) that

$$\begin{aligned} F &= F' + H \\ G &= G' + H \end{aligned}$$

where $F' + G' + H \leq I$. Hence, any two effects $F, G \in \overline{S}^\sigma$ are coexistent, that is, $\overline{S}^\sigma$ is pairwise coexistent. □

An implication of Theorem 1 is that a set of pairwise coexistent effects or its convex hull cannot be σ-dense in $\mathcal{E}(\mathcal{H})$ (provided that $\dim \mathcal{H} \geq 2$). In [13] this fact is proved differently and some physical consequences are discussed.

3. Compatibility

It is well known that the set $\partial_e\mathcal{E}(\mathcal{H})$ of the sharp effects is, with the order relation of $\mathcal{B}_s(\mathcal{H})$, a complete orthomodular orthocomplemented lattice. One can prove that the greatest lower and the least upper bound of $P, Q \in \partial_e\mathcal{E}(\mathcal{H})$ taken in $\partial_e\mathcal{E}(\mathcal{H})$ coincide with the corresponding bounds taken in $\mathcal{E}(\mathcal{H})$, but they do not exist when taken in $\mathcal{B}_s(\mathcal{H})$ unless P and Q are comparable. Furthermore, $\mathcal{E}(\mathcal{H})$ is not a lattice (for the lattice properties of $\mathcal{E}(\mathcal{H})$, see [16] and some references quoted therein). We use only the lattice structure of $\partial_e\mathcal{E}(\mathcal{H})$ and the fact—which we shall prove—that the infimum of an arbitrary subset of $\partial_e\mathcal{E}(\mathcal{H})$ taken in $\partial_e\mathcal{E}(\mathcal{H})$ is the same as taken in $\mathcal{E}(\mathcal{H})$.

We give a simple proof of the following known result (cf. [2]).

Theorem 2. *Two sharp effects $P, Q \in \partial_e\mathcal{E}(\mathcal{H})$ are coexistent if and only if $PQ = QP$. If P and Q are coexistent, then, in the representation*

$$\begin{aligned} P &= F_1 + H \\ Q &= F_2 + H \end{aligned} \tag{7}$$

where $F_1, F_2, H \in \mathcal{E}(\mathcal{H})$ satisfying $F_1 + F_2 + H \leq I$, F_1, F_2, and H are uniquely determined and given by the orthogonal projections $F_1 = P(I - Q) = P \wedge Q^\perp$, $F_2 = Q(I - P) = Q \wedge P^\perp$, and $H = PQ = P \wedge Q$.

Proof. If $PQ = QP$, then

$$\begin{aligned} P &= P(I - Q) + PQ \\ Q &= Q(I - P) + PQ \end{aligned}$$

and $P(I - Q) + Q(I - P) + PQ = P + Q(I - P) \leq P + (I - P) = I$. Thus, P and Q are coexistent.

Conversely, assume (7) holds with $F_1, F_2, H \in \mathcal{E}(\mathcal{H})$ and $F_1 + F_2 + F_3 \leq I$. From $P = F_1 + H$ it follows that $F_1 \leq P$ and from this, according to Lemma 1, $F_1 = PF_1$ and

$$PF_1 = F_1P. \tag{8}$$

Journal of Physics: Conference Series **1638** (2020) 012004 doi:10.1088/1742-6596/1638/1/012004

The relation $F_1 + F_2 + H = F_1 + Q \leq I$ implies $F_1 \leq I - Q$; therefore, analogously to (8), we have that $F_1(I - Q) = (I - Q)F_1$ and in consequence

$$QF_1 = F_1Q. \tag{9}$$

Furthermore, $H \leq Q$ and

$$QH = HQ. \tag{10}$$

Eqs. (9), (10), and $P = F_1 + H$ entail

$$PQ = QP. \tag{11}$$

Since $F_1 + Q \leq I$, $(I - Q) - F_1 \geq 0$. By (8) and (11), it follows that $PF_1 \leq P(I - Q)$. Taking account of $F_1 = PF_1$ we obtain

$$F_1 \leq P(I - Q). \tag{12}$$

Moreover, $H \leq P, Q$ implies $H \leq P \wedge Q$ which is, by (11), equivalent to

$$H \leq PQ. \tag{13}$$

From (12) and (13) we conclude that

$$P = F_1 + H \leq P(I - Q) + PQ = P.$$

Hence,

$$F_1 + H = P(I - Q) + PQ,$$

or, equivalently,

$$F_1 - P(I - Q) = PQ - H. \tag{14}$$

By (12), $F_1 - P(I - Q) \leq 0$ whereas, by (13), $PQ - H \geq 0$. Thus, we obtain from (14) that

$$F_1 - P(I - Q) = PQ - H = 0;$$

consequently, $F_1 = P(I - Q)$, $H = PQ$, and $F_2 = Q - H = Q(I - P)$. □

Two coexistent sharp effects are often called *compatible* (*commensurable* in the terminology of Ludwig), and the compatibility of $P, Q \in \partial_e\mathcal{E}(\mathcal{H})$ is frequently denoted by $P \leftrightarrow Q$. The denotation $P \leftrightarrow S$ where $S \subseteq \partial_e\mathcal{E}(\mathcal{H})$ means that $P \leftrightarrow Q$ for all $Q \in S$. The compatibility of a set of sharp effects is understood in the pairwise sense, that is, $S \subseteq \partial_e\mathcal{E}(\mathcal{H})$ is *compatible* if $P \leftrightarrow Q$ for all $P, Q \in \partial_e\mathcal{E}(\mathcal{H})$ (in other words, S is pairwise coexistent). We make use of these conventions.

It is well known that a set of sharp effects is compatible if and only if it is contained in a Boolean sublattice of $\partial_e\mathcal{E}(\mathcal{H})$. To prove an even stronger result, the concept of the compatant of a set of sharp effects is useful (cf. [1], [8, pp. 53–129]). The *compatant* S^c of a set $S \in \partial_e\mathcal{E}(\mathcal{H})$ is the set of all sharp effects that are compatible with all effects of S, i.e., $S^c := \{Q \in \partial_e\mathcal{E}(\mathcal{H}) \mid Q \leftrightarrow S\}$. We notice the following simple statements:

(a) If $S_1 \subseteq S_2 \subseteq \partial_e\mathcal{E}(\mathcal{H})$, then $S_2^c \subseteq S_1^c$.

(b) For every $S \subseteq \partial_e\mathcal{E}(\mathcal{H})$, $S \subseteq S^{cc}$ where $S^{cc} := (S^c)^c$ is the *bicompatant* of S.

(c) For every $S \subseteq \partial_e\mathcal{E}(\mathcal{H})$, $S^c = S^{ccc}$.

(d) A set S of sharp effects is compatible if and only if $S \subseteq S^c$.

(e) If $S \subseteq \partial_e\mathcal{E}(\mathcal{H})$ is compatible, then S^{cc} is compatible.

Statements (a), (b), and (d) are obvious. From (b) it follows that $S^c \subseteq (S^c)^{cc} = S^{ccc}$, whereas (b) and (a) imply that $S^{ccc} = (S^{cc})^c \subseteq S^c$, hence, (c) holds. If S is compatible, then, by (d), (a), and (c), $S^{cc} \subseteq S^c = (S^c)^{cc}$. Using (d) again, we obtain from $S^{cc} \subseteq (S^c)^{cc} = (S^{cc})^c$ that S^{cc} is compatible, thus proving (e). The next lemma is in particular necessary to prove some further statements about S^c.

Lemma 2. *Let $P \in \partial_e\mathcal{E}(\mathcal{H})$ and $S \subseteq \partial_e\mathcal{E}(\mathcal{H})$ nonempty. If $P \leftrightarrow S$, then $P \leftrightarrow \bigvee_{Q\in S} Q$ and $P \leftrightarrow \bigwedge_{Q\in S} Q$, and the distributive laws*

$$\begin{aligned} P \wedge \bigvee_{Q\in S} Q &= \bigvee_{Q\in S} P \wedge Q \\ P \vee \bigwedge_{Q\in S} Q &= \bigwedge_{Q\in S} P \vee Q \end{aligned} \tag{15}$$

hold.

Proof. Since $P \leftrightarrow Q$ for all $Q \in S$, we have, by Theorem 2,

$$\bigvee_{Q\in S} Q = \bigvee_{Q\in S} ((Q \wedge P^\perp) \vee (P \wedge Q)) = \left[\bigvee_{Q\in S} (Q \wedge P^\perp)\right] \vee \left[\bigvee_{Q\in S} (P \wedge Q)\right], \tag{16}$$

and since $\bigvee_{Q\in S}(P \wedge Q) \leq P$,

$$P = \left[P \wedge \left(\bigvee_{Q\in S} (P \wedge Q)\right)^\perp\right] \vee \left[\bigvee_{Q\in S} (P \wedge Q)\right]. \tag{17}$$

Obviously, the two bracketed projection-valued expressions of (16) are orthogonal to each other, and the same is true for (17). Due to

$$\bigvee_{Q\in S} (Q \wedge P^\perp) \leq P^\perp \leq P^\perp \vee \bigvee_{Q\in S} (P \wedge Q) = \left[P \wedge \left(\bigvee_{Q\in S} (P \wedge Q)\right)^\perp\right]^\perp$$

all the three different expressions inside the brackets of (16) and (17) are mutually orthogonal, so P and $\bigvee_{Q\in S} Q$ commute. Hence, by Theorem 2, $P \leftrightarrow \bigvee_{Q\in S} Q$. Now, from the uniqueness statement of Theorem (2) it follows that

$$P \wedge \bigvee_{Q\in S} Q = \bigvee_{Q\in S} P \wedge Q.$$

If $P \leftrightarrow E$, $E \in \partial_e\mathcal{E}(\mathcal{H})$, then $P \leftrightarrow E^\perp = I - E$. Thus, $P \leftrightarrow Q^\perp$ for all $Q \in S$, and $P \leftrightarrow \left(\bigvee_{Q\in S} Q^\perp\right)^\perp = \bigwedge_{Q\in S} Q$. Finally, $P^\perp \leftrightarrow Q^\perp$ for all $Q \in S$ so that

$$P \vee \bigwedge_{Q\in S} Q = \left(P^\perp \wedge \bigvee_{Q\in S} Q^\perp\right)^\perp = \left(\bigvee_{Q\in S} P^\perp \wedge Q^\perp\right)^\perp = \bigwedge_{Q\in S} P \vee Q.$$

□

We recall that an orthocomplemented sublattice $\mathcal{L}$ of $\partial_e\mathcal{E}(\mathcal{H})$ is Boolean if the distributive laws

$$\begin{aligned} E \wedge (P \vee Q) &= (E \wedge P) \vee (E \wedge Q) \\ E \vee (P \wedge Q) &= (E \vee P) \wedge (E \vee Q), \end{aligned} \tag{18}$$

$E, P, Q \in \mathcal{L}$, hold. As a consequence, any two $P, Q \in \mathcal{L}$ are compatible, i.e., $\mathcal{L}$ is a compatible set. In fact, because

$$P = (P \wedge Q^{\perp}) \vee (P \wedge Q)$$
$$Q = (Q \wedge P^{\perp}) \vee (P \wedge Q)$$

and $(P \wedge Q^{\perp})$, $(Q \wedge P^{\perp})$, and $(P \wedge Q)$ are mutually orthogonal, P and Q commute, i.e., $P \leftrightarrow Q$. If $\mathcal{L}$ is complete, then, according to Lemma 2, the validity of the general distributive laws (15) in $\mathcal{L}$ is implied. Thus, Eqs. (18) entails compatibility, and the completeness of $\mathcal{L}$ and its compatibility yield Eqs. (15).

Now we can prove the following simple, but important theorem about compatants.

Theorem 3. *Let $S \subseteq \partial_e\mathcal{E}(\mathcal{H})$.*

(a) The compatant S^c is a complete orthocomplemented sublattice of $\partial_e\mathcal{E}(\mathcal{H})$. Moreover, S^c is closed in the topology $\sigma \cap \partial_e\mathcal{E}(\mathcal{H})$.

(b) If S is compatible, the bicompatant S^{cc} is a complete Boolean sublattice of $\partial_e\mathcal{E}(\mathcal{H})$ which, moreover, is $\sigma \cap \partial_e\mathcal{E}(\mathcal{H})$-closed.

Proof. The first statement of (a) is a consequence of $0, I \in S^c$, $Q \in S^c \Rightarrow Q^{\perp} \in S^c$, and Lemma 2. To prove the second statement of (a), recall that the topology $\sigma \cap \mathcal{E}(\mathcal{H})$ is second countable, which implies the second countability of $\sigma \cap \partial_e\mathcal{E}(\mathcal{H})$. Consequently, if $E \in \overline{S^c}^{\sigma \cap \partial_e\mathcal{E}(\mathcal{H})}$, then there exists a sequence $\{Q_n\}_{n\in\mathbb{N}}$ in S^c such that $Q_n \xrightarrow{\sigma} E$ as $n \to \infty$. Taking account of $Q_n \leftrightarrow P$, $P \in S$, we obtain, since the product of operators of $\mathcal{B}_s(\mathcal{H})$ is sequentially σ-continuous separately in each factor,

$$PE = P\sigma\text{-}\lim_{n\to\infty} Q_n = \sigma\text{-}\lim_{n\to\infty} PQ_n = \sigma\text{-}\lim_{n\to\infty} Q_nP = (\sigma\text{-}\lim_{n\to\infty} Q_n)P = EP.$$

Hence, $E \in S^c$, and $S^c = \overline{S^c}^{\sigma \cap \partial_e\mathcal{E}(\mathcal{H})}$.

Concerning (b), we only have to prove that in the complete orthocomplemented sublattice S^{cc} the distributive laws are satisfied. We already remarked that S^{cc} is compatible if S is compatible. Thus, by Lemma 2, S^{cc} is distributive. □

Corollary. *If a set S of sharp effects is contained in a Boolean sublattice $\mathcal{L}$ of $\partial_e\mathcal{E}(\mathcal{H})$, then S is compatible. Conversely, any compatible set S of sharp effects is contained in a complete Boolean sublattice of $\partial_e\mathcal{E}(\mathcal{H})$.*

Proof. The first statement of the corollary is clear since $\mathcal{L}$ is compatible as shown above. The second follows from $S \subseteq S^{cc}$ and part (b) of the theorem since S is compatible. □

In Section 2 we noticed that $\partial_e\mathcal{E}(\mathcal{H})$ is σ-dense in $\mathcal{E}(\mathcal{H})$. So one cannot expect that the compatant of S is σ-closed in $\mathcal{E}(\mathcal{H})$ or, equivalently, in $\mathcal{B}_s(\mathcal{H})$; S^c is closed in the σ-topology on $\partial_e\mathcal{E}(\mathcal{H})$.

4. Some crucial properties of countably complete Boolean sublattices

According to Lemma 2, the bicompatant of a compatible set is a $\sigma \cap \partial_e\mathcal{E}(\mathcal{H})$-closed complete Boolean sublattice of $\partial_e\mathcal{E}(\mathcal{H})$. We are going to prove that every countably complete Boolean sublattice of $\partial_e\mathcal{E}(\mathcal{H})$ is $\sigma \cap \partial_e\mathcal{E}(\mathcal{H})$-closed; we understand a *countably complete lattice* to be a partially ordered set in which the greatest lower and the least upper bound of every countable subset exist[3]. To prove the indicated property of countably complete Boolean sublattices of $\partial_e\mathcal{E}(\mathcal{H})$, we need the following two lemmata.

Lemma 3. *Let $\{P_n\}_{n\in\mathbb{N}}$ be a sequence of mutually compatible sharp effects. Then*

$$\bigvee_{n\in\mathbb{N}} P_n = \bigvee_{n\in\mathbb{N}} Q_n$$

where $\{Q_n\}_{n\in\mathbb{N}}$ is a sequence in $\partial_e\mathcal{E}(\mathcal{H})$ whose members are mutually orthogonal.

Proof. Define $Q_1 := P_1$ and

$$Q_n := P_n \wedge \left(\bigvee_{j=1}^{n-1} P_j\right)^\perp \tag{19}$$

for $n \geq 2$. It follows that, for $m < n$,

$$Q_n \leq \left(\bigvee_{j=1}^{n-1} P_j\right)^\perp = \bigwedge_{j=1}^{n-1} P_j^\perp \leq P_m^\perp \leq P_m^\perp \vee \bigvee_{j=1}^{m-1} P_j = \left(P_m \wedge \left(\bigvee_{j=1}^{m-1} P_j\right)^\perp\right)^\perp = Q_m^\perp;$$

therefore, the projections Q_n are mutually orthogonal. By induction, we now show that

$$\bigvee_{j=1}^{n} P_j = \bigvee_{j=1}^{n} Q_j. \tag{20}$$

The case $n = 1$ is clear. Assume (20) holds for $n = k$. Then, for $n = k + 1$,

$$\begin{aligned}\bigvee_{j=1}^{k+1} Q_j &= \left(\bigvee_{j=1}^{k} Q_j\right) \vee Q_{k+1} = \left(\bigvee_{j=1}^{k} P_j\right) \vee \left(P_{k+1} \wedge \left(\bigvee_{j=1}^{k} P_j\right)^\perp\right) \\ &= \left(\bigvee_{j=1}^{k+1} P_j\right) \wedge \left(\left(\bigvee_{j=1}^{k} P_j\right) \vee \left(\bigvee_{j=1}^{k} P_j\right)^\perp\right) \\ &= \bigvee_{j=1}^{k+1} P_j\end{aligned}$$

where we used one of the distributive laws, which, by Lemma 2, is possible since $\bigvee_{j=1}^{k} P_j \leftrightarrow \left(\bigvee_{j=1}^{k} P_j\right)^\perp$ and, again by Lemma 2, $\bigvee_{j=1}^{k} P_j \leftrightarrow P_{k+1}$. Thus, (20) has been proved.

[3] A countably complete Boolean lattice is often called a Boolean σ-algebra. The author would like to call a countably complete lattice *σ-complete*; but this is misleading because we could understand a σ-complete subset of $\mathcal{B}_s(\mathcal{H})$ to be set that is complete w.r.t. the uniform structure induced by our σ-topology.

Eq. (19) implies $Q_n \leq P_n$, consequently,

$$Q_n \leq \bigvee_{n\in\mathbb{N}} P_n. \tag{21}$$

Let $E \in \partial_e\mathcal{E}(\mathcal{H})$ be any upper bound of all Q_n. Then, by (20), $E \geq \bigvee_{j=1}^n Q_j = \bigvee_{j=1}^n P_j$ holds for all $n \in \mathbb{N}$. Hence, $E \geq P_n$ for all $n \in \mathbb{N}$, which entails that $E \geq \bigvee_{n\in\mathbb{N}} P_n$. From this and (21) we conclude that $\bigvee_{n\in\mathbb{N}} Q_n = \bigvee_{n\in\mathbb{N}} P_n$. □

The first statement of the next lemma is of general functional analytic character.

Lemma 4.

(a) Let $(\mathcal{V}, \|\,.\,\|)$ be a separable Banach space and $(\mathcal{V}', \|\,.\,\|)$ its dual. Then there exists a second norm $\|\,.\,\|_\sigma$ on $\mathcal{V}'$ such that the $\|\,.\,\|_\sigma$-topology on the $\|\,.\,\|$-closed unit ball $B_{\mathcal{V}'}$ coincides with the weak-∗ topology $\sigma(\mathcal{V}', \mathcal{V})$ on $B_{\mathcal{V}'}$, that is, the topology on $B_{\mathcal{V}'}$ induced by the norm $\|\,.\,\|_\sigma$ equals $\sigma(\mathcal{V}', \mathcal{V}) \cap B_{\mathcal{V}'}$.

(b) In the case $\langle\mathcal{V}, \mathcal{V}'\rangle = \langle\mathcal{T}_s(\mathcal{H}), \mathcal{B}_s(\mathcal{H})\rangle$ the norm $\|\,.\,\|_\sigma$ can be chosen such that, in addition,

$$\|A_1\|_\sigma \leq \|A_2\|_\sigma$$

holds for $A_1, A_2 \in \mathcal{B}_s(\mathcal{H})$ whenever $0 \leq A_1 \leq A_2$.

Proof. Choose a countable dense subset $D := \{v_1, v_2, \ldots\}$ of the $\|\,.\,\|$-closed unit ball $B_\mathcal{V}$ of $\mathcal{V}$ being $\|\,.\,\|$-dense in $B_\mathcal{V}$ and define

$$\|\ell\|_\sigma := \sum_{j=1}^{\infty} \lambda_j |\ell(v_j)| \tag{22}$$

where $\ell \in \mathcal{V}'$ and $\lambda_j > 0$, $\sum_{j=1}^{\infty} \lambda_j = 1$. It is easily shown that D separates $\mathcal{V}'$ and $\|\,.\,\|_\sigma$ is a norm on $\mathcal{V}'$. Moreover, the weak topology $\sigma(\mathcal{V}', D)$ is Hausdorff and coincides on $B_{\mathcal{V}'}$ with the $\|\,.\,\|_\sigma$-topology, as we are going to show. To that end, observe that the sets

$$U(\ell; v_1, \ldots, v_n; \varepsilon) := \{\tilde{\ell} \in \mathcal{V}' \mid |\tilde{\ell}(v_j) - \ell(v_j)| < \varepsilon, j = 1, \ldots, n\},$$

$n \in \mathbb{N}$, $\varepsilon > 0$, constitute a neighborhood base of $\ell \in \mathcal{V}'$ (compare to Eq. (1)).

Let $\ell_1, \ell_2 \in \mathcal{V}'$ with $\ell_1 \neq \ell_2$. Since D separates $\mathcal{V}'$, there exists a vector $v_N \in D$ such that $\ell_1(v_N) \neq \ell_2(v_N)$. Let $|\ell_1(v_N) - \ell_2(v_N)| =: \varepsilon$ and consider the neighborhoods

$$U\left(\ell_k; v_N; \frac{\varepsilon}{2}\right) := \left\{\tilde{\ell} \in \mathcal{V}' \,\middle|\, |\tilde{\ell}(v_N) - \ell_k(v_N)| < \frac{\varepsilon}{2}\right\}$$

of ℓ_1 and ℓ_2, respectively, $k = 1, 2$. For $\tilde{\ell} \in U(\ell_1; v_N; \frac{\varepsilon}{2})$ we have that

$$\begin{aligned} |\tilde{\ell}(v_N) - \ell_2(v_N)| &= |(\tilde{\ell}(v_N) - \ell_1(v_N)) - (\ell_2(v_N) - \ell_1(v_N))| \\ &\geq \Big||\tilde{\ell}(v_N) - \ell_1(v_N)| - |\ell_2(v_N) - \ell_1(v_N)|\Big| \\ &= \varepsilon - |\tilde{\ell}(v_N) - \ell_1(v_N)| \\ &> \frac{\varepsilon}{2}. \end{aligned}$$

In consequence, $\tilde{\ell} \notin U(\ell_2; v_N; \frac{\varepsilon}{2})$; so $U(\ell_1; v_N; \frac{\varepsilon}{2})$ and $U(\ell_2; v_N; \frac{\varepsilon}{2})$ are disjoint, and $\sigma(\mathcal{V}', D)$ is Hausdorff.

Using (22), we obtain from $\|\tilde{\ell} - \ell\|_\sigma < \varepsilon$ that $|(\tilde{\ell} - \ell)(v_j)| < \varepsilon$ for all $j \in \mathbb{N}$. Therefore, $\tilde{\ell} \in U(\ell; v_1, \ldots, v_n; \varepsilon)$ for all $n \in \mathbb{N}$. Hence, on $\mathcal{V}'$ and in particular on $B_{\mathcal{V}'}$, $\sigma(\mathcal{V}', D)$ is weaker than the $\|\,.\,\|_\sigma$-topology. On $B_{\mathcal{V}'}$, the converse is also true. Namely, let $\ell, \tilde{\ell} \in B_{\mathcal{V}'}$, $\varepsilon > 0$, $\sum_{j=n+1}^\infty \lambda_j < \frac{\varepsilon}{4}$, and $\tilde{\ell} \in U(\ell; v_1, \ldots, v_n; \frac{\varepsilon}{2})$. It follows that

$$\begin{aligned}\|\tilde{\ell} - \ell\|_\sigma &= \sum_{j=1}^n \lambda_j |(\tilde{\ell} - \ell)(v_j)| + \sum_{j=n+1}^\infty \lambda_j |(\tilde{\ell} - \ell)(v_j)| < \sum_{j=1}^n \lambda_j \frac{\varepsilon}{2} + \sum_{j=n+1}^\infty \lambda_j \|\tilde{\ell} - \ell\| \\ &< \frac{\varepsilon}{2} + \frac{\varepsilon}{4} \cdot 2 \\ &= \varepsilon;\end{aligned}$$

in the estimation of the infinite sums $|(\tilde{\ell} - \ell)(v_j)| \leq \|\tilde{\ell} - \ell\| \leq 2$ has been used where $\|\tilde{\ell} - \ell\|$ is the usual norm of $\tilde{\ell} - \ell$ and $\|v_j\|, \|\tilde{\ell}\|, \|\ell\| \leq 1$ has been taken into account. Thus, $\tilde{\ell} \in U(\ell; v_1, \ldots, v_n; \frac{\varepsilon}{2}) \cap B_{\mathcal{V}'}$ implies $\tilde{\ell} \in \{\hat{\ell} \in \mathcal{B}_s(\mathcal{H}) \,|\, \|\hat{\ell} - \ell\|_\sigma < \varepsilon\} \cap B_{\mathcal{V}'}$, and on $B_{\mathcal{V}'}$ the $\|\,.\,\|_\sigma$-topology and $\sigma(\mathcal{V}', D)$ are equal.

Now, the topology $\sigma(\mathcal{V}', D)$ is Hausdorff and weaker than $\sigma(\mathcal{V}', \mathcal{V})$, and $\sigma(\mathcal{V}', \mathcal{V}) \cap B_{\mathcal{V}'}$ is compact. Hence, $\sigma(\mathcal{V}', D) \cap B_{\mathcal{V}'} = \sigma(\mathcal{V}', \mathcal{V}) \cap B_{\mathcal{V}'}$. Thus, on $B_{\mathcal{V}'}$ the $\|\,.\,\|_\sigma$-topology and $\sigma(\mathcal{V}', \mathcal{V})$ coincide.

In the case $\langle \mathcal{V}, \mathcal{V}' \rangle = \langle \mathcal{T}_s(\mathcal{H}), \mathcal{B}_s(\mathcal{H}) \rangle$, Eq. (22) reads

$$\|A\|_\sigma = \sum_{j=1}^\infty \lambda_j |\mathrm{tr}\, V_j A|$$

where $A \in \mathcal{B}_s(\mathcal{H})$ and the $V_j \in \mathcal{T}_s(\mathcal{H})$ form, w.r.t. the trace norm, a dense subset of the unit ball of $\mathcal{T}_s(\mathcal{H})$. We can replace the set $D = \{V_1, V_2, \ldots\}$ by a set $\widetilde{D} := \{W_1, W_2, \ldots\}$ of density operators being dense in the set $\mathcal{S}(\mathcal{H})$ of all density operators. Since $\widetilde{D}$ separates $\mathcal{B}_s(\mathcal{H})$, $\sigma(\mathcal{B}_s(\mathcal{H}), \widetilde{D})$ is Hausdorff, and since $\|W_j\| = 1$, the $\|\,.\,\|_\sigma$-topology corresponding to $\widetilde{D}$ coincides on $B_{\mathcal{B}_s(\mathcal{H})}$ with $\sigma(\mathcal{B}_s(\mathcal{H}), \widetilde{D})$. Again, the Hausdorff topology $\sigma(\mathcal{B}_s(\mathcal{H}), \widetilde{D}) \cap B_{\mathcal{B}_s(\mathcal{H})}$ is weaker than the compact topology $\sigma(\mathcal{B}_s(\mathcal{H}), \mathcal{T}_s(\mathcal{H})) \cap B_{\mathcal{B}_s(\mathcal{H})}$, so they must be equal; thus, on $B_{\mathcal{B}_s(\mathcal{H})}$ the $\|\,.\,\|_\sigma$-topology corresponding to $\widetilde{D}$ and our σ-topology coincide. Finally, from

$$\|A\|_\sigma = \sum_{j=1}^\infty \lambda_j |\mathrm{tr}\, W_j A|$$

and $0 \leq A_1 \leq A_2$ it follows that

$$\|A_1\|_\sigma = \sum_{j=1}^\infty \lambda_j |\mathrm{tr}\, W_j A_1| \leq \sum_{j=1}^\infty \lambda_j |\mathrm{tr}\, W_j A_2| = \|A_2\|_\sigma.$$

□

The following important theorem and its corollary can be proved on the basis of the preceding two lemmata (cf. [2, pp. 95–96]).

Theorem 4. *Let $\{P_j\}_{j \in \mathbb{N}}$ be a sequence of compatible sharp effects converging to $P \in \partial_e \mathcal{E}(\mathcal{H})$ in the σ-topology.*

(a) If $Q \in \partial_e \mathcal{E}(\mathcal{H})$ is compatible with all P_j, then $Q \leftrightarrow P$. The limit P is compatible with all P_j.

(b) There exists a subsequence $\{P_{j_k}\}_{k\in\mathbb{N}}$ *such that*

$$P = \bigwedge_{n=1}^{\infty} \bigvee_{k=n}^{\infty} P_{j_k}.$$

Proof. From $P_jQ = QP_j$ for all $j \in \mathbb{N}$, $P_j \xrightarrow{\sigma} P$ as $j \to \infty$, and the sequential σ-continuity of the product of operators of $\mathcal{B}_s(\mathcal{H})$ separately in each factor, it follows that $PQ = QP$. So the first statement of (a) has been shown. Setting $Q = P_k$, $P_k \leftrightarrow P_j$ implies $P_k \leftrightarrow P$.

We are now going to prove statement (b). From $P_j \xrightarrow{\sigma} P$ as $j \to \infty$ we obtain that

$$P_j(I-P) \xrightarrow{\sigma} P(I-P) = 0.$$

Using a norm according to part (b) of Lemma 4, we have that

$$\|P_j(I-P)\|_\sigma \to 0$$

as $j \to \infty$. We can choose a subsequence $\{P_{j_k}\}_{k\in\mathbb{N}}$ of $\{P_j\}_{j\in\mathbb{N}}$ such that

$$\|P_{j_k}(I-P)\|_\sigma < \frac{1}{2^k}. \tag{23}$$

Observe that, for every $n \in \mathbb{N}$, the infinite sum $\sum_{k=n}^{\infty} P_{j_k}(I-P)$ exists in the σ-topology because $\sum_{k=n}^{\infty} \|P_{j_k}(I-P)\|_\sigma < \sum_{k=n}^{\infty} \frac{1}{2^k} \leq 1$.

Since $\{\bigvee_{k=n}^{\infty} P_{j_k}\}_{n\in\mathbb{N}}$ is a decreasing sequence of orthogonal projections, it is σ-convergent and its limit is equal to $\widetilde{P} := \bigwedge_{n=1}^{\infty} \bigvee_{k=n}^{\infty} P_{j_k} \in \partial_e\mathcal{E}(\mathcal{H})$. From $\bigvee_{k=n}^{\infty} P_{j_k} \geq P_{j_n}$ and $P_{j_k} \xrightarrow{\sigma} P$ it then follows that $\widetilde{P} \geq P$; hence,

$$\bigvee_{k=n}^{\infty} P_{j_k} \geq \widetilde{P} \geq P. \tag{24}$$

Because the sharp effects P_{j_k} are compatible, Lemma 3 implies that

$$\bigvee_{k=n}^{\infty} P_{j_k} = \bigvee_{k=n}^{\infty} Q_{j_k} = \sum_{k=n}^{\infty} Q_{j_k} \tag{25}$$

where the Q_{j_k} are mutually orthogonal. From (24) and (25) we obtain that

$$\left(\bigvee_{k=n}^{\infty} P_{j_k}\right) - P = \left(\bigvee_{k=n}^{\infty} P_{j_k}\right)(I-P) = \sum_{k=n}^{\infty} Q_{j_k}(I-P). \tag{26}$$

Choosing Q_{j_k} according to (19), we have that $P_l \leftrightarrow Q_{j_k}$ for all $k, l \in \mathbb{N}$. Consequently, $Q_{j_k} \leftrightarrow P$ and $Q_{j_k}P = PQ_{j_k}$; furthermore, $P_{j_k} \leftrightarrow P$ and $P_{j_k}P = PP_{j_k}$. Thus, $Q_{j_k} \leq P_{j_k}$ implies $Q_{j_k}(I-P) \leq P_{j_k}(I-P)$. Now, from (26) it follows that

$$\left(\bigvee_{k=n}^{\infty} P_{j_k}\right) - P = \sum_{k=n}^{\infty} Q_{j_k}(I-P) \leq \sum_{k=n}^{\infty} P_{j_k}(I-P) \tag{27}$$

where the last sum exists in the σ-topology, as we already remarked. Finally, from (27) and (23) we conclude that

$$\left\|\left(\bigvee_{k=n}^{\infty} P_{j_k}\right) - P\right\|_\sigma \leq \left\|\sum_{k=n}^{\infty} P_{j_k}(I-P)\right\|_\sigma \leq \sum_{k=n}^{\infty} \|P_{j_k}(I-P)\|_\sigma < \sum_{k=n}^{\infty} \frac{1}{2^k} \to 0$$

as $n \to \infty$. Hence, $P = \sigma\text{-}\lim_{n\to\infty} \bigvee_{k=n}^{\infty} P_{j_k} = \widetilde{P} = \bigwedge_{n=1}^{\infty} \bigvee_{k=n}^{\infty} P_{j_k}$. □

Corollary. *Every countably complete Boolean sublattice $\mathcal{L}$ of $\partial_e\mathcal{E}(\mathcal{H})$*[4] *is σ-closed in $\partial_e\mathcal{E}(\mathcal{H})$, i.e., closed in the topology $\sigma \cap \partial_e\mathcal{E}(\mathcal{H})$.*

Proof. Let $P \in \overline{\mathcal{L}}^{\sigma\cap\partial_e\mathcal{E}(\mathcal{H})}$. Then, as a consequence of the second countability of $\sigma \cap \partial_e\mathcal{E}(\mathcal{H})$, there exists a sequence $\{P_j\}_{j\in\mathbb{N}}$ in $\mathcal{L}$ converging to P in the σ-topology. Recall that $\mathcal{L}$ is a compatible set. Then, according to the theorem, there is a subsequence $\{P_{j_k}\}_{j\in\mathbb{N}}$ such that $P = \bigwedge_{n=1}^{\infty}\bigvee_{k=n}^{\infty} P_{j_k}$. Hence, $P \in \mathcal{L}$, and $\mathcal{L}$ is $\sigma \cap \partial_e\mathcal{E}(\mathcal{H})$-closed. □

For our purpose, the theorem is important because of its corollary. The latter enables a short proof of the next lemma on which the lemma afterwards is based.

Lemma 5. *Every countably complete Boolean sublattice $\mathcal{L}$ of $\partial_e\mathcal{E}(\mathcal{H})$ is countably generated, i.e., there exists a countable set $\{P_1, P_2, \ldots\} \subseteq \mathcal{L}$ such that $\mathcal{L}$ is the smallest countably complete Boolean sublattice of $\partial_e\mathcal{E}(\mathcal{H})$ containing $\{P_1, P_2, \ldots\}$.*

Proof. Since the topology $\sigma\cap\mathcal{L}$ on $\mathcal{L}$ is second countable as well, $\mathcal{L}$ is separable in this topology. That is, there exists a countable set $\{P_1, P_2, \ldots\} \subseteq \mathcal{L}$ being $\sigma \cap \mathcal{L}$-dense in $\mathcal{L}$. In consequence, $\{P_1, P_2, \ldots\}$ is also $\sigma \cap \partial_e\mathcal{E}(\mathcal{H})$-dense in $\mathcal{L}$ or, equivalently, $\overline{\{P_1, P_2, \ldots\}}^{\sigma\cap\partial_e\mathcal{E}(\mathcal{H})} = \overline{\mathcal{L}}^{\sigma\cap\partial_e\mathcal{E}(\mathcal{H})}$. Thus, by the corollary, $\mathcal{L} = \overline{\{P_1, P_2, \ldots\}}^{\sigma\cap\partial_e\mathcal{E}(\mathcal{H})}$.

Let $\widetilde{\mathcal{L}}$ be the smallest countably complete Boolean sublattice containing $\{P_1, P_2, \ldots\}$. Clearly, $\widetilde{\mathcal{L}} \subseteq \mathcal{L}$. Again, by the corollary, $\widetilde{\mathcal{L}} = \overline{\widetilde{\mathcal{L}}}^{\sigma\cap\partial_e\mathcal{E}(\mathcal{H})} \supseteq \overline{\{P_1, P_2, \ldots\}}^{\sigma\cap\partial_e\mathcal{E}(\mathcal{H})} = \mathcal{L}$. Hence, $\mathcal{L} = \widetilde{\mathcal{L}}$. □

The final lemma of this section is the basis of our conclusions in the next section. The author took the idea of its proof from [10, pp. 268–270], and [18, pp. 64–65]; however, the idea obviously originates from [19, proof of Satz 10].

Lemma 6. *Let $\mathcal{L}$ be a countably complete Boolean sublattice of $\partial_e\mathcal{E}(\mathcal{H})$. Then there exists an increasing family $\{E_\lambda\}_{\lambda\in\mathbb{R}}$ of orthogonal projections that is σ-continuous from the right, satisfies $E_\lambda = 0$ for $\lambda < 0$, $E_\lambda = I$ for $\lambda \geq I$, and generates $\mathcal{L}$.*

Proof. Choose a set $\{P_1, P_2, \ldots\} \subseteq \mathcal{L}$ that generates $\mathcal{L}$. The sharp effects P_j are mutually compatible. We consider the case of an infinite set $\{P_1, P_2, \ldots\}$; if the generating set is finite (i.e., if $\mathcal{L}$ is finite), the following argumentation can easily be modified. By induction, we construct the following increasing sequence of increasing families of orthogonal projections:

(i) $0 \leq P_1 \leq I$.

(ii) $0 \leq P_1 \wedge P_2 \leq P_1 \leq P_1 \vee P_2 \leq I$ or, what is the same, $0 \leq P_1P_2 \leq P_1 \leq P_1 + P_2(I - P_1) \leq I$.

(iii) Rename the family of (ii) as $0 \leq Q_1 \leq Q_2 \leq Q_3 \leq I$ and define

$$0 \leq Q_1 \wedge P_3 \leq Q_1 \leq Q_1 \vee (Q_2 \wedge P_3) \leq Q_2 \leq Q_2 \vee (Q_3 \wedge P_3) \leq Q_3 \leq Q_3 \vee P_3 \leq I$$

or

$$0 \leq Q_1P_3 \leq Q_3 \leq Q_1 + (Q_2 - Q_1)P_3 \leq Q_2 \leq Q_2 + (Q_3 - Q_2)P_3 \leq Q_3 \leq Q_3 + (I - Q_3)P_3 \leq I.$$

(vi) In general, if the nth family is $0 \leq E_1 \leq E_2 \leq \ldots \leq E_{2^n-1} \leq I$, define

$$0 \leq E_1 \wedge P_{n+1} \leq E_1 \leq E_1 \vee (E_2 \wedge P_{n+1}) \leq E_2 \leq E_2 \vee (E_3 \wedge P_{n+1}) \leq E_3 \leq \ldots$$
$$\leq E_{2^n-1} \leq E_{2^n-1} \vee P_{n+1} \leq I.$$

[4] It should be remarked that, due to the separability of the Hilbert space $\mathcal{H}$, every countably complete Boolean lattice is already complete. We do not use this fact; it is proved in [17, pp. 303–304], where the well-ordering principle is used.

Each of these families $0, E_1, \ldots, E_{2^n-1}, I$ where $n \in \mathbb{N}$ contains the preceding one as a subfamily. The members of each such family are mutually compatible with each other and with all $P_1, P_2, \ldots$. Moreover, the Boolean lattice $\mathcal{L}_n$ generated by the family $0, E_1, \ldots, E_{2^n-1}, I$ contains $P_1, \ldots, P_n$. This is true for $n = 1$. Assume $\mathcal{L}_n$ contains $P_1, \ldots, P_n$. Since $\mathcal{L}_n \subseteq \mathcal{L}_{n+1}$, $P_1, \ldots, P_n$ also belong to $\mathcal{L}_{n+1}$, and we only have to show that $P_{n+1} \in \mathcal{L}_{n+1}$. From

$$(E_1 \wedge P_{n+1}) \vee [(E_1 \vee (E_2 \wedge P_{n+1})) \wedge E_1^{\perp}] = E_2 \wedge P_{n+1}$$

it follows that $E_2 \wedge P_{n+1} \in \mathcal{L}_{n+1}$, and

$$(E_2 \wedge P_{n+1}) \vee [(E_2 \vee (E_3 \wedge P_{n+1})) \wedge E_2^{\perp}] = E_3 \wedge P_{n+1}$$

then implies $E_3 \wedge P_{n+1} \in \mathcal{L}_{n+1}$. Continuing this way, we obtain $E_{2^n-1} \wedge P_{n+1} \in \mathcal{L}_{n+1}$ and finally

$$(E_{2^n-1} \wedge P_{n+1}) \vee [(E_{2^n-1} \vee P_{n+1}) \wedge E_{2^n-1}^{\perp}] = P_{n+1}.$$

Thus, $P_{n+1} \in \mathcal{L}_{n+1}$.

As a consequence of $P_1, \ldots, P_n \in \mathcal{L}_n$, the Boolean sublattice $\widetilde{\mathcal{L}}_n$ generated by $P_1, \ldots, P_n$ is contained in $\mathcal{L}_n$. By the construction of the family $0, E_1, \ldots, E_{2^n-1}, I$, we have that $0, E_1, \ldots, E_{2^n-1}, I \in \widetilde{\mathcal{L}}_n$ which entails $\mathcal{L}_n \subseteq \widetilde{\mathcal{L}}_n$. Hence, $\mathcal{L}_n = \widetilde{\mathcal{L}}_n$.

Let S be the set of all members of all families $0, E_1, \ldots, E_{2^n-1}, I$, $n \in \mathbb{N}$. Because the projections of each family are increasing and the families themselves increase, S is a countable totally ordered set of orthogonal projections. Let $\mathcal{L}(S)$ be the Boolean sublattice generated by S and $\mathcal{L}_c(S)$ the countably complete Boolean sublattice generated by S. From $\mathcal{L}(S) \supseteq \mathcal{L}_n$ it follows that $\mathcal{L}(S) \supseteq \bigcup_{n\in\mathbb{N}} \mathcal{L}_n$. Since $\mathcal{L}_n \subseteq \mathcal{L}_{n+1}$, $\bigcup_{n\in\mathbb{N}} \mathcal{L}_n$ itself is a Boolean lattice containing S, so $\mathcal{L}(S) \subseteq \bigcup_{n\in\mathbb{N}} \mathcal{L}_n$. Hence,

$$\mathcal{L}(S) = \bigcup_{n\in\mathbb{N}} \mathcal{L}_n = \bigcup_{n\in\mathbb{N}} \widetilde{\mathcal{L}}_n. \tag{28}$$

The original countably complete Boolean sublattice $\mathcal{L}$ is generated by $\{P_1, P_2, \ldots\}$. Consequently, $\mathcal{L} \supseteq \widetilde{\mathcal{L}}_n$ and $\mathcal{L} \supseteq \bigcup_{n\in\mathbb{N}} \widetilde{\mathcal{L}}_n$; further, by (28), $\mathcal{L} \supseteq \mathcal{L}(S)$ and in particular $\mathcal{L} \supseteq S$. Therefore,

$$\mathcal{L} \supseteq \mathcal{L}_c(S) \supseteq \mathcal{L}(S) \supseteq \{P_1, P_2, \ldots\} \tag{29}$$

where the latter inclusion is a consequence of $\widetilde{\mathcal{L}}_n \supseteq \{P_1, \ldots, P_n\}$ and (28) again. Now, from (29) we obtain $\mathcal{L} \supseteq \mathcal{L}_c(S) \supseteq \{P_1, P_2, \ldots\}$ which implies

$$\mathcal{L} \supseteq \mathcal{L}_c(S) \supseteq \mathcal{L}.$$

Thus, $\mathcal{L} = \mathcal{L}_c(S)$.

We have proved that S is a countable totally ordered set of orthogonal projections generating $\mathcal{L}$. Now, to avoid confusion, we rename the families $0, E_1, \ldots, E_{2^n-1}, I$ as $0, Q_1, \ldots, Q_{2^n-1}, I$ and use them to construct the family $\{E_\lambda\}_{\lambda\in\mathbb{R}}$ occurring in the statement of the lemma. Consider the interval $[0, 1]$ of real numbers and define $E_\lambda = P_1$ for $\lambda \in]\frac{1}{3}, \frac{2}{3}[$. Next define

$$E_\lambda := \begin{cases} Q_1 & \text{if } \lambda \in]\frac{1}{9}, \frac{2}{9}[\\ Q_2 & \text{if } \lambda \in]\frac{1}{3}, \frac{2}{3}[\\ Q_3 & \text{if } \lambda \in]\frac{7}{9}, \frac{8}{9}[\end{cases}$$

where Q_1, Q_2, Q_3 are the projections of step (ii) above and $Q_2 = P_1$. The considered intervals are just the open intervals that occur in the first two steps of the construction of the Cantor set. If the values of E_λ on the open intervals $I_1, \ldots, I_{2^n-1}$ occurring in the n-th step of the construction of the Cantor set are defined to be $Q_1, \ldots, Q_{2^n-1}$, respectively, then the respective values of E_λ on the open intervals $J_1, \ldots, J_{2^{n+1}-1}$ of the step $n+1$ are to be $Q_1 \wedge P_{n+1} \leq Q_1 \leq Q_1 \vee (Q_2 \wedge P_{n+1}) \leq Q_2 \leq Q_2 \vee (Q_3 \wedge P_{n+1}) \leq Q_3 \leq \ldots \leq Q_{2^n-1} \leq Q_{2^n-1} \vee P_{n+1}$. So E_λ is consistently defined for all $\lambda \in [0,1] \setminus C$ where C is the Cantor set. Defining $E_\lambda := 0$ for $\lambda < 0$ and $E_\lambda := I$ for $\lambda \geq 1$, $\{E_\lambda\}_{\lambda \in (\mathbb{R} \setminus C) \cup \{1\}}$ is an increasing family of orthogonal projections whose members constitute the set S.

To define E_λ on the Cantor set, let $\lambda \in C$, $\lambda \neq 1$. Since $[0,1] \setminus C$ is dense in $[0,1]$, there exist a sequence $\{\lambda_n\}_{n \in \mathbb{N}}$ in $[0,1] \setminus C$ satisfying $\lambda < \lambda_{n+1} \leq \lambda_n$ for all n and converging to λ. For every such sequence, the decreasing sequence $\{E_{\lambda_n}\}_{n \in \mathbb{N}}$ is σ-convergent. Given two different decreasing sequences in $[0,1] \setminus C$ converging to λ from the right, one can construct a third such sequence containing subsequences the first two ones. From this it follows that $\sigma\text{-}\lim_{n \to \infty} E_{\lambda_n}$ does not depend on the chosen sequence $\{\lambda_n\}_{n \in \mathbb{N}}$ and $E_\lambda := \sigma\text{-}\lim_{n \to \infty} E_{\lambda_n}$ for $\lambda \in C \setminus \{0,1\}$ is well defined.

Since $\{E_\lambda\}_{\lambda \in \mathbb{R}}$ increases on $(\mathbb{R} \setminus C) \cup \{1\}$ and $E_\lambda = \sigma\text{-}\lim_{n \to \infty} E_{\lambda_n} = \bigwedge_{n \in \mathbb{N}} E_{\lambda_n} = \bigwedge_{\mu > \lambda} E_\mu$ is valid for all $\lambda \in C \setminus \{1\}$, $\{E_\lambda\}_{\lambda \in \mathbb{R}}$ is increasing on $\mathbb{R}$. It remains to show its σ-continuity from the right. This is trivially the case for $\lambda < 0$ and $\lambda \geq 1$. It is also true for $\lambda \in [0,1] \setminus C$ because then λ belongs to one of the open boundary intervals of C and E_λ is constant on such an interval. So we still have to consider $\lambda \in C \setminus \{1\}$.

Let $\lambda \in C \setminus \{1\}$ and $\{\lambda_n\}_{n \in \mathbb{N}}$ be a sequence used in the definition of E_λ, i.e., $\lambda_n \in [0,1] \setminus C$, $\lambda < \lambda_n \leq \lambda_{n+1}$ for all $n \in \mathbb{N}$ and $\lambda_n \to \lambda$ as $n \to \infty$. We then have $E_\lambda = \sigma\text{-}\lim_{n \to \infty} E_{\lambda_n}$ and, for all states $W \in \mathcal{S}(\mathcal{H})$, $\operatorname{tr} W E_{\lambda_n} \to \operatorname{tr} W E_\lambda$. Now consider an arbitrary sequence $\{\mu_n\}_{n \in \mathbb{N}}$ converging to λ from the right, i.e., $\mu_n \in \mathbb{R}$, $\mu < \mu_n$, and $\mu_n \to \lambda$ as $n \to \infty$. Given $W \in \mathcal{S}(\mathcal{H})$ and $\varepsilon > 0$, choose n_0 such that $\operatorname{tr} W E_\lambda \leq \operatorname{tr} W E_{\lambda_{n_0}} < \operatorname{tr} W E_\lambda + \varepsilon$ and $N \in \mathbb{N}$ such that $\lambda < \mu_n < \lambda_{n_0}$ for all $n \geq N$. It follows that $E_\lambda \leq E_{\mu_n} \leq E_{\lambda_{n_0}}$ and

$$\operatorname{tr} W E_\lambda \leq \operatorname{tr} W E_{\mu_n} \leq \operatorname{tr} W E_{\lambda_{n_0}} < \operatorname{tr} W E_\lambda + \varepsilon$$

hold for all $n \geq N$. Hence, $\operatorname{tr} W E_{\mu_n} \to \operatorname{tr} W E_\lambda$ for all $W \in \mathcal{S}(\mathcal{H})$ as $n \to \infty$, and in consequence $E_\lambda = \sigma\text{-}\lim_{n \to \infty} E_{\mu_n}$. Thus, for all $\lambda \in C \setminus \{1\}$, $\{E_\lambda\}_{\lambda \in \mathbb{R}}$ is σ-continuous from the right.

Summarizing, $\{E_\lambda\}_{\lambda \in \mathbb{R}}$ is increasing, σ-continuous from the right, and equal to 0 and I for $\lambda < 0$ and $\lambda \geq 1$, respectively. Moreover, since $\{E_\lambda \mid \lambda \in (\mathbb{R} \setminus C) \cup \{1\}\} = S$, $\{E_\lambda\}_{\lambda \in \mathbb{R}}$ generates $\mathcal{L}$. □

5. Sharp observables and countably complete Boolean sublattices

An increasing family $\{E_\lambda\}_{\lambda \in \mathbb{R}}$ of orthonormal projections being σ-continuous from the right and fulfilling $\sigma\text{-}\lim_{\lambda \to -\infty} E_\lambda = 0$ and $\sigma\text{-}\lim_{\lambda \to \infty} E_\lambda = I$ is usually called a spectral family. So the family of Lemma 6 is a particular spectral family. Given any spectral family $\{E_\lambda\}_{\lambda \in \mathbb{R}}$, it is well known that there is a unique spectral measure $\boldsymbol{E}$ on the Borel sets $\mathcal{B}(\mathbb{R})$ of $\mathbb{R}$ (in physical terms, a sharp observable on $(\mathbb{R}, \mathcal{B}(\mathbb{R}))$ satisfying

$$\boldsymbol{E}(]\lambda, \mu]) = E_\mu - E_\lambda \tag{30}$$

for all $\lambda, \mu \in \mathbb{R}$, $\lambda < \mu$. To present a self-contained treatise, we prove this statement.

Let $\{E_\lambda\}_{\lambda \in \mathbb{R}}$ be a spectral family and $\varphi \in \mathcal{H}$. The function $\lambda \mapsto \langle \varphi | E_\lambda \varphi \rangle$ is, except for normalization, a cumulative distribution function, and there is exactly one (finite) measure m_φ on $\mathcal{B}(\mathbb{R})$ such that

$$m_\varphi(]\lambda, \mu]) = \langle \varphi | E_\mu \varphi \rangle - \langle \varphi | E_\lambda \varphi \rangle = \langle \varphi | \boldsymbol{E}(]\lambda, \mu]) \varphi \rangle \tag{31}$$

where $\boldsymbol{E}(]\lambda,\mu]) := E_\mu - E_\lambda$, $\lambda<\mu$ (see, e.g., [20]). Defining the complex measures

$$m_{\varphi\psi} := \frac{1}{4}\sum_{j=0}^{3} i^j m_{\psi+i^j\varphi} \tag{32}$$

we obtain, by polarization,

$$m_{\varphi\psi}(]\lambda,\mu]) = \langle\varphi|\boldsymbol{E}(]\lambda,\mu])\psi\rangle.$$

It follows that

$$m_{\varphi,\varphi+\psi}(]\lambda,\mu]) = m_{\varphi\psi_1}(]\lambda,\mu]) + m_{\varphi,\psi_2}(]\lambda,\mu]).$$

Since the measures $m_{\varphi,\varphi+\psi}$ and $m_{\varphi\psi_1}+m_{\varphi\psi_2}$ coincide on all half-open intervals, they must be equal. Therefore, adding some analogous arguments, $(\varphi,\psi)\mapsto m_{\varphi\psi}$ is linear in ψ and conjugate linear in φ.

By means of the limit properties of $\{E_\lambda\}_{\lambda\in\mathbb{R}}$ for $\lambda\to\pm\infty$ and the continuity properties of a measure, Eq. (31) implies that $m_\varphi(\mathbb{R}) = \|\varphi\|^2$. Thus, it follows from (32) that, for any Borel set $B\in\mathcal{B}(\mathbb{R})$,

$$|m_{\varphi\psi}(B)| \le \frac{1}{4}\sum_{j=0}^{3} m_{\psi+i^j\varphi}(B) \le \frac{1}{4}\sum_{j=0}^{3} m_{\psi+i^j\varphi}(\mathbb{R}) = \frac{1}{4}\sum_{j=0}^{3}\|\psi+i^j\varphi\|^2 \le (\|\varphi\|+\|\psi\|)^2.$$

For unit vectors $\varphi,\psi\in\mathcal{H}$, we have $|m_{\varphi\psi}(B)|\le 4$; by linearity, this entails that $|m_{\varphi\psi}(B)|\le 4\|\varphi\|\|\psi\|$ for arbitrary vectors $\varphi,\psi\in\mathcal{H}$. Hence, for every B, $(\varphi,\psi)\mapsto m_{\varphi\psi}$ is a bounded sesquilinear functional and there exists a bounded linear operator $\boldsymbol{E}(B)$ such that

$$m_{\varphi\psi}(B) = \langle\varphi|\boldsymbol{E}(B)\psi\rangle. \tag{33}$$

Since $0\le\langle\varphi|\boldsymbol{E}(B)\varphi\rangle = m_\varphi(B)\le m_\varphi(\mathbb{R}) = \|\varphi\|^2$ we obtain that $\boldsymbol{E}(B)=\boldsymbol{E}(B)^*$ and $0\le\boldsymbol{E}(B)\le I$, i.e., $\boldsymbol{E}(B)\in\mathcal{E}(\mathcal{H})$.

We still have to show that, for all Borel sets $B\in\mathcal{B}(\mathbb{R})$, $\boldsymbol{E}(B)$ is a projection. To do this, we use a representation of the measure m_φ that is related to the construction of m_φ by means of the cumulative distribution function $\lambda\mapsto\langle\varphi|E_\lambda\varphi\rangle$. In fact, we have, for all $B\in\mathcal{B}(\mathbb{R})$,

$$m_\varphi(B) = m_\varphi^*(B) = \inf\sum_{j=1}^{\infty}(\langle\varphi|E_{\mu_j}\varphi\rangle - \langle\varphi|E_{\lambda_j}\varphi\rangle) = \inf\sum_{j=1}^{\infty}\langle\varphi|\boldsymbol{E}(]\lambda_j,\mu_j])\varphi\rangle, \tag{34}$$

where m_φ^* is the outer measure occurring in the construction of m_φ and the infimum is taken over all sequences of half-open intervals $]\lambda_j,\mu_j]$ with $B\subseteq\bigcup_{j=1}^\infty]\lambda_j,\mu_j]$. Defining $D_1 :=]\lambda_1,\mu_1]$ and $D_j :=]\lambda_j,\mu_j]\setminus\bigcup_{k=1}^{j-1}]\lambda_k,\mu_k]$ for $j\ge 2$, we observe that $\bigcup_{j=1}^\infty]\lambda_j,\mu_j] = \bigcup_{j=1}^\infty D_j$ and $D_j\cap D_k=\emptyset$ for $j\ne k$ (note the analogy with Lemma 3 and Eq. (19)). Further, each D_j is the union of finitely many disjoint half-open intervals. Hence, there is a sequence of pairwise disjoint intervals $]a_k,b_k]$ so that $B\subseteq\bigcup_{j=1}^\infty]\lambda_j,\mu_j] = \bigcup_{k=1}^\infty]a_k,b_k]$ and $\sum_{k=1}^\infty(\langle\varphi|E_{b_k}\varphi\rangle - \langle\varphi|E_{a_k}\varphi\rangle)\le\sum_{j=1}^\infty(\langle\varphi|E_{\mu_j}\varphi\rangle - \langle\varphi|E_{\lambda_j}\varphi\rangle)$. In consequence, it suffices to take the infimum in (34) over all sequences of disjoint intervals $]a_k,b_k]$ with $B\subseteq\bigcup_{k=1}^\infty]a_k,b_k]$.

For the disjoint pairwise intervals $]a_k,b_k]$ the projections $\boldsymbol{E}(]a_k,b_k])$ are mutually orthogonal, and we obtain from (33) and (34) that

$$\langle\varphi|\boldsymbol{E}(B)\varphi\rangle = \inf\sum_{j=1}^{\infty}\langle\varphi|\boldsymbol{E}(]a_k,b_k])\varphi\rangle = \inf\left\langle\varphi\left|\left(\sum_{j=1}^{\infty}\boldsymbol{E}(]a_k,b_k])\right)\varphi\right.\right\rangle, \tag{35}$$

where the latter sum converges in the σ-topology, for instance. Let S be the set of all orthogonal projections $\sum_{j=1}^{\infty} \boldsymbol{E}(]a_k, b_k])$ occurring in (35). So we can write

$$\langle \varphi | \boldsymbol{E}(B) \varphi \rangle = \inf_{P \in S} \langle \varphi | P \varphi \rangle. \tag{36}$$

It follows that $\boldsymbol{E}(B) \leq P$ for all $P \in S$ and thus $\boldsymbol{E}(B) \leq \inf_{P \in S} P$. But note that this infimum must be taken in $\mathcal{E}(\mathcal{H})$, provided it exists. Fortunately, the infimum of a subset of $\partial_e \mathcal{E}(\mathcal{H})$ taken in $\mathcal{E}(\mathcal{H})$ is the same as the infimum taken in $\partial_e \mathcal{E}(\mathcal{H})$, as we shall prove below. So $\inf_{P \in S} P = \bigwedge_{P \in S} P$ exists and belongs to $\partial_e \mathcal{E}(\mathcal{H})$. Now, from $\bigwedge_{P \in S} P \leq P$ for all $P \in S$ and (36) we conclude that

$$\left\langle \varphi \middle| \left(\bigwedge_{P \in S} P \right) \varphi \right\rangle \leq \inf_{P \in S} \langle \varphi | P \varphi \rangle = \langle \varphi | \boldsymbol{E}(B) \varphi \rangle.$$

Hence, $\bigwedge_{P \in S} P \leq \boldsymbol{E}(B)$. Thus, $\boldsymbol{E}(B) = \bigwedge_{P \in S} P$, and $\boldsymbol{E}(B)$ is an orthogonal projection.

Let $S \subseteq \partial_e \mathcal{E}(\mathcal{H})$ and $\bigwedge_{P \in S} P$ be the greatest lower bound of S taken in $\partial_e \mathcal{E}(\mathcal{H})$. It could be that the infimum of S taken in $\mathcal{E}(\mathcal{H})$ does not exist or is greater than $\bigwedge_{P \in S} P$. However, we now show that, for $A \in \mathcal{E}(\mathcal{H})$, $A \leq P$ for all $P \in S$ implies $A \leq \bigwedge_{P \in S} P$. Then $\bigwedge_{P \in S} P$ is also the greatest lower bound of S taken in $\mathcal{E}(\mathcal{H})$.

For $E := \bigwedge_{P \in S} P$ we have that $R(E) = \bigcap_{P \in S} R(P)$ and

$$R(E)^{\perp} = \left(\bigcap_{P \in S} R(P) \right)^{\perp} = \bigvee_{P \in S} R(P)^{\perp} = \overline{\operatorname{span} \bigcup_{P \in S} R(P)^{\perp}}, \tag{37}$$

where $R(E)$ and $R(P)$ denote the ranges of E and P, respectively, and $\bigvee_{P \in S} R(P)^{\perp}$ is the smallest closed subspace of $\mathcal{H}$ containing all the subspaces $R(P)^{\perp}$. Let $A \in \mathcal{E}(\mathcal{H})$ and suppose $A \leq P$ for all $P \in S$. Considering any $P \in S$, it follows that, for $\varphi \in R(E) \subseteq R(P)$,

$$\langle \varphi | A \varphi \rangle \leq \langle \varphi | P \varphi \rangle = \langle \varphi | \varphi \rangle = \langle \varphi | E \varphi \rangle \tag{38}$$

and, for $\chi \in R(P)^{\perp}$,

$$\| A^{\frac{1}{2}} \chi \|^2 = \langle \chi | A \chi \rangle \leq \langle \chi | P \chi \rangle = 0.$$

Thus, $A\chi = 0$ for χ belonging to any subspace $R(P)^{\perp}$, and, by (37), $A\chi = 0$ for $\chi \in R(E)^{\perp}$. Hence, decomposing $\psi \in \mathcal{H}$ according to $\psi = \varphi + \chi$, $\varphi \in R(E)$, $\chi \in R(E)^{\perp}$, and taking account of (38), we obtain that

$$\langle \psi | A \psi \rangle = \langle \psi | A \varphi \rangle = \langle \varphi | A \varphi \rangle \leq \langle \varphi | E \varphi \rangle = \langle \psi | E \psi \rangle,$$

i.e., $A \leq E = \bigwedge_{P \in S} P$.

Summarizing, we have constructed a map $\boldsymbol{E} : \mathcal{B}(\mathbb{R}) \to \partial_e \mathcal{E}(\mathcal{H})$, $B \mapsto \boldsymbol{E}(B)$, such that $\boldsymbol{E}(]\lambda, \mu]) = E_\mu - E_\lambda$, $\lambda < \mu$, and for every $\varphi \in \mathcal{H}$

$$B \mapsto \langle \varphi | \boldsymbol{E}(B) \varphi \rangle$$

is a measure, namely, m_φ. From $m_\varphi(\emptyset) = 0$ and $m_\varphi(\mathbb{R}) = \|\varphi\|^2$ it follows that $\boldsymbol{E}(\emptyset) = 0$ and $\boldsymbol{E}(\mathbb{R}) = I$, and the σ-additivity of m_φ implies that $\boldsymbol{E}(\bigcup_{j=1}^{\infty} B_j) = \sum_{j=1}^{\infty} \boldsymbol{E}(B_j)$ where $B_j \cap B_k = \emptyset$ for $j \neq k$ and the sum converges in the σ-topology, for instance. That is, $\boldsymbol{E}$ is a spectral measure satisfying (30). Two spectral measures satisfying (30) for a given spectral family must be the same since every measure m_φ is uniquely determined by its values on the half-open intervals and all the m_φ determine $\boldsymbol{E}$ uniquely according to $m_\varphi(B) = \langle \varphi | \boldsymbol{E}(B) \varphi \rangle$. Finally, $\boldsymbol{E}(B_1 \cup B_2) = \boldsymbol{E}(B_1) + \boldsymbol{E}(B_2)$ for $B_1 \cap B_2 = \emptyset$ entails $\boldsymbol{E}(B_1)\boldsymbol{E}(B_2) = 0$ because all $\boldsymbol{E}(B)$ are orthogonal projections.

The following lemma states important properties of any sharp observable.

Lemma 7. *Let $\boldsymbol{E}$ be a sharp observable on any measurable space (M, Ξ).*

(a) The range $R(\boldsymbol{E}) = \boldsymbol{E}(\Xi) = \{\boldsymbol{E}(B) | B \in \Xi\}$ is a countably complete Boolean sublattice of $\partial_e \mathcal{E}(\mathcal{H})$. Moreover, $\boldsymbol{E}$ satisfies (i) $\boldsymbol{E}(\emptyset) = 0$, $\boldsymbol{E}(M) = I$, (ii) $\boldsymbol{E}(CB) = \boldsymbol{E}(B)^\perp$ where CB denotes the complement of B, and (iii) $\boldsymbol{E}(\bigcup_{j=1}^\infty B_j) = \bigvee_{j=1}^\infty \boldsymbol{E}(B_j)$ and $\boldsymbol{E}(\bigcap_{j=1}^\infty B_j) = \bigwedge_{j=1}^\infty \boldsymbol{E}(B_j)$ where $\{B_j\}_{j\in\mathbb{N}}$ is any sequence of measurable sets. That is, $\boldsymbol{E}\colon \mathcal{B}(\mathbb{R}) \to \partial_e \mathcal{E}(\mathcal{H})$ is a σ-homomorphism.

(b) If the σ-algebra Ξ is generated by $\Gamma \subseteq \Xi$, then $R(\boldsymbol{E})$ is generated by $\boldsymbol{E}(\Gamma) = \{\boldsymbol{E}(B) \mid B \in \Gamma\}$.

Proof. Clearly, $\boldsymbol{E}(\emptyset) = 0 \in R(\boldsymbol{E})$ and $\boldsymbol{E}(M) = I \in R(\boldsymbol{E})$. By the additivity of $\boldsymbol{E}$, we have the orthogonal decompositions

$$\begin{aligned}\boldsymbol{E}(B_1) &= \boldsymbol{E}(B_1 \setminus B_2) + \boldsymbol{E}(B_1 \cap B_2) \\ \boldsymbol{E}(B_2) &= \boldsymbol{E}(B_2 \setminus B_1) + \boldsymbol{E}(B_1 \cap B_2),\end{aligned} \tag{39}$$

where $B_1, B_2 \in \Xi$. So any two sharp effects of $R(\boldsymbol{E})$ are compatible and

$$\boldsymbol{E}(B_1)\boldsymbol{E}(B_2) = \boldsymbol{E}(B_1 \cap B_2) = \boldsymbol{E}(B_2)\boldsymbol{E}(B_1).$$

This implies that $\boldsymbol{E}(B_1) \wedge \boldsymbol{E}(B_2) = \boldsymbol{E}(B_1 \cap B_2) \in R(\boldsymbol{E})$ and further, by induction,

$$\bigwedge_{j=1}^{n} \boldsymbol{E}(B_j) = \boldsymbol{E}\left(\bigcap_{j=1}^{n} B_j\right) \in R(\boldsymbol{E}), \tag{40}$$

where $B_1, \ldots, B_n \in \Xi$. To show that, for a sequence of sets $B_j \in \Xi$, the equality

$$\bigwedge_{j=1}^{\infty} \boldsymbol{E}(B_j) = \boldsymbol{E}\left(\bigcap_{j=1}^{\infty} B_j\right) \in R(\boldsymbol{E}) \tag{41}$$

holds, observe that $\boldsymbol{E}(\bigcap_{j=1}^\infty B_j) \leq \boldsymbol{E}(B_j)$ is valid for every B_j, and let $P \in \partial_e \mathcal{E}(\mathcal{H})$ with $P \leq \boldsymbol{E}(B_j)$ for all B_j. Then, by (40), $P \leq \bigwedge_{j=1}^n \boldsymbol{E}(B_j) = \boldsymbol{E}\left(\bigcap_{j=1}^n B_j\right)$. Using one of the continuity properties of the measure $B \mapsto \langle\varphi|\boldsymbol{E}(B)\varphi\rangle$, $\varphi \in \mathcal{H}$, we obtain that

$$\langle\varphi|P\varphi\rangle \leq \left\langle \varphi \middle| \boldsymbol{E}\left(\bigcap_{j=1}^{n} B_j\right)\varphi \right\rangle \longrightarrow \left\langle \varphi \middle| \boldsymbol{E}\left(\bigcap_{j=1}^{\infty} B_j\right)\varphi \right\rangle$$

as $n \to \infty$. Therefore, $P \leq \boldsymbol{E}(\bigcap_{j=1}^\infty B_j)$ and thus $\bigwedge_{j=1}^\infty \boldsymbol{E}(B_j) = \boldsymbol{E}(\bigcap_{j=1}^\infty B_j)$.

From (39) it follows that

$$\begin{aligned}\boldsymbol{E}(B_1) \vee \boldsymbol{E}(B_2) &= (\boldsymbol{E}(B_1 \setminus B_2) \vee \boldsymbol{E}(B_1 \cap B_2)) \vee (\boldsymbol{E}(B_2 \setminus B_1) \vee \boldsymbol{E}(B_1 \cap B_2)) \\ &= \boldsymbol{E}(B_1 \setminus B_2) \vee \boldsymbol{E}(B_2 \setminus B_1) \vee \boldsymbol{E}(B_1 \cap B_2) \\ &= \boldsymbol{E}(B_1 \setminus B_2) + \boldsymbol{E}(B_2 \setminus B_1) + \boldsymbol{E}(B_1 \cap B_2) \\ &= \boldsymbol{E}(B_1 \cup B_2)\end{aligned}$$

and further

$$\bigvee_{j=1}^{n} \boldsymbol{E}(B_j) = \boldsymbol{E}\left(\bigcup_{j=1}^{n} B_j\right) \in R(\boldsymbol{E}). \tag{42}$$

Since $\boldsymbol{E}(CB) = I - \boldsymbol{E}(B)$ where $B \in \Xi$,

$$\boldsymbol{E}(B)^{\perp} = \boldsymbol{E}(CB) \in R(\boldsymbol{E}). \tag{43}$$

According to (40), (42), and (43), $R(\boldsymbol{E})$ is an orthocomplemented sublattice of $\partial_e\mathcal{E}(\mathcal{H})$, and because of (41) this sublattice is countably complete. Namely,

$$\bigvee_{j=1}^{\infty} \boldsymbol{E}(B_j) = \left(\bigwedge_{j=1}^{\infty} \boldsymbol{E}(B_j)^{\perp}\right)^{\perp} = \left(\boldsymbol{E}\left(\bigcap_{j=1}^{\infty} CB_j\right)\right)^{\perp} = \boldsymbol{E}\left(\bigcup_{j=1}^{\infty} B_j\right),$$

so

$$\bigvee_{j=1}^{\infty} \boldsymbol{E}(B_j) = \boldsymbol{E}\left(\bigcup_{j=1}^{\infty} B_j\right) \in R(\boldsymbol{E})$$

also holds.

Finally, according to Lemma 2 the sublattice $R(\boldsymbol{E})$ is Boolean because its elements are mutually compatible.

To prove part (b) of the lemma, let $\mathcal{L}(\boldsymbol{E}(\Gamma))$ be the countably complete Boolean sublattice of $\partial_e\mathcal{E}(\mathcal{H})$ generated by $\boldsymbol{E}(\Gamma)$. Since $\boldsymbol{E}(\Gamma) \subseteq \boldsymbol{E}(\Xi) = R(\boldsymbol{E})$ and $R(\boldsymbol{E})$ is a countably complete sublattice, we have that

$$\mathcal{L}(\boldsymbol{E}(\Gamma)) \subseteq R(\boldsymbol{E}). \tag{44}$$

To show the converse, let $\mathcal{L}$ be any countably complete Boolean sublattice contained in $R(\boldsymbol{E})$. Then $\boldsymbol{E}^{-1}(\mathcal{L}) = \{B \in \Xi \,|\, \boldsymbol{E}(B) \in \mathcal{L}\}$ is a σ-subalgebra of Ξ. If $B \in \Gamma$, then $\boldsymbol{E}(B) \in \mathcal{L}(\boldsymbol{E}(\Gamma))$ and so $B \in \boldsymbol{E}^{-1}(\mathcal{L}(\boldsymbol{E}(\Gamma)))$. Hence,

$$\Gamma \subseteq \boldsymbol{E}^{-1}(\mathcal{L}(\boldsymbol{E}(\Gamma))) \subseteq \Xi.$$

In virtue of the preceding remark, $\boldsymbol{E}^{-1}(\mathcal{L}(\boldsymbol{E}(\Gamma)))$ is a σ-subalgebra of Ξ. Since Γ generates Ξ, we obtain $\Xi = \boldsymbol{E}^{-1}(\mathcal{L}(\boldsymbol{E}(\Gamma)))$. Consequently,

$$R(\boldsymbol{E}) = \boldsymbol{E}(\Xi) = \boldsymbol{E}[\boldsymbol{E}^{-1}(\mathcal{L}(\boldsymbol{E}(\Gamma)))] \subseteq \mathcal{L}(\boldsymbol{E}(\Gamma))$$

and thus, by (44), $R(\boldsymbol{E}) = \mathcal{L}(\boldsymbol{E}(\Gamma))$. □

After these extensive preparations, we are now able to prove the following main theorem which asserts the converse of the first statement of part (a) of the preceding lemma.

Theorem 5. *Let $\mathcal{L}$ be a countably complete Boolean sublattice of $\partial_e\mathcal{E}(\mathcal{H})$. Then there exists a sharp observable $\boldsymbol{E}$ on the Borel space $(\mathbb{R}, \mathcal{B}(\mathbb{R}))$ (in other words, a spectral measure) such that $\mathcal{L} = R(\boldsymbol{E})$.*

Proof. According to Lemma 6, there exists a spectral family $\{E_\lambda\}_{\lambda\in\mathbb{R}}$ that generates $\mathcal{L}$. The family $\{E_\lambda\}_{\lambda\in\mathbb{R}}$ determines uniquely a spectral measure $\boldsymbol{E} : \mathcal{B}(\mathbb{R}) \to \mathcal{E}(\mathcal{H})$ satisfying $\boldsymbol{E}(]\lambda,\mu]) = E_\mu - E_\lambda$. One of the continuity properties of the measures $B \mapsto \langle\varphi|\boldsymbol{E}(B)\varphi\rangle$, $\varphi \in \mathcal{H}$, and $\sigma\text{-}\lim_{\lambda\to-\infty} E_\lambda = 0$ imply that $\langle\varphi|\boldsymbol{E}(]-\infty,\mu])\varphi\rangle = \langle\varphi|E_\mu\varphi\rangle$ which entails $\boldsymbol{E}(]-\infty,\lambda]) = E_\lambda$ for all $\lambda \in \mathbb{R}$.

The σ-algebra $\mathcal{B}(\mathbb{R})$ is generated by the half-open intervals $]\lambda,\mu]$; since $]\lambda,\mu] =]-\infty,\mu]\setminus]-\infty,\lambda]$ and every interval $]-\infty,\lambda]$ can be represented as a countable union of intervals of the form $]\lambda,\mu]$, $\mathcal{B}(\mathbb{R})$ is generated by the intervals $]-\infty,\lambda]$, $\lambda \in \mathbb{R}$, as well. Then, by part (b) of Lemma 7, $R(\boldsymbol{E})$ is generated by

$$\{\boldsymbol{E}(]-\infty,\lambda]) \,|\, \lambda \in \mathbb{R}\} = \{E_\lambda \,|\, \lambda \in \mathbb{R}\}.$$

Hence, $R(\boldsymbol{E}) = \mathcal{L}$. □

Corollary. *Any set of mutually compatible (i.e., pairwise coexistent) sharp effects is contained in the range of a sharp observable on* $(\mathbb{R}, \mathcal{B}(\mathbb{R}))$ *and is therefore coexistent.*

Proof. An arbitrary set $S \subseteq \partial_e \mathcal{E}(\mathcal{H})$ is contained in its bicompatant S^{cc}. If S is compatible, then S^{cc} is compatible also. According to Theorem 3, S^{cc} is a countably complete Boolean sublattice of $\partial_e \mathcal{E}(\mathcal{H})$. Taking account of the last theorem, we obtain that

$$S \subseteq S^{cc} = R(\boldsymbol{E})$$

where $\boldsymbol{E}$ is some sharp observable on $(\mathbb{R}, \mathcal{B}(\mathbb{R}))$. □

Within the framework of Hilbert space and bounded linear operators acting in Hilbert space, we proved that, for sharp effects, the concepts of pairwise coexistence and coexistence are equivalent.

References

[1] Gudder S 1979 *Stochastic Methods in Quantum Mechanics* (New York: North Holland)
[2] Ludwig G 1983 *Foundations of Quantum Mechanics I* (New York: Springer)
[3] Busch P, Lahti P, Pellonpää J-P and Ylinen K 2016 *Quantum Measurement* (Cham: Springer)
[4] Ludwig G 1964 Versuch einer axiomatischen Grundlegung der Quantenmechanik und allgemeinerer physikalischer Theorien *Z. Phys.* **181** 233–60
[5] Ludwig G 1967 Attempt of an axiomatic foundation of quantum mechanics and more general theories II *Commun. Math. Phys.* **4** 331–48
[6] Ludwig G 1968 Attempt of an axiomatic foundation of quantum mechanics and more general theories III *Commun. Math. Phys.* **9** 1–12
[7] Ludwig G 1970 *Deutung des Begriffs "physikalische Theorie" und axiomatische Grundlegung der Hilbertraumstruktur der Quantenmechanik durch Hauptsätze des Messens* (Lecture Notes in Physics vol 4) (Berlin: Springer)
[8] Gudder S 1970 Axiomatic quantum mechanics and generalized probability theory *Probabilistic Methods in Applied Mathematics 2* ed A Bharucha-Reid (New York: Academic Press) pp 53–129
[9] Riesz F and Szőkefalvi-Nagy B 1972 *Functional Analysis* (New York: Ungar)
[10] Achieser N I and Glasmann I M 1968 *Theorie der linearen Operatoren im Hilbert-Raum* (Berlin: Akademie-Verlag)
[11] Varadarajan V S 1968 *Geometry of Quantum Theory 1* (Princeton: Van Nostrand)
[12] Loomis L 1947 On the representation of σ-complete Boolean algebras *Bull. Amer. Math. Soc.* **53** 757–60
[13] Stulpe W 2014 From the attempt of certain classical reformulations of quantum mechanics to quasi-probability representations *J. Math. Phys.* **55** 012109
[14] Haapasalo E, Pellonpää J-P and Uola R 2015 Compatibility properties of extreme quantum observables *Lett. Math. Phys.* **105** 661–73
[15] Davies E B 1976 *Quantum Theory of Open Systems* (London: Academic Press)
[16] Gudder S 1996 Sharp and unsharp quantum effects *Adv. Appl. Math.* **20** 169–87
[17] Nikodým O M 1966 *The Mathematical Apparatus for Quantum-Theories* (Berlin: Springer)
[18] Ludwig G 1954 *Die Grundlagen der Quantenmechanik* (Berlin: Springer)
[19] von Neumann J 1929 Zur Algebra der Funktionaloperationen und Theorie der normalen Operatoren *Math. Ann.* **102** 370–427
[20] Royden H L 1988 *Real Analysis* (New York: Macmillan)

Journal of Physics: Conference Series **1638** (2020) 012005 doi:10.1088/1742-6596/1638/1/012005

Time-energy uncertainty does not create particles

Bryan W Roberts[1] **and Jeremy Butterfield**[2]

[1] Philosophy, Logic & Scientific Method, Centre for Philosophy of Natural and Social Sciences, London School of Economics & Political Science, London, WC2A 2AE, United Kingdom

[2] Trinity College, University of Cambridge, Cambridge, CB2 1TQ, United Kingdom

E-mail: b.w.roberts@lse.ac.uk, jb56@cam.ac.uk

Abstract. In this contribution in honour of Paul Busch, we criticise the claims of many expositions that the time-energy uncertainty principle allows both a violation of energy conservation and particle creation, provided that this happens for a sufficiently short time. But we agree that there are grains of truth in these claims: which we make precise and justify using perturbation theory.

1. Introduction

In expositions of quantum theory, it is often said that energy conservation can be violated, and that particles can 'pop in to existence' out of nowhere, thanks to a time-energy uncertainty principle. Time-energy uncertainty was, of course, one of Paul Busch's areas of expertise. His insightful analyses of it began already in his two 1990 works [1, 2] and the topic remained an abiding interest of his, shown for example in his 2008 review article about time in quantum physics [3]. It seems to us very likely that Busch, with his clarity and precision of thought, would have had misgivings about this 'folklore'. Accordingly, we propose to commemorate his life and work by criticising it—in a way that we hope is worthy of his memory. But we will also argue that the folklore contains grains of truth: which we will make precise and justify, using perturbation theory.

We begin with some illustrative quotations. Thus Jones proposes that

> *a consequence of the Heisenberg Uncertainty Principle is that we can take seriously the possibility of the existence of energy non-conserving processes—provided the amount by which energy is not conserved, $E_{violation}$, exists for a time less than $t = \hbar/2E_{violation}$'.* [4, p 226]

This folklore is not just a myth of pedagogy or popular exposition. Many excellent textbooks[1] make similar comments; such as the classic quantum field theory textbook of Peskin and Schroeder:

[1] As Fermat might say: we have discovered a marvellous number of comments of this kind, which this footnote is too small to contain. Here are a couple from books about quantum field theory; though one of our main points will be that the issues are not specific to quantum field theory, but arise already in quantum mechanics. The first book is about quantum field theory's philosophical interpretation, the second about its mathematics. Thus Teller [5, p 148] reports—and we agree—that violation of local energy conservation is 'customarily excused' by the time-energy uncertainty principle; while Folland [6, p 133] writes that, 'the uncertainty principle allows the particle and/or quantum to be temporarily 'off mass-shell' between the times of emission and absorption or vice versa'.

Journal of Physics: Conference Series **1638** (2020) 012005 doi:10.1088/1742-6596/1638/1/012005

> *Even when there is not enough energy for pair creation, multiparticle states appear, for example, as intermediate states in second-order perturbation theory. We can think of such states as existing only for a very short time, according to the uncertainty principle* $\Delta E \cdot \Delta t = \hbar$. *As we go to higher orders in perturbation theory, arbitrarily many such 'virtual' particles can be created.* [7, p 13]

Similarly, in their textbook on general relativity, Hobson *et al* [8] say that time-energy uncertainty is responsible not just for virtual particles, but for concrete physical predictions like Hawking radiation[2]:

> *Hawking's original calculation uses the techniques of quantum field theory, but we can derive the main results very simply from elementary arguments. ... Pair creation violates the conservation of energy and so is classically forbidden. In quantum mechanics, however, one form of Heisenberg's uncertainty principle is* $\Delta t \Delta E = \hbar$, *where* ΔE *is the minimum uncertainty in the energy of a particle that resides in a quantum mechanical state for a time* Δt. *Thus, provided the pair annihilates in a time less than* $\Delta t = \hbar/\Delta E$, *where* ΔE *is the amount of energy violation, no physical law has been broken.* [8, §11.11]

Agreed: not everyone endorses this use of time-energy uncertainty to justify particle creation. In particular, Griffiths [9] is unimpressed:

> *It is often said that the uncertainty principle means that energy is not strictly conserved in quantum mechanics-that you're allowed to 'borrow' energy* ΔE, *as long as you 'pay it back' in a time* $\Delta t \sim \hbar/2\Delta E$; *the greater the violation, the briefer the period over which it can occur. There are many legitimate readings of the energy-time uncertainty principle, but this is not one of them. Nowhere does quantum mechanics license violation of energy conservation.* [9, p 115]

We agree with Griffiths.[3] Of course, mass-energy is exactly conserved in an isolated physical system, in quantum physics no less than classical physics. In quantum theory, the unitary propagator U_t (a strongly continuous unitary representation of the reals under addition) can be written $U_t = e^{-itH}$, where the self-adjoint generator H is the energy. And, for any initial state ρ that evolves unitarily according to $\rho_t = U_t \rho U_t^*$, the energy expectation value does not change over time, since H and U_t commute: $\mathrm{Tr}(\rho_t H) = \mathrm{Tr}(\rho U_t^* H U_t) = \mathrm{Tr}(\rho H)$ for all $t \in \mathbb{R}$.

Nevertheless, what the folklore says about time, energy, and energy conservation contains some grains of truth. We will focus on three ideas:

(i) (*non-conservation*) There is some sense in which 'energy' associated with a perturbed system is not conserved;

(ii) (*particle creation*) There is some sense in which that non-conservation allows the non-conservation of particle-number; and

(iii) (*shorter times*) There is some sense in which more particle creation occurs during shorter times.

So our aim is to do the exercise of making these statements precise, and verifying them. Our lesson will be that, in each case, it is not a time-energy uncertainty relation that provides the wiggle-room to create particles: hence our slogan, 'time-energy uncertainty does not create

[2] There is a sense in which Hawking radiation [10], when viewed as a comparison between a quantum field theory constructed at past null infinity and one constructed at future null infinity, is indeed associated with particle creation. However, this is a matter of inequivalent vacua, and so spontaneous symmetry breaking, not of time-energy uncertainty.

[3] Another dissenter is Bunge [11], who gives a short but scathing criticism of understanding virtual particles in terms of time-energy uncertainty.

particles.' The particles are rather best viewed as artefacts of the shifted perspective one adopts when approximating a physical system using perturbation theory. We will discuss the statements (i), (ii) and (iii) in Sections 3, 4 and 5, respectively. But we precede this with an overview, Section 2, about perturbation theory, and the way it justifies the idea of a virtual state.

2. The perturbation view of virtual states

For our purposes, a *quantum system* is a triple $(\mathcal{H}, \mathcal{A}, t \mapsto U_t)$, where $\mathcal{H}$ is a separable complex Hilbert space, $\mathcal{A}$ is a von Neumann algebra of linear operators on $\mathcal{H}$, and $t \mapsto U_t$ is a strongly continuous one-parameter unitary representation in $\mathcal{A}$ of the group $\mathbb{R}$ under addition. The Hilbert space represents a collection of states; the algebra contains a collection of operators that represent observables; and the unitary representation provides the dynamics. By Stone's theorem, there exists a unique self-adjoint operator $H \in \mathcal{A}$ (the 'energy' operator or 'Hamiltonian') such that the dynamics can be written in the form, $U_t = e^{-itH}$, for all $t \in \mathbb{R}$.

Perturbation theory uses a quantum system that is in some way tractable to approximate a quantum system that is not. For example: the energy H of a Helium atom, which contains two mutually repulsive electrons, can be approximated by a system whose energy H_0 ignores the mutual repulsion of the electrons. In such examples, one begins with a quantum system $(\mathcal{H}, \mathcal{A}, t \mapsto U_t)$ and a set of operators $\{A_\lambda\} \subset \mathcal{A}$ parametrised by a real number λ, one value of which gives the correct, physically real, or at least appropriate, operator of interest:

$$A_\lambda = A_0 + V_\lambda. \tag{1}$$

This set is constructed in such a way that as $\lambda \to 0$, we get $A_\lambda \to A_0$ in the operator norm. A_0 is called the 'unperturbed' operator, and the set A_λ is called a 'perturbation'. The hope is both that a physical system can be correctly described by A_λ for some value of λ, and also that its properties can be accurately approximated using known facts about the more tractable operator A_0. Agreed, the appropriate operator A_λ does not always exist in quantum field theory: most notoriously, when associated with a series expansion that does not converge. But to keep matters simple we will set aside this issue, since our main points do not depend on it.

Perturbation theory allows one to approximate various aspects of the operator A_λ when it can be represented in a power series expansion around $\lambda = 0$. That is, one seeks an exact expression of A_λ of the form,

$$A_\lambda = A_0 + \lambda \left(\frac{d}{d\lambda} V_\lambda \right) \Big|_{\lambda=0} + \frac{\lambda^2}{2!} \left(\frac{d^2}{d\lambda^2} V_\lambda \right) \Big|_{\lambda=0} + \cdots \tag{2}$$

The nth-order approximation of A_λ is by definition the sum of the first n terms in this series. As $n \to \infty$, the series approaches A_λ in the operator norm. A wide class of problems can be solved by adopting the simple approximation where only the first two terms are calculated. Defining $V := \frac{d}{d\lambda} V_\lambda \Big|_{\lambda=0}$ this gives what is called a 'linear' perturbation:

$$A_\lambda \approx A_0 + \lambda V. \tag{3}$$

The eigenvalue problem for an operator expressed as a linear perturbation can typically be given an approximate analysis thanks to classic results in perturbation theory.[4]

A typical way for virtual states to arise in perturbation theory is in its application to the dynamics. Again: for simplicity we set aside the issue of non-convergent expansions, so that

[4] For example, if it is the case that for all λ, $A_\lambda = A_0 + \lambda V$ has a discrete spectrum and commutes with its adjoint ($A_\lambda A_\lambda^* = A_\lambda^* A_\lambda$), then by the Mitzkin-Taussky theorem, the eigenvalues of a_λ^i of A_λ are linear in λ (in that $a_\lambda^i = a_0^i + \lambda k^i$), and its eigenvectors are analytic functions of λ [12, §2.5]. It is then possible to expand these eigenvectors around $\lambda = 0$ to give an approximation in terms of (usually already-known) eigenvectors of A_0.

the following discussion is straightforward. (Thus we forego discussion of which results can be recovered for asymptotic series.) Let $(\mathcal{H}, \mathcal{A}, t \mapsto U_t)$ be a quantum system, and consider a second one-parameter unitary representation $t \mapsto U_t^0$. We write H and H_0 for their respective Hamiltonian generators, and refer to the former as the 'perturbed' (or 'interacting') Hamiltonian, while the the latter is the 'unperturbed' Hamiltonian. We define $V := H - H_0$ and refer to it as the 'potential'. Writing $U_t = U_t^0(U_{-t}^0 U_t)$, we expand the term in parentheses as a power series around $t = 0$,

$$U_t = U_t^0 \left(I + t\frac{d}{dt}(U_{-t}^0 U_t)\Big|_{t=0} + \frac{t^2}{2!}\frac{d^2}{dt^2}(U_{-t}^0 U_t)\Big|_{t=0} + \cdots \right). \quad (4)$$

To the extent that t is close to zero and $V = H - H_0$ is small (in the operator norm), the operator U_t is approximated by summing the first N terms in this series and ignoring the rest [12, §II.3]. Note however that this partial sum of the first N terms is in general not unitary.

Writing $U_t^{(n)}$ to denote the nth term in the series, the first few orders of approximation can be calculated by applying the Leibniz rule to the derivatives:

$$\begin{aligned} U_t^{(0)} &= U_t^0; \\ U_t^{(1)} &= -itU_t^0 V; \\ U_t^{(2)} &= \tfrac{t^2}{2!}U_t^0\left([V, H_0] - V^2\right). \end{aligned} \quad (5)$$

The virtual states picture arises from imagining that each contribution to the series is the amplitude of a "scattering event" in its own right. For example, suppose ψ_a and ψ_b are orthogonal eigenvectors of the unperturbed Hamiltonian H_0, and that we wish to approximate the amplitude $\langle\psi_b, U_t\psi_a\rangle$ associated with a transition during time t from ψ_a to ψ_b. In a first-order approximation, we replace U_t with the sum of the terms $U_t^{(0)} + U_t^{(1)}$. In a more accurate second-order approximation, we replace it with $U_t^{(0)} + U_t^{(1)} + U_t^{(2)}$, and so on. With a little calculation[5], the terms in these approximations are found to be,

$$\begin{aligned} \langle\psi_b, U_t^{(0)}\psi_a\rangle &= 0; \\ \langle\psi_b, U_t^{(1)}\psi_a\rangle &= -ite^{ibt}\langle\psi_b, V\psi_a\rangle; \\ \langle\psi_b, U_t^{(2)}\psi_b\rangle &= e^{ibt}(a-b)\langle\psi_b, V\psi_a\rangle - e^{ibt}\langle\psi_b, V^2\psi_a\rangle; \end{aligned} \quad (6)$$

where a and b are the H_0 eigenvalues associated with ψ_a and ψ_b, respectively. The zeroth-order contribution $\langle\psi_b, U^{(0)}\psi_a\rangle$, viewed as a transition amplitude in its own right, describes an initial state ψ_a that never transitions to ψ_b. The first-order contribution might lead us to say that the presence of the potential V 'deflects' the initial state ψ_a to ψ_b with some probability. The second-order contribution has two terms, the first of which will just get collected together with the first-order one in the series, and the second of which is a 'deflection' by the potential V^2. Of course, we have not given any reason to view this language as anything more than short-hand; strictly speaking, each is simply a contribution to a series that approximates $\langle\psi_b, U_t\psi_a\rangle$.

Virtual states arise as 'intermediate states' in this kind of analysis. In the example just given of approximating the transition amplitude $\langle\psi_b, U_t\psi_a\rangle$, they begin to appear at the level of the second-order contribution $U_t^{(2)}$. Instead of thinking of the second-order term $e^{ibt}\langle\psi_b, V^2\psi_a\rangle$ as a deflection in the potential V^2, let us rewrite it in the form,

$$e^{ibt}\langle\psi_b, V^2\psi_a\rangle = \langle\psi_b, Ve^{ib(t-t')}\psi'\rangle\langle\psi', Ve^{ibt'}\psi_a\rangle \quad (7)$$

[5] Zeroth order: $\langle\psi_b, U_t^{(0)}\psi_a\rangle = e^{iat}\langle\psi_b, \psi_a\rangle = 0$. First-order: $\langle\psi_b, U_t^{(1)}\psi_a\rangle = -it\langle\psi_b, U_t^0 V\psi_a\rangle = -ite^{ibt}\langle\psi_b, V\psi_a\rangle$. Second-order: $\langle\psi_b, U^{(2)}\psi_b\rangle = \langle\psi_b, U_t^0[V, H_0]\psi_a\rangle - \langle\psi_b, U_t^0 V^2\psi_a\rangle = e^{ibt}(a-b)\langle\psi_b, V\psi_a\rangle + e^{ibt}\langle\psi_b, V^2\psi_a\rangle$.

where we define $\psi' := \frac{1}{|V\psi_a|}V\psi_a$, and choose any t' such that $0 < t' < t$. Now, instead of viewing the transition as going from ψ_a to ψ_b in the potential V^2 during a time t, we can view it as consisting of an intermediate transition from ψ_a to ψ' in V during time t', followed by a transition from ψ' to ψ_b in V during the later time-interval $t - t'$. The intermediate state ψ' is an example of a *virtual state*. As expected, the third-order transitions have a V^3 term that gives rise to a pair of virtual states, and so on up the series.

One can draw a Feynman diagram for each contribution $\langle\psi_b, U_t^{(n)}\psi_a\rangle$ in this series, illustrated for $U^{(0)}$, $U^{(1)}$, and $U^{(2)}$ in Figure 1. Virtual states appear, beginning in the second-order diagrams, as states without open endpoints, such as ψ' in the right-most diagram of the Figure. This encodes the fact that they are not associated with the measured in-state or out-state in the scattering experiment associated with this amplitude. However, again: we have given no reason to view each diagram as anything other than shorthand for a term $\langle\psi_b, U^{(n)}\psi_a\rangle$ in a series approximation of the amplitude $\langle\psi_a, U_t\psi_b\rangle \approx \langle\psi_a, U_t^{(0)}\psi_b\rangle + \langle\psi_a, U_t^{(1)}\psi_b\rangle + \langle\psi_a, U_t^{(2)}\psi_b\rangle + \cdots$.

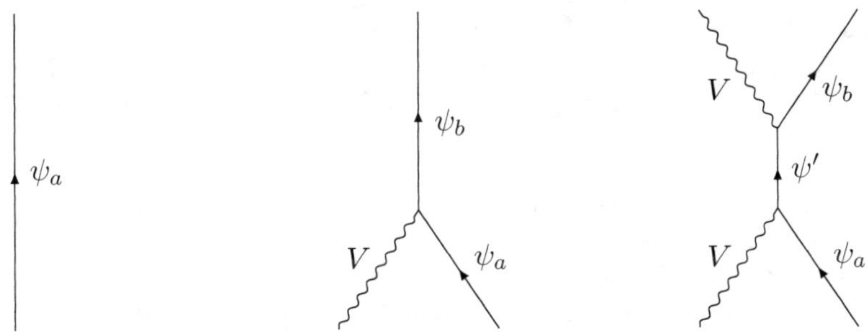

Figure 1.
Feynman diagrams for the first three terms in a perturbation series for $\langle\psi_b, U_t\psi_a\rangle$, with $U_t = e^{-it(H_0+V)}$ and ψ_a, ψ_b eigenvectors of H_0. The second-order state ψ' is 'virtual'.

3. The appearance of energy non-conservation

To sum up: the perturbation view is one of shifting perspectives. We describe a quantum system $(\mathcal{H}, \mathcal{A}, t \mapsto U_t)$ from the perspective of the nth-order approximation in a perturbation series, recognising that on this perspective, the system will sometimes appear to deviate from its true 'perturbed' dynamics, as well as from the idealised 'unperturbed' dynamics.

As it happens, one such deviation is that energy conservation can appear to fail. Thus we have our first claim:

(i) (*non-conservation*) There is some sense in which 'energy' associated with a perturbed system is not conserved.

In the context of the virtual states described in the previous section, the 'energy' that fails to be conserved is associated with the idealised Hamiltonian H_0 of the unperturbed dynamics. In the zeroth-order approximation $U^{(0)}$, it is conserved. However, in the first-order and second-order approximations, it is not, in that there are eigenstates of H_0 that are not stationary. This is due to the fact that for interacting systems with $[H_0, V] \neq 0$, the unperturbed Hamiltonian is not stationary, $U_t H_0 U_t^* \neq 0$, or equivalently, $[H, H_0] \neq 0$. So this is one thing that could be meant by saying 'energy' is not conserved in the presence of virtual states.

However, in a quantum system $(\mathcal{H}, \mathcal{A}, t \mapsto U_t)$, it is the generator H of the true i.e. perturbed dynamics U_t that represents the 'true energy' of the system, not the idealised Hamiltonian H_0. So, a more physically interesting question is whether H commutes with the approximate dynamics given by a partial sum, up to say the Nth term, of the perturbation series, eq. 4 and 5.

Of course, it may happen that for some finite N, we find that $\sum_n^N U^{(n)} = U_t$, so that $\sum_n^N U^{(n)}$ is an exact description rather than an approximate one, and the energy H is indeed conserved. But in general, the approximate dynamics $\sum_n^N U^{(n)}$ is not equal to U_t, indeed is not even unitary; and H will not be conserved under it.

On the other hand, there is a sense in which energy H for the perturbed dynamics U_t comes closer to being conserved by the approximate dynamics, the higher the order of approximation. For as we add more terms, the resulting approximate dynamics $\sum_n U_t^{(n)}$ better approximates the true dynamics U_t, at least under appropriate conditions (cf. [12]). More generally, if $\sum_{n=0}^N U_t^{(n)} \to U_t$ in the operator norm as N becomes arbitrarily large, then:[6]

$$\left[\sum_{n=0}^N U_t^{(n)}, H\right] \to 0 \text{ as } N \to +\infty, \tag{8}$$

giving perfect conservation of energy in the limit. This is ironic, in that Peskin and Schroeder [7] write: 'As we go to higher orders in perturbation theory, arbitrarily many such 'virtual' particles can be created', suggesting that energy conservation gets worse with higher order terms [7, p 13]. Agreed: there is 'more room' for virtual states in higher-order terms. But this does not mean that energy conservation gets worse. In general, it gets better as the 'true' perturbed energy becomes more distant (in the operator norm) from the idealised, unperturbed description of the system.

4. Particle creation

We now turn to the statement of (ii), the particle-number claim. For this we need to add some notion of particle number to our description. This will consist in a representation of annihilation (a_i) and creation (a_i^*) operators on $\mathcal{H}$ for $i \in \mathbb{Z}^+$, which satisfy $[a_i, a_j^*] = \delta_{ij}$ and $[a_i, a_j] = [a_i^*, a_j^*] = 0$. Let $N = \sum_i a_i^* a_i$ be the 'particle number' operator. Taking U_t^0 as the unperturbed dynamics associated with no interactions, we will here assume for simplicity that $[N, U_t^0] = 0$, and hence that the unperturbed system is one in which this (unperturbed) particle-number is conserved. (Although much more could be said about number operators in this context, we here confine ourselves to this simple situation.)

However, if $[N, U_t] \neq 0$, then particle number will in general *not* be conserved by the 'true' dynamics; and in general, it will not be conserved under the dynamics given by any of the partial-sum approximations $\sum_n U_t^{(n)}$. Agreed, N *might* be conserved by U_t and-or by an approximate dynamics. But this will not hold true in general. That is, we have:

(ii) (*particle-creation*) The dynamics generated by the partial-sum approximation $\sum_n U_t^{(n)}$ does not in general conserve the unperturbed particle number N.

5. Shorter times

We now turn to our final claim:

(iii) (*shorter times*) There is some sense in which more particle creation occurs during shorter times.

Here we enter the realm of the time-energy uncertainty principle—or, rather, principles—that are invoked by the cavalier textbook tradition with which we began. These principles are surveyed by Busch [1–3]. For us, there are two main points to make, corresponding to two broad

[6] In general, $\|A_n - B\| \to 0$ implies the commutator $\|[A_n, B]\| \to 0$, since $\|[A_n, B]\| = \|(A_n - B)B + B(B - A_n)\| \leq \|A_n - B\|\|B\| + \|B\|\|B - A_n\| = 2\|B\|\|A_n - B\|$.

Journal of Physics: Conference Series **1638** (2020) 012005 doi:10.1088/1742-6596/1638/1/012005

understandings of time-energy uncertainty. The first point is not connected to perturbation theory: it is really a warning against an untenable understanding of time-energy uncertainty. The second point will be more positive, in that it will vindicate the shorter-time claim (iii).

The first point concerns what Busch [3] suggests one should call 'external time' (or in [1]: 'pragmatic time'): namely, time as measured by clocks that are not coupled to the objects studied in the experiment. So in this role, time specifies a parameter or parameters of the experiment: e.g. an instant or duration of preparation or of measurement, or the time-interval between preparation and measurement. In this role, there seems to be no scope for uncertainty about time. And indeed, our first point here is a warning—following Busch [1]. For as Busch discusses, there is a tradition (deriving from the founding fathers of quantum theory) of an uncertainty principle between:

(1) the duration of an energy measurement; and

(2) either the range of an uncontrollable change of the measured system's energy, or the resolution of the energy measurement, or the statistical spread of the system's energy.

To give a little more detail: Busch [1, §1-2] describes how various authorities (including Landau and Peierls [13] and Landau and Lifshitz [14, §44]) endorse at least one of the following claims:

(P) An energy measurement of duration Δt leads to an uncontrollable and unpredictable change of the (previously sharply defined) energy by an amount of the order ΔE such that $\Delta E.\Delta t \geq \hbar$; so that there is no short-time reproducible (first kind) energy measurement.

(P') An energy measurement of duration Δt must carry an inaccuracy ΔE such that the uncertainty relation $\Delta E.\Delta t \geq \hbar$ is satisfied.

Busch argues, and we agree, that Aharonov and Bohm [15] refute this tradition; (see [1], especially §4, and [3, §3.1]). They give a simple model of an arbitrarily accurate and arbitrarily rapid energy measurement, where in short: two particles are confined to a line and are both free, except for an impulsive measurement of the momentum and so energy of the first by the second, with the momentum of the second being the pointer-quantity.[7] This leaves little room for 'energy uncertainty' to develop during brief (in external time) energy measurements, either in terms of uncontrollable changes in energy, or in terms of measurement inaccuracy.

Our second, more positive, point concerns what Busch [3] suggests one should call 'intrinsic time' (or in his [1]: 'dynamical time'): namely, a dynamical variable of the target system itself whose function is to measure the time, such as the position of a clock's dial relative to its face. Busch suggests that in principle, every non-stationary quantity $A \in \mathcal{A}$ and density operator state ρ defines a characteristic time interval, $\tau_\rho(A)$, in which the expectation value changes 'significantly'. For example: in the Schrödinger representation on the space of $L^2(\mathbb{R})$ functions, if $A\psi(x) := Q\psi(x) = x\psi(x)$ for all $\psi(x)$ in the domain of Q and if $\rho = E_\psi$ is the projection associated with a wave packet $\psi(x)$, then $\tau_\rho(A)$ could be defined as the time interval required for the bulk of the wave packet to shift by its width—in some sense of 'width'.

Various definitions of such a time interval are available. We will choose what is sometimes called the 'characteristic time' associated with the dispersion of an operator-state pair. This obeys what is probably the best-known time-energy uncertainty principle for intrinsic times: the *Mandelstam-Tamm uncertainty principle.*

[7] Note that Busch argues that a proper analysis and vindication of Aharonov and Bohm's refutation uses positive operator-valued measures (POVMs) to describe measurement outcomes, a notion of physical quantity that generalises projection-valued measures (PVMs); this follows the tradition of Ludwig [16,17], and is developed at book-length by Busch *et al* [18].

Journal of Physics: Conference Series **1638** (2020) 012005 doi:10.1088/1742-6596/1638/1/012005

One arrives at it for a quantum system $(\mathcal{H}, \mathcal{A}, t \mapsto U_t)$ by combining three ideas: (a) the Heisenberg equation of motion for an operator $A \in \mathcal{A}$ with $A(t) := U_t A U_t^*$,

$$i\hbar \tfrac{d}{dt} A(t) = [H, A(t)]; \tag{9}$$

(b) the Heisenberg-Robertson uncertainty principle[8], that for any quantities $A, B \in \mathcal{A}$, and density operator (quantum state) ρ,

$$\Delta_\rho A \Delta_\rho B \geq \frac{1}{2} |\operatorname{Tr}(i[A, B]\rho)| ; \tag{10}$$

and (c) the definition of a characteristic time for an operator A that does not commute with the unitary dynamics $t \mapsto U_t$, and for a time-independent state ρ that is not an eigenstate of A:

$$\tau_\rho(A) := \frac{\Delta_\rho A}{|\frac{d}{dt} \operatorname{Tr}(A(t)\rho)|}, \tag{11}$$

i.e. the time it takes for the expectation value of A to change by its standard deviation. From these definitions it immediately follows that,

$$\tau_\rho(A) \Delta_\rho(H) \geq \frac{1}{2} \hbar, \tag{12}$$

whenever $[A, H] \neq 0$ and ρ is not an eigenstate of H.

This principle can be applied in a perturbation theory analysis of a quantum system $(\mathcal{H}, \mathcal{A}, t \mapsto U_t)$ with Hamiltonian generator H and an unperturbed Hamiltonian H_0, in two ways, so as to give two construals of the shorter times claims (iii). The first way puts H_0 for A in eq. 12, while the second way puts a number operator N for A.

So first set $A := H_0$ in eq. 12, to get: $\tau_\rho(H_0)\Delta_\rho(H) \geq \frac{1}{2}\hbar$. Suppose that $\Delta_\rho H_0$ is very small, corresponding to a state ρ that is 'peaked' in unperturbed energy, so that the characteristic time $\tau_\rho(H_0)$ is comparatively small. The Mandelstam-Tamm uncertainty principle then implies that $\Delta_\rho H$ is comparatively large, and hence that the spread of the system's true (unperturbed) energy is large. In this sense, a short characteristic time and a peaked distribution of the unperturbed energy is associated with a large uncertainty of the system's true energy. Thus we have one general way to construe the shorter-times claim (iii) above. Agreed, this is a construal that uses the unperturbed Hamiltonian H_0, not an (unperturbed) number operator N.

And finally: a related construal of claim (iii) can be given, about a particle number operator N associated with the unperturbed dynamics i.e. that satisfies $[N, H_0] = 0$; so that under unperturbed dynamics, this particle number is conserved. Such an operator will typically not be conserved by the true dynamics, in that $[H, N] \neq 0$. Thus let us suppose now that ρ is a state for which the true i.e. perturbed energy H is peaked, in that $\Delta_\rho H$ is small, and hence (eq. 12) that the characteristic time (under the true i.e. perturbed dynamics) $\tau_\rho(N)$ is large. Then, the larger $|\frac{d}{dt} \operatorname{Tr}(N(t)\rho)|$ is, corresponding to a fast rate of change of the expectation value of N in the true dynamics, the larger the spread $\Delta_p N$ must be, in order to guarantee that $\tau_\rho(N) := \Delta_\rho N / |\frac{d}{dt} \operatorname{Tr}(N(t)\rho)|$ (cf. eq. 11) is sufficiently large. So this construal of claim (iii) amounts to: 'If the perturbed energy H is peaked, and the expectation value of N changes fast ('short times'), then N has large spread ('non-negligible amplitudes for values far from the expectation value')'.

To conclude: we hope to have shown that with a little *Buschian wisdom*, one can recover some grains of truth from some cavalier statements in the textbook tradition. May Paul's legacy continue to inspire our community to emulate his craftsmanship and creativity.

[8] Cf. [19, Theorem 8.1.2].

Acknowledgments
We thank Klaas Landsman, Michael Miller, Jos Uffink and Reinhard Werner for advice and encouragement; Erik Curiel and two anonymous referees for helpful comments and corrections; and the editors, not least for their patience. Bryan W. Roberts was supported by the Leverhulme Trust's Philip Leverhulme Prize and as a Visiting Fellow Commoner at Trinity College, Cambridge.

References

[1] Busch P 1990 On the energy-time uncertainty relation. Part I: Dynamical time and time indeterminacy *Found. Phys.* **20** 1–32
[2] Busch P 1990 On the energy-time uncertainty relation. Part II: Pragmatic time versus energy indeterminacy *Found. Phys.* **20** 33–43
[3] Busch P 2008 The time-energy uncertainty relation *Time In Quantum Mechanics I* (Lecture Notes in Physics vol 734) ed J G Muga *et al* (Berlin: Springer) pp 73–105
[4] Jones G T 2002 The uncertainty principle, virtual particles and real forces *Phys. Educ.* **37** 223–33
[5] Teller P 1995 *An Interpretive Introduction to Quantum Field Theory* (Princeton: Princeton University Press)
[6] Folland G B 2008 *Quantum Field Theory: A Tourist Guide for Mathematicians* (Mathematical Surveys and Monographs vol 149) (Providence, RI: American Mathematical Society)
[7] Peskin M E and Schroeder D V 1995 *An Introduction to Quantum Field Theory* (Reading, MA: Perseus)
[8] Hobson M P, Efstathiou G P and Lasenby A N 2006 *General Relativity: An Introduction for Physicists* (Cambridge: Cambridge University Press)
[9] Griffiths D J 1995 *Introduction to Quantum Mechanics* (Upper Saddle River, NJ: Prentice Hall)
[10] Hawking S W 1975 Particle creation by black holes *Comm. Math. Phys.* **43** 199–220
[11] Bunge M 1970 Virtual processes and virtual particles: real or fictitious? *Int. J. Theor. Phys.* **3** 507–8
[12] Kato T 1995 *Perturbation Theory for Linear Operators* (Berlin: Springer)
[13] Landau L and Peierls R 1931 Erweiterung des Unbestimmtheitsprinzips für die relativistische Quantentheorie *Z. Phys.* **69** 56–69
[14] Landau L and Lifshitz E M 1977 *Quantum Mechanics: Non-Relativistic Theory* (Oxford: Pergamon)
[15] Aharonov Y and Bohm D 1961 Time in the quantum theory and the uncertainty relation for time and energy *Phys. Rev.* **122** 1649–58
[16] Ludwig G 1983 *Foundations of Quantum Mechanics I* (Berlin: Springer)
[17] Ludwig G 1985 *Foundations of Quantum Mechanics II* (Berlin: Springer)
[18] Busch P, Grabowski M and Lahti P J 1995 *Operational Quantum Physics* (Berlin: Springer)
[19] Blank J, Exner P and Havlíček M 2008 *Hilbert Space Operators in Quantum Physics* (Dordrecht: Springer)

Journal of Physics: Conference Series **1638** (2020) 012006 doi:10.1088/1742-6596/1638/1/012006

Notes on Naimark's dilation theorem

Roberto Beneduci

Departimento di Fisica, Universitá della Calabria and Istituto Nazionale di Fisica Nucleare via P. Bucci, cubo 30/B, 87036 Rende, Italy

E-mail: roberto.beneduci@unical.it

Abstract. After a short review of the main properties of unsharp observables we provide examples of commutative unsharp observables obtained as the projection of sharp observables. In particular, we consider a family of PVMs defined on the tensor product Hilbert space $\mathcal{H}\otimes\mathcal{K}$ which include the one introduced by Ozawa in his modification of the von Neumann position measurement model and show that their projections onto $\mathcal{H}$ are commutative POVMs. Then, we make some observations on the concept of Naimark dilation and its connections with the integral representation of commutative POVMs. Finally, we introduce a partial order relation in the set of Naimark dilations, show that there is a minimal element and that it is unique. Minimality is a consequence of the partial ordered structure and coincides with the usual definition of minimal Naimark dilation. In passing, we use Naimark's theorem in order to prove some well known properties of positive operator valued measures.

Dedication

The present notes are dedicated to the memory of my friend Paul Busch. In 2010 we started a scientific collaboration which soon turned into a deep connection. He was a wise and precious friend. Paul's death leaves an unbridgeable loss for the community. Some of the observations I present here are answers to questions that were raised during our conversations.

1. Introduction

Positive operator valued measures are fruitfully used in quantum mechanics in order to generalize the concept of quantum observable which is otherwise represented by a self-adjoint operator. Such a generalization is rooted in the statistical character of quantum physics and opened new perspectives [1–9]. Just to recall some examples, it allowed a phase space representation of quantum mechanics where the distribution functions describing the states of the system are positive-definite [4, 7, 8, 10, 11], it made possible a rigorous description of the Heisenberg error-disturbance relations [9, 12, 13], it made clear that commutativity is not equivalent to compatibility [2, 7], and it made it possible to define a photon localization observable [4–6, 14–19]. As a relevant physical example that illustrates the problem of joint measurability in the framework of POVMs, one can consider the case of the position and momentum operators, Q, P, in the Hilbert space $\mathcal{H} = L^2(\mathbb{R})$. Although they are incompatible (they do not commute) they can be smeared to two commutative POVMs, F^Q, F^P which are non-commuting but compatible, i.e., they are the marginals of a joint POVM [2, 7]. In turn, F^Q and F^P provide examples of commutative POVMs obtained as the randomization of a spectral measure. That is a general feature of commutative POVMs, i.e., they are random versions of spectral measures [20–24]. To be more precise, they can be represented as the integral of a Markov kernel with respect to a spectral measure [22–24]. There are other integral representations of POVMs [25, 26]. In the commutative case, all of them are equivalent [23, 27]. A different kind of connection between POVMs and projection valued measures (PVMs) is established by the Naimark dilation theorem which ensures that every POVM is the projection of a PVM in an extended Hilbert space [28–31]. From the physical viewpoint, the Naimark theorem ensures the existence of a measurement for any observable [7].

In the present note we make some observations on Naimark's dilation theorem. For example, it is interesting to ask under what conditions a POVM obtained as the projection of a PVM is commutative. As an example, we analyze a family of PVMs which includes the one introduced by Ozawa [32] in his modification of the von Neumann position measurement model, and show that their projection is a commutative POVM. Then, we review some of the main results on the connection between the Naimark dilation theorem and integral representation of commutative POVMs, discuss the relevance of the Naimark theorem to the compatibility problem and derive some well known properties of POVMs by means of the Naimark theorem. Finally we analyze the concept of minimality of the Naimark dilation in the framework of partially ordered sets (posets) where minimality is a consequence of the poset structure and prove the existence of a unique minimal Naimark dilation.

In what follows, we denote by $\mathcal{B}(X)$ the Borel σ-algebra of a topological space X and by $\mathcal{L}_s(\mathcal{H})$ the space of all bounded self-adjoint linear operators acting in a Hilbert space $\mathcal{H}$ with scalar product $\langle\cdot,\cdot\rangle$. The subspace of positive operators is denoted by $\mathcal{L}_s^+(\mathcal{H})$.

Definition 1.1. *A Positive Operator Valued Measure (for short, POVM) is a map* $F : \mathcal{B}(X) \to \mathcal{L}_s^+(\mathcal{H})$ *such that:*

$$F(\bigcup_{n=1}^{\infty}\Delta_n) = \sum_{n=1}^{\infty} F(\Delta_n),$$

where $\{\Delta_n\}$ *is a countable family of disjoint sets in* $\mathcal{B}(X)$ *and the series converges in the weak operator topology. It is said to be normalized if*

$$F(X) = \mathbf{1},$$

where $\mathbf{1}$ *is the identity operator.*

Definition 1.2. *A POVM is said to be commutative if*

$$\big[F(\Delta_1), F(\Delta_2)\big] = \mathbf{0}, \quad \forall\, \Delta_1\,, \Delta_2 \in \mathcal{B}(X). \tag{1}$$

Definition 1.3. *A POVM is said to be orthogonal if* $\Delta_1 \cap \Delta_2 = \emptyset$ *implies*

$$F(\Delta_1)F(\Delta_2) = \mathbf{0}, \tag{2}$$

where $\mathbf{0}$ *is the null operator.*

Definition 1.4. *A Projection Valued Measure (for short, PVM) is an orthogonal, normalized POVM. A spectral measure is a real PVM* ($X = \mathbb{R}$).

Let E be a PVM. By equation (2),

$$\mathbf{0} = E(\Delta)E(X - \Delta) = E(\Delta)[\mathbf{1} - E(\Delta)] = E(\Delta) - E(\Delta)^2.$$

We can then restate Definition 1.4 as follows.

Definition 1.5. *A PVM E is a POVM such that $E(\Delta)$ is a projection operator for each $\Delta \in \mathcal{B}(X)$.*

In quantum mechanics, non-orthogonal normalized POVMs are also called **generalised** or **unsharp** observables while PVMs are called **standard** or **sharp** observables.

Definition 1.6. *The von Neumann algebra (or W^*-algebra) $\mathcal{A}^W(F)$ generated by the POVM F is the von Neumann algebra generated by the set $\{F(\Delta)\}_{\Delta\in\mathcal{B}(X)}$. It is the smallest $*$-algebra of bounded operators closed in the weak operator topology and generated by $\{F(\Delta)\}_{\Delta\in\mathcal{B}(X)}$.*

In what follows, we shall always refer to normalized POVMs and we shall use the term "measurable" for the Borel measurable functions. For any unit vector $\psi \in \mathcal{H}$, the map

$$\langle F(\cdot)\psi, \psi\rangle \,:\, \mathcal{B}(X) \to [0,1], \qquad \Delta \mapsto \langle F(\Delta)\psi, \psi\rangle,$$

is a measure. In the following, we shall use the symbol $d\langle F_x\psi, \psi\rangle$ to mean integration with respect to $\langle F(\cdot)\psi, \psi\rangle$.

For any real, bounded and measurable function f and for any POVM F, there is a unique [33] bounded self-adjoint operator $B \in \mathcal{L}_s(\mathcal{H})$ such that

$$\langle B\psi, \psi\rangle = \int f(x) d\langle F_x\psi, \psi\rangle, \quad \text{for each} \quad \in \mathcal{H}. \tag{3}$$

If equation (3) is satisfied, we write $B = \int f(x)dF_x$ or $B = \int f(x)F(dx)$ equivalently.

Definition 1.7. *The spectrum $\sigma(F)$ of a POVM F is the closed set*

$$\{x \in X : F(\Delta) \neq \mathbf{0},\, \forall\Delta \textit{ open},\, x \in \Delta\}.$$

By the spectral Theorem [34], there is a one-to-one correspondence between spectral measures, E, and self-adjoint operators, B, the correspondence being given by

$$B = \int \lambda dE^B_\lambda.$$

Notice that the spectrum of E^B coincides with the spectrum of the corresponding self-adjoint operator B. Moreover, in this case a functional calculus can be developed. Indeed, if $f : \mathbb{R} \to \mathbb{R}$ is a measurable real-valued function, we can define the self-adjoint operator [34]

$$f(B) = \int f(\lambda) dE^B_\lambda. \tag{4}$$

If f is bounded, then $f(B)$ is bounded [34]. In particular,

$$E[i] := \int t^i \, dE_t = \Big(\int t\, dE_t\Big)^i = A^i \tag{5}$$

and $A = \int t\, dE_t$ is the generator of the von Neumann algebra generated by E. We point out that if F is not projection valued, equations (5) and (4) do not hold [35] and, in order to recover the generator of $\mathcal{A}^W(F)$, we need all the moments of F. In particular, in the case of a real commutative POVM F with bounded spectrum and such that $F(\Delta)$ is discrete for any Δ, we have [36, 37]

$$A = \sum_{i=0}^{\infty} \alpha_i\, F[i], \quad \alpha_i \geq 0, \quad \sum_{i=0}^{\infty} \alpha_i < \infty,$$

where A is a generator of the von Neumann algebra $\mathcal{A}^W(F)$.

In the following we assume X to be Hausdorff, locally compact and second countable.

2. Naimark's dilation theorem

A procedure to derive a POVM is to start from a PVM E in a Hilbert space $\mathcal{H}$ and to project to a subspace $\mathcal{H}_0 \subset \mathcal{H}$, i.e., $F(\Delta) = PE(\Delta)P$ where $P\mathcal{H} = \mathcal{H}_0$. In this context it is worth analyzing the properties of the POVM F as for example its commutativity. Here we provide a family of PVMs whose projection onto a subspace gives a commutative POVM. In particular, we consider a PVM of the kind

$$\bar{Q}(\tau) = cQ(0) \otimes I + dI \otimes \bar{Q}(0), \tag{6}$$

where $\bar{Q}(0)$ is a self-adjoint operator in the Hilbert space $\mathcal{K}$, $Q(0)$ is a self-adjoint operator in the Hilbert space $\mathcal{H}$, $\bar{Q}(\tau)$ is a self-adjoint operator in the Hilbert space $\mathcal{H} \otimes \mathcal{K}$ and c and b are functions depending on τ. We suppose $\mathcal{H}$ and $\mathcal{K}$ are infinite-dimensional. An observable of this kind has been derived by Ozawa [32] as the solution of the Heisenberg equation in his model for the measurement of position. In the Ozawa model, $\bar{Q}(\tau)$ denotes the meter position observable at time τ and $Q(0)$ the position observable of the system at time $t = 0$. The outcome of the measurement is obtained by measuring the meter observable $\bar{Q}$ at time Δt[1]. Analogous equations can be derived for the momentum. Now we show that the projection of $\bar{Q}(\tau)$ onto the subspace $\mathcal{H} \subset \mathcal{H} \otimes \mathcal{K}$ is a commutative POVM. First we recall that the tensor product $\mathcal{H} \otimes \mathcal{K}$ can be represented as the direct sum[2]

$$\mathcal{H} \otimes \mathcal{K} = \mathcal{H} \oplus \mathcal{H} \oplus \cdots$$

Note also that $\mathcal{H}$ is isomorphic to $\mathcal{H} \otimes \phi$ where ϕ is a unit vector in $\mathcal{K}$. Now, let $\{\psi_i\}_i$ and $\{\phi_j\}_j$ be bases for $\mathcal{H}$ and $\mathcal{K}$ respectively. Then

$$\mathcal{H} \otimes \mathcal{K} = \bigoplus_j (\mathcal{H} \otimes \phi_j)$$

so that $\mathcal{H}$ is embedded in $\mathcal{H} \otimes \mathcal{K}$ as

$$(\mathcal{H} \otimes \phi_1) \oplus \mathbf{0} \oplus \mathbf{0} \cdots$$

Let P be the projection operator from $\mathcal{H} \otimes \mathcal{K}$ onto $\mathcal{H}$ defined by $P(\xi_1 \otimes \phi_1, \dots, \xi_i \otimes \phi_i, \dots) = \xi_1 \otimes \phi_1$ where $\xi_i \in \mathcal{H}$, $i \in \mathbb{N}$. We have

$$P(\psi \otimes \phi_j) = (\psi \otimes \phi_j)\delta_{1j}.$$

Let $E^{Q(0)}$ and $E^{\bar{Q}(0)}$ denote respectively the spectral measures of $Q(0)$ and $\bar{Q}(0)$ in equation (6). Then, the spectral measure of $\bar{Q}(\tau)$ is $E^{Q(0)} \otimes E^{\bar{Q}(0)}$. Indeed, we have

$$cQ(0) \otimes I + dI \otimes \bar{Q}(0) = \int \int (c\,\lambda + d\,\beta)\, dE^{Q(0)}_\lambda \otimes dE^{\bar{Q}(0)}_\beta.$$

What about the projection of $E^{Q(0)} \otimes E^{\bar{Q}(0)}$ onto $\mathcal{H}$? For any $\psi_i \in \mathcal{H}$, we have

$$\begin{aligned} P(E^{Q(0)}(\Delta_1) \otimes E^{\bar{Q}(0)}(\Delta_2))P(\psi_i \otimes \phi_1) &= P(E^{Q(0)}(\Delta_1) \otimes E^{\bar{Q}(0)}(\Delta_2))(\psi_i \otimes \phi_1) \\ &= P(E^{Q(0)}(\Delta_1)\psi_i \otimes E^{\bar{Q}(0)}(\Delta_2)\phi_1) \\ &= (E^{Q(0)}(\Delta_1)\psi_i \otimes \alpha_1\phi_1), \end{aligned}$$

[1] τ is chosen such that $0 \le \tau \le \Delta t$ with $K\Delta t = 1$ where K is the coupling constant which is supposed to be very large, $K \gg 1$.
[2] We recall that the direct sum of a countable family of Hilbert spaces $\{(\mathcal{H}_i, \langle\cdot,\cdot\rangle_i)\}_{i\in\mathbb{N}}$ is the space $\bigoplus_{i=1}^{\infty} \mathcal{H}_i = \{(\psi^1, \dots, \psi^i, \dots) \,|\, \psi^i \in \mathcal{H}_i, \sum_{i=1}^{\infty} \|\psi^i\|^2 < \infty\}$ with the inner product $\langle(\psi^1, \dots, \psi^i, \dots), (\phi^1, \dots, \phi^i, \dots)\rangle = \sum_{i=1}^{\infty} \langle\psi^i, \phi^i\rangle_i$.

Journal of Physics: Conference Series **1638** (2020) 012006 doi:10.1088/1742-6596/1638/1/012006

where α_1 comes from the expansion $E^{\bar{Q}(0)}(\Delta_2)\phi_1 = \sum_l \alpha_l \phi_l$. Moreover, $\forall \psi_i \in \mathcal{H}$,

$$
\begin{aligned}
P(E^{Q(0)}(\Delta_1') \otimes E^{\bar{Q}(0)}(\Delta_2'))P(E^{Q(0)}(\Delta_1)\psi_i \otimes \alpha_1\phi_1) &= \\
&= P\big(E^{Q(0)}(\Delta_1')E^{Q(0)}(\Delta_1)\psi_i \otimes E^{\bar{Q}(0)}(\Delta_2')\alpha_1\phi_1\big) \\
&= (E^{Q(0)}(\Delta_1')E^{Q(0)}(\Delta_1)\psi_i \otimes \alpha_1'\alpha_1\phi_1),
\end{aligned}
$$

where α_1' comes from the expansion $E^{\bar{Q}(0)}(\Delta_2')\phi_1 = \sum_l \alpha_l' \phi_l$. We can now calculate

$$
\begin{aligned}
P(E^{Q(0)}(\Delta_1') \otimes E^{\bar{Q}(0)}(\Delta_2'))P(\psi_i \otimes \phi_1) &= P(E^{Q(0)}(\Delta_1') \otimes E^{\bar{Q}(0)}(\Delta_2'))(\psi_i \otimes \phi_1) \\
&= P(E^{Q(0)}(\Delta_1')\psi_i \otimes E^{\bar{Q}(0)}(\Delta_2')\phi_1) \\
&= (E^{Q(0)}(\Delta_1')\psi_i \otimes \alpha_1'\phi_1)
\end{aligned}
$$

and

$$
\begin{aligned}
&P(E^{Q(0)}(\Delta_1) \otimes E^{\bar{Q}(0)}(\Delta_2))P((E^{Q(0)}(\Delta_1')\psi_i \otimes \alpha_1'\phi_1)) \\
=&P(E^{Q(0)}(\Delta_1)E^{Q(0)}(\Delta_1')\psi_i \otimes E^{\bar{Q}(0)}(\Delta_2)\alpha_1'\phi_1) \\
=&(E^{Q(0)}(\Delta_1')E^{Q(0)}(\Delta_1)\psi_i \otimes \alpha_1'\alpha_1\phi_1),
\end{aligned}
$$

which show that the POVM $P(E^{Q(0)}(\Delta_1') \otimes E^{\bar{Q}(0)}(\Delta_2'))P$ is commutative. Thus $P(E^{Q(0)}(\Delta_1') \otimes E^{\bar{Q}(0)}(\Delta_2'))P$ should be the randomization of a sharp position observable and it would be interesting to analyze the properties of the Markov kernel that realizes the randomization. Moreover, the sharp version should be equivalent to the projection of the pointer observable $cQ(0) \otimes I + dI \otimes \bar{Q}(0)$ (see Theorem 2.7 below). Those are problems deserving further work. Anyway, we have a relevant physical example (perhaps a family of physical examples) where the projection of a pointer observable is a commutative POVM.

An important result concerning the connections between PVMs and POVMs is due to Naimark. He proved that not only the projection of a PVM provides a POVM but that every non-orthogonal POVM is the projection of a PVM.

Theorem 2.1 (Naimark [3, 8, 28–31]). *Let F be a POVM. Then, there exist an extended Hilbert space $\mathcal{H}^+$ and a PVM E^+ on $\mathcal{H}^+$ such that*

$$F(\Delta)\psi = PE^+(\Delta)\psi, \quad \forall \psi \in \mathcal{H}$$

where P is the operator of projection onto $\mathcal{H}$. The dilation $E^+ : \mathcal{H}^+ \to \mathcal{H}^+$ can be chosen such that the extended Hilbert space $\mathcal{H}^+$ coincides with the closure of $\mathrm{span}\{E^+(\Delta)\psi : \psi \in \mathcal{H},\ \Delta \in \mathcal{B}(X)\}$. *In this case the dilation is said to be minimal and is unique up to unitary transformations.*

Therefore, up to isomorphisms, there is a unique minimal Naimark dilation of F. In subsection 2.3 we reformulate the concept of minimality of the Naimark dilation in the language of partially ordered sets (posets) where the existence of a minimal dilation can be proved by Zorn's lemma. In this framework, the definition of minimal dilation (see Theorem 2.1) becomes a consequence of the poset structure.

In the case of a commutative POVM F, Naimark's theorem can be used to define a self-adjoint operator A^+ in $\mathcal{H}^+$ whose projection onto $\mathcal{H}$ gives the sharp reconstruction of F [36–41]. Let us first recall the integral representation of commutative POVMs. In the following Λ denotes a measurable space.

Definition 2.2. *A Markov kernel is a map* $\mu : \Lambda \times \mathcal{B}(X) \to [0,1]$ *such that,*

1. $\mu_\Delta(\cdot)$ *is a measurable function for each* $\Delta \in \mathcal{B}(X)$,
2. $\mu_{(\cdot)}(\lambda)$ *is a probability measure for each* $\lambda \in \Lambda$.

Definition 2.3. *A Feller Markov kernel is a Markov kernel* $\mu_{(\cdot)}(\cdot) : \Lambda \times \mathcal{B}(X) \to [0,1]$ *such that the function*

$$G(\lambda) = \int_X f(x)\, \mu_{dx}(\lambda), \quad \lambda \in \Lambda$$

is continuous and bounded whenever f *is continuous and bounded.*

Theorem 2.4 ([24]). *A POVM* $F : \mathcal{B}(X) \to \mathcal{L}_s^+(\mathcal{H})$ *is commutative if and only if there exists a bounded self-adjoint operator* $A = \int \lambda\, dE_\lambda$ *with spectrum* $\sigma(A) \subset [0,1]$, *a subset* $\Gamma \subset \sigma(A)$, $E(\Gamma) = \mathbf{1}$, *a ring* $\mathcal{R}$ *which generates* $\mathcal{B}(X)$ *and a Feller Markov Kernel* $\mu : \Gamma \times \mathcal{B}(X) \to [0,1]$ *such that*

1) $F(\Delta) = \int_\Gamma \mu_\Delta(\lambda)\, dE_\lambda, \quad \Delta \in \mathcal{B}(X)$.
2) $\mathcal{A}^W(F) = \mathcal{A}^W(A)$.
3) μ *separates the points in* Γ.
4) μ_Δ *is continuous for each* $\Delta \in \mathcal{R}$.
5) A *is called the sharp version of* F *and is unique up to bijections.*

An equivalence relation $\sim_a$ such that $A \sim_a F$ whenever A is discrete can be introduced. That can be interpreted as the equivalence of A and F with respect to their informational content [40].

Definition 2.5. *Let* A *and* B *be self-adjoint operators. Whenever there exists a one-to-one measurable function* f *such that* $A = f(B)$, *we say that* A *is equivalent to* B *and write* $A \leftrightarrow B$.

Theorem 2.6 ([38]). *Let* $F : \mathcal{B}(\mathbb{R}) \to \mathcal{L}_s^+(\mathcal{H})$ *be a real POVM. Let* f *be a bounded measurable real valued function. Let* A^+ *be the self-adjoint operator corresponding to the Naimark dilation* E^+ *of* F. *Then,*

$$F(f) := \int f(t)\, dF_t = P f(A^+) P.$$

In the case f is unbounded, the domain of definition of the operators must be taken into account [9,35].

Theorem 2.7 ([36,37]). *Let* $F : \mathcal{B}(\mathbb{R}) \to \mathcal{L}_s^+(\mathcal{H})$ *be a commutative POVM such that the operators in the range of* F *are discrete.*[3] *Let* A *be the sharp version of* F *and* $A^+ = \int \lambda\, dE_\lambda^+$ *the Naimark operator corresponding to the Naimark dilation* E^+. *Then, there are two bounded, one-to-one functions* f *and* h *such that*

$$h(A) = \int f(t)\, dF_t = P^+ f(A^+) P.$$

Theorem 2.7 establishes that $h(A)$ is the projection of $f(A^+)$ with h and f one-to-one. According to Definition 2.5 we can say that a correspondence between A and A^+ is established. We denote such a correspondence by $A \leftrightarrow PrA^+$.

[3] $F(\Delta)$ is discrete if it has a complete set of eigenvectors.

Journal of Physics: Conference Series **1638** (2020) 012006 doi:10.1088/1742-6596/1638/1/012006

2.1. Conditions for the joint measurability

In the present section we recall the definition and some of the main theorems on the joint measurability of two POVMs. Then, we show how Naimark's dilation theorem can be used in order to give necessary and sufficient conditions for joint measurability.

Definition 2.8. *Two POVMs $F_1 : \mathcal{B}(X_1) \to \mathcal{L}_s^+(\mathcal{H})$, $F_2 : \mathcal{B}(X_2) \to \mathcal{L}_s^+(\mathcal{H})$ are compatible (or jointly measurable) if they are marginals of a joint POVM $F : \mathcal{B}(X_1 \times X_2) \to \mathcal{L}_s^+(\mathcal{H})$.*

We recall that the symbol $\mathcal{B}(X_1 \times X_2)$ denotes the product σ-algebra generated by the family of sets $\{\Delta_1 \times \Delta_2 \,:\, \Delta_1 \in \mathcal{B}(X_1),\, \Delta_2 \in \mathcal{B}(X_2)\}$.

Two POVMs F_1 and F_2 commute if $[F_1(\Delta_1), F_2(\Delta_2)] = \mathbf{0}$, for each $\Delta_1 \in \mathcal{B}(X_1)$ and $\Delta_2 \in \mathcal{B}(X_2)$. In the following, the commutativity of two POVMs F_1 and F_2 is denoted by the symbol $[F_1, F_2] = \mathbf{0}$.

As the following theorem shows, in the case of two POVMs, commutativity implies compatibility but the converse is not true, i.e, commutativity is not a necessary condition for the compatibility. That is one of the main advantage in using POVMs in order to represent quantum observables.

Theorem 2.9 ([42])**.** *Two commuting POVMs are compatible.*

Now, we use Naimark's dilation theorem in order to characterize the compatibility of two POVMs. The proof is provided for completeness.

Theorem 2.10 ([43])**.** *Two POVMs $F_1 : \mathcal{B}(X_1) \to \mathcal{L}_s^+(\mathcal{H})$ and $F_2 : \mathcal{B}(X_2) \to \mathcal{L}_s^+(\mathcal{H})$ are compatible if and only if there are two Naimark dilations $E_1^+ : \mathcal{B}(X_1) \to \mathcal{L}_s^+(\mathcal{H})$ and $E_2^+ : \mathcal{B}(X_2) \to \mathcal{L}_s^+(\mathcal{H})$ such that $[E_1^+, E_2^+] = \mathbf{0}$.*

Proof. Suppose F_1 and F_2 are compatible. Then, there is a POVM F of which F_1 and F_2 are the marginals; i.e., $F_1(\Delta_1) = F(\Delta_1 \times X_2)$, $F_2(\Delta_2) = F(X_1 \times \Delta_2)$. Let E^+ be a Naimark dilation of F and consider the PVMs $E_1^+(\Delta_1) = E^+(\Delta_1 \times X_2)$ and $E_2^+(\Delta_2) = E^+(X_1 \times \Delta_2)$. We have, $PE_1^+(\Delta_1)P = PE^+(\Delta_1 \times X_2)P = F(\Delta_1 \times X_2) = F_1(\Delta_1)$ and $PE_2^+(\Delta_2)P = PE^+(X_1 \times \Delta_2)P = F(X_1 \times \Delta_2) = F_2(\Delta_2)$. Moreover, E_1^+ and E_2^+ commute since they are the marginals of the PVM E^+.

Conversely, suppose there are two Naimark dilations E_1^+ and E_2^+ such that $[E_1^+, E_2^+] = \mathbf{0}$. Thanks to the commutativity $[E_1^+, E_2^+] = \mathbf{0}$, there is a joint PVM E^+; i.e., $E_1^+(\Delta_1) = E^+(\Delta_1 \times X_2)$, $E_2^+(\Delta_2) = E^+(X_1 \times \Delta_2)$. We have,

$$F_1(\Delta_1) = PE_1^+(\Delta_1)P = PE^+(\Delta_1 \times X_2)P = F(\Delta_1 \times X_2);$$
$$F_2(\Delta_2) = PE_2^+(\Delta_2)P = PE^+(X_1 \times \Delta_2)P = F(X_1 \times \Delta_2),$$

where $F := PE^+P$. Therefore, F is a joint POVM for F_1 and F_2. □

Note that if F_1 and F_2 are PVMs, theorems 2.10 and 2.9 imply that F_1 and F_2 are compatible if and only if they commute. Indeed, if PE_1^+P and PE_2^+P are projections then $[P, E_i^+] = \mathbf{0}$, $i = 1, 2$. Hence $[F_1, F_2] = \mathbf{0}$. Moreover [35], two PVMs are compatible if and only if they are the marginals of a PVM.

Theorem 2.10 is illustrated in the following diagram.

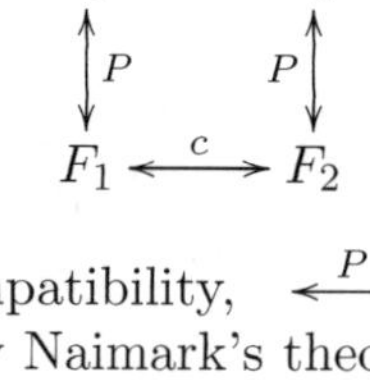

where the arrow $\xleftrightarrow{c}$ denotes compatibility, $\xleftrightarrow{P}$ denotes the relationship between a POVM and its dilation as expressed by Naimark's theorem.

Journal of Physics: Conference Series **1638** (2020) 012006 doi:10.1088/1742-6596/1638/1/012006

2.2. On some properties of POVMs

In the present subsection, we use the Naimark dilation theorem and the results in section 2.1 to prove some well known theorems about POVMs and compatibility of POVMs.

Theorem 2.11. *Let $F : \mathcal{B}(X) \to \mathcal{L}_s^+(\mathcal{H})$ be a POVM and suppose there is $\Delta \in \mathcal{B}(X)$ such that $F(\Delta)^2 = F(\Delta)$. Then, $[F(\Delta), F(\Delta')] = \mathbf{0}$, $\forall \Delta' \in \mathcal{B}(X)$.*

Proof. Let E^+ be a Naimark dilation of F. Then $PE^+(\Delta)_{|\mathcal{H}} = F(\Delta)$. Since by hypothesis $F(\Delta)$ is a projection, it must be $[P, E^+(\Delta)] = \mathbf{0}$. Then, for any $\psi \in \mathcal{H}$ and $\Delta' \in \mathcal{B}(X)$,

$$\begin{aligned} F(\Delta)F(\Delta')\psi &= PE^+(\Delta)PE^+(\Delta')\psi \\ &= PE^+(\Delta)E^+(\Delta')\psi \\ &= PE^+(\Delta')E^+(\Delta)\psi \\ &= PE^+(\Delta')E^+(\Delta)P\psi \\ &= PE^+(\Delta')PE^+(\Delta)\psi \\ &= F(\Delta')F(\Delta)\psi. \end{aligned}$$

□

Theorem 2.12. *Let $E : \mathcal{B}(X_1) \to \mathcal{L}_s^+(\mathcal{H})$ and $F : \mathcal{B}(X_2) \to \mathcal{L}_s^+(\mathcal{H})$ be a PVM and a POVM respectively. They are compatible if and only if they commute.*

Proof. If they commute they are compatible by Theorem 2.9. If they are compatible there are commuting dilations $E^+ : \mathcal{B}(X_1) \to \mathcal{L}_s^+(\mathcal{H}^+)$ and $\widetilde{E}^+ : \mathcal{B}(X_2) \to \mathcal{L}_s^+(\mathcal{H}^+)$ of E and F respectively. Since for any $\Delta_1 \in \mathcal{B}(X_1)$, $E(\Delta_1)$ is a projection, it must be $[P, E^+(\Delta_1)] = \mathbf{0}$. Therefore, for any $\in \mathcal{H}$,

$$\begin{aligned} F(\Delta_2)E(\Delta_1)\psi &= P\widetilde{E}^+(\Delta_2)PE^+(\Delta_1)\psi = P\widetilde{E}^+(\Delta_2)E^+(\Delta_1)P\psi \\ &= PE^+(\Delta_1)\widetilde{E}^+(\Delta_2)\psi = PE^+(\Delta_1)P\widetilde{E}^+(\Delta_2)\psi = E(\Delta_1)F(\Delta_2)\psi. \end{aligned}$$

□

2.3. Minimal dilation

In the present subsection we analyze the problem of the existence of a minimal Naimark dilation in the framework of partially ordered sets (posets). We introduce a preorder in the set of all Naimark dilations, then we consider the partial order it induces and prove the existence of a minimal element as well as its uniqueness. In this approach the definition of the minimal Naimark dilation becomes a consequence of the partial order structure. First we reformulate the definition of Naimark's dilation in a language which is more appropriate to the partial order sets framework.

Definition 2.13. *Let $F : \mathcal{B}(X) \to \mathcal{L}_s^+(\mathcal{H})$ be a POVM on the Hilbert space $\mathcal{H}$. A Naimark dilation is a couple $(E^+, \mathcal{H}^+)$, where $\mathcal{H}^+$ is a Hilbert space containing $\mathcal{H}$ and E^+ is a PVM, such that*

$$F(\Delta)\psi = PE^+(\Delta)\psi, \quad \forall \psi \in \mathcal{H} \text{ and } \forall \Delta \in \mathcal{B}(X),$$

where P is the operator of projection onto $\mathcal{H}$.

Let $\mathcal{N}_d(F)$ be the set of all the Naimark dilations of a POVM F. We can introduce a preorder in $\mathcal{N}_d(F)$.

Definition 2.14. *Let F be a POVM on the Hilbert space $\mathcal{H}$. Let $(E^+, \mathcal{H}^+), (E^\sim, \mathcal{H}^\sim) \in \mathcal{N}_d(F)$. We say that $(E^\sim, \mathcal{H}^\sim)$ extends $(E^+, \mathcal{H}^+)$, and write $(E^\sim, \mathcal{H}^\sim) \sqsubseteq (E^+, \mathcal{H}^+)$ if there exists a unitary transformation $U : \mathcal{H}^+ \to U\mathcal{H}^+ \subset \mathcal{H}^\sim$, $\mathcal{H} \subset U\mathcal{H}^+$, such that $UE^+U^\dagger = E^\sim|_{U\mathcal{H}^+}$.*

Journal of Physics: Conference Series **1638** (2020) 012006 doi:10.1088/1742-6596/1638/1/012006

Now we use the preorder to define an equivalence relation between the elements of $\mathcal{N}_d(F)$.

Definition 2.15. *We say that* $(E^+,\mathcal{H}^+),(E^\sim,\mathcal{H}^\sim)\in\mathcal{N}_d(F)$ *are equivalent and write* $(E^+,\mathcal{H}^+)\sim(E^\sim,\mathcal{H}^\sim)$ *if* $(E^\sim,\mathcal{H}^\sim)\sqsubseteq(E^+,\mathcal{H}^+)$ *and* $(E^+,\mathcal{H}^+)\sqsubseteq(E^\sim,\mathcal{H}^\sim)$. *The set* $\mathcal{N}_d(F)$ *modulo the equivalence relation is denoted* $\mathcal{N}_d(F)^\sim$.

Now the relation $\sqsubseteq$ induces a partial order in the set $(\mathcal{N}_d(F))^\sim$. Indeed, let $[a]$ denote the equivalence class of $a\in\mathcal{N}_d(F)$. Let $[a],[b]\in\mathcal{N}_d(F)^\sim$. We say that $[a]\sqsubseteq[b]$ if $a\sqsubseteq b$. It is easy to see that the $(\mathcal{N}_d(F)^\sim,\sqsubseteq)$ is a poset.

Proposition 2.16. $(E^+,\mathcal{H}^+)\sim(E^\sim,\mathcal{H}^\sim)$ *if and only if there is a unitary transformation* $U:\mathcal{H}^+\to\mathcal{H}^\sim$ *such that* $UE^+U^\dagger=E^\sim$.

Proof. Since $(E^\sim,\mathcal{H}^\sim)\sqsubseteq(E^+,\mathcal{H}^+)$, there is an isometry $U_1:\mathcal{H}^+\to U_1(\mathcal{H}^+)\subset\mathcal{H}^\sim$ such that $U_1E^+U_1^\dagger=E^\sim|_{U_1\mathcal{H}^+}$. Since $(E^+,\mathcal{H}^+)\sqsubseteq(E^\sim,\mathcal{H}^\sim)$, there is an isometry $U_2:\mathcal{H}^\sim\to U_2(\mathcal{H}^\sim)\subset\mathcal{H}^+$ such that $U_2E^\sim U_2^\dagger=E^+|_{U_2\mathcal{H}^\sim}$. Since the map $U_2U_1:\mathcal{H}^+\to\mathcal{H}^+$ preserves the inner product, it must be unitary (surjective and isometric). Hence U_1 and U_2 are unitary as well. In particular $U_1(\mathcal{H}^+)=\mathcal{H}^+$ and $U_1E^+U_1^\dagger=E^\sim$. It is easy to prove the converse. □

Now, we want to prove that $\mathcal{N}_d(F)^\sim$ has a $\sqsubseteq$-maximal element.

Theorem 2.17. *Let F be a POVM on the Hilbert space $\mathcal{H}$. Then, there is a Naimark dilation* $(E^+,\mathcal{H}^+)\in\mathcal{N}_d^\sim(F)$ *which is $\sqsubseteq$-maximal.*

Proof. We prove that every chain in $\mathcal{N}_d(F)^\sim$ has an upper bound. Then, an application of Zorn's lemma will prove the thesis. Let $\{(E^\alpha,\mathcal{H}^\alpha)\}_{\alpha\in A}$ be a chain in $\mathcal{N}_d(F)^\sim$ (i.e., it is a non-empty subset $\mathcal{N}$ of $\mathcal{N}_d(F)^\sim$ such that for all $[a],[b]\in\mathcal{N}$, either $[a]\sqsubseteq[b]$ or $[b]\sqsubseteq[a]$). Let $\overline{\alpha}\in A$ and set $\overline{A}:=\{\alpha\in A:\ (E^{\overline{\alpha}},\mathcal{H}^{\overline{\alpha}})\sqsubseteq(E^\alpha,\mathcal{H}^\alpha)\}$. By definition of the order relation, for each $\alpha\in\overline{A}$, $\mathcal{H}^\alpha$ can be embedded isometrically into $\mathcal{H}^{\overline{\alpha}}$, i.e., there exists a unitary isomorphism $U^\alpha:\mathcal{H}^\alpha\to U^\alpha\mathcal{H}^\alpha\subset\mathcal{H}^{\overline{\alpha}}$ such that $U^\alpha E^\alpha(U^\alpha)^\dagger=E^{\overline{\alpha}}|_{U^\alpha\mathcal{H}^\alpha}$. Now, we claim that $(\overline{E},\overline{\mathcal{H}})$ with $\overline{\mathcal{H}}:=\bigcap_{\alpha\in\overline{A}}U^\alpha\mathcal{H}^\alpha\supseteq\mathcal{H}$ and $\overline{E}=E^{\overline{\alpha}}|_{\overline{\mathcal{H}}}$ is an upper bound of the collection $\{(E^\alpha,\mathcal{H}^\alpha)\}_{\alpha\in A}$.
First we prove that $\overline{E}(\overline{\mathcal{H}})\subset\overline{\mathcal{H}}$. Suppose $\in\overline{\mathcal{H}}$. Then, $\in U^\alpha\mathcal{H}^\alpha$, for any $\alpha\in\overline{A}$. Hence, $E^{\overline{\alpha}}\psi\in U^\alpha\mathcal{H}^\alpha$, for any $\alpha\in\overline{A}$ and $\overline{E}\psi=E^{\overline{\alpha}}\psi\in\bigcap_{\alpha\in\overline{A}}U^\alpha\mathcal{H}^\alpha=\overline{\mathcal{H}}$.
Now, we prove that $(E^\alpha,\mathcal{H}^\alpha)\sqsubseteq(\overline{E},\overline{\mathcal{H}})$ for each $\alpha\in\overline{A}$. The map $\overline{U}^\alpha:=(U^\alpha)^\dagger:\overline{\mathcal{H}}\to(U^\alpha)^\dagger(\overline{\mathcal{H}})\subset\mathcal{H}^\alpha$ is unitary with inverse $(\overline{U}^\alpha)^\dagger:\overline{U}^\alpha\overline{\mathcal{H}}\to\overline{\mathcal{H}}$ such that $(\overline{U}^\alpha)^\dagger=U^\alpha_{|\overline{U}^\alpha\overline{\mathcal{H}}}$. Moreover, $\overline{U}^\alpha\overline{E}(\overline{U}^\alpha)^\dagger=E^\alpha_{|\overline{U}^\alpha\overline{\mathcal{H}}}$. Indeed, $(E^{\overline{\alpha}},\mathcal{H}^{\overline{\alpha}})\sqsubseteq(E^\alpha,\mathcal{H}^\alpha)$ implies

$$(U^\alpha)E^\alpha(U^\alpha)^\dagger=E^{\overline{\alpha}}_{|U^\alpha\mathcal{H}^\alpha}$$

and then,

$$E^\alpha=(U^\alpha)^\dagger E^{\overline{\alpha}}_{|U^\alpha\mathcal{H}^\alpha}U^\alpha=(U^\alpha)^\dagger E^{\overline{\alpha}}U^\alpha.$$

Therefore,

$$E^\alpha_{|\overline{U}^\alpha\overline{\mathcal{H}}}=(U^\alpha)^\dagger E^{\overline{\alpha}}U^\alpha_{|\overline{U}^\alpha\overline{\mathcal{H}}}=(U^\alpha)^\dagger E^{\overline{\alpha}}_{|\overline{\mathcal{H}}}U^\alpha_{|\overline{U}^\alpha\overline{\mathcal{H}}}=(U^\alpha)^\dagger\overline{E}(\overline{U}^\alpha)^\dagger=\overline{U}^\alpha\overline{E}(\overline{U}^\alpha)^\dagger$$

so that $(E^\alpha,\mathcal{H}^\alpha)\sqsubseteq(\overline{E},\overline{\mathcal{H}})$, for each $\alpha\in\overline{A}$. Thus $(\overline{E},\overline{\mathcal{H}})$ is an upper bound for the chain. By the Zorn lemma, there is a $\sqsubseteq$-maximal element $(E^+,\mathcal{H}^+)$ in $\mathcal{N}_d(F)^\sim$.

□

Now, we can introduce a partial order $\sqsubseteq^*$ in $\mathcal{N}_d(F)^\sim$ which is the dual of $\sqsubseteq$ and such that $(E^+,\mathcal{H}^+)$ is $\sqsubseteq^*$-minimal.

Definition 2.18. *We say that* $(E^{\sim}, \mathcal{H}^{\sim}) \sqsubseteq^{*} (E^{+}, \mathcal{H}^{+})$ *if* $(E^{+}, \mathcal{H}^{+}) \sqsubseteq (E^{\sim}, \mathcal{H}^{\sim})$.

Corollary 2.19. *There is a* $\sqsubseteq^{*}$*-minimal element in* $\mathcal{N}_d(F)^{\sim}$.

Proof. Since $(E^{+}, \mathcal{H}^{+})$ is $\sqsubseteq$-maximal in $\mathcal{N}_d(F)^{\sim}$, it is $\sqsubseteq^{*}$-minimal in $\mathcal{N}_d(F)^{\sim}$. □

The minimal Naimark dilation with respect to the relation $\sqsubseteq^{*}$ is exactly the minimal dilation that is usually introduced by definition.

Proposition 2.20. *Let* $(E^{+}, \mathcal{H}^{+})$ *be a* $\sqsubseteq^{*}$*-minimal Naimark dilation. Then,*

$$\mathcal{H}^{+} = \widetilde{\mathcal{H}} = \overline{\operatorname{span}\{E^{+}(\Delta)\psi \,|\, \Delta \in \mathcal{B}(X),\, \psi \in \mathcal{H}\}}.$$

Proof. First we prove that $(E^{+}_{|\widetilde{\mathcal{H}}}, \widetilde{\mathcal{H}})$ is a Naimark dilation. Indeed, $\mathcal{H} \subset \widetilde{\mathcal{H}}$ and for any

$$\psi^{+} \in \{E^{+}(\Delta)\psi \,|\, \Delta \in \mathcal{B}(X),\, \psi \in \mathcal{H}\},$$

we have

$$\begin{aligned} E^{+}_{|\widetilde{\mathcal{H}}}(\Delta)\psi^{+} &= E^{+}(\Delta)[E^{+}(\Delta')\psi] \\ &= E^{+}(\Delta \cap \Delta')\psi \in \{E^{+}(\Delta)\psi \,|\, \Delta \in \mathcal{B}(X),\, \psi \in \mathcal{H}\}. \end{aligned}$$

Then, by linearity and continuity $E^{+}_{|\widetilde{\mathcal{H}}}(\Delta)(\widetilde{\mathcal{H}}) \subset \widetilde{\mathcal{H}}$ and then

$$F(\Delta)\psi = PE^{+}(\Delta)\psi = PE^{+}_{|\widetilde{\mathcal{H}}}(\Delta)\psi.$$

Moreover, $(E^{+}_{|\widetilde{\mathcal{H}}}, \widetilde{\mathcal{H}}) \sqsubseteq^{*} (E^{+}, \mathcal{H}^{+})$ which, by the minimality of $(E^{+}, \mathcal{H}^{+})$, implies $(E^{+}_{|\widetilde{\mathcal{H}}}, \widetilde{\mathcal{H}}) = (E^{+}, \mathcal{H}^{+})$. □

Corollary 2.21. *A* $\sqsubseteq^{*}$*- minimal Naimark dilation is minimal in the sense of Theorem 2.1.*

Theorem 2.22. *A* $\sqsubseteq^{*}$*-minimal Naimark dilation is unique.*

Proof. Let $(E_1^{+}, \mathcal{H}_1^{+})$ and $(E_2^{+}, \mathcal{H}_2^{+})$ be two minimal Naimark dilations of F. By proposition 2.20,

$$\mathcal{H}_1^{+} = \overline{\operatorname{span}\{E_1^{+}(\Delta)\psi \,|\, \Delta \in \mathcal{B}(X),\, \psi \in \mathcal{H}\}},$$

$$\mathcal{H}_2^{+} = \overline{\operatorname{span}\{E_2^{+}(\Delta)\psi \,|\, \Delta \in \mathcal{B}(X),\, \psi \in \mathcal{H}\}}.$$

By proposition 2.16, it is sufficient to prove that there is a unitary map $U : \mathcal{H}_1^{+} \to \mathcal{H}_2^{+}$ such that $UE_1^{+}U^{\dagger} = E_2^{+}$. Let us prove that the linear map

$$U : E_1^{+}(\Delta)\psi \longmapsto E_2^{+}(\Delta)\psi$$

is an isomorphism. It is sufficient to show that it is surjective and isometric. First we verify that U preserves the norm, for any choice of $\Delta \in \mathcal{B}(X)$ and $\psi \in \mathcal{H}$. We have

$$\begin{aligned} \|E_1^{+}(\Delta)\psi\|^2 &= \langle E_1^{+}(\Delta)\psi, E_1^{+}(\Delta)\psi\rangle = \langle E_1^{+}(\Delta)\psi, \psi\rangle \\ &= \langle F(\Delta)\psi, \psi\rangle \\ &= \langle E_2^{+}(\Delta)\psi, \psi\rangle = \langle E_2^{+}(\Delta)\psi, E_2^{+}(\Delta)\psi\rangle = \|E_2^{+}(\Delta)\psi\|^2. \end{aligned}$$

By the linearity of the inner product and the continuity of the norm, U can be extended to a linear isometry from $\mathcal{H}_1^{+}$ onto $\mathcal{H}_2^{+}$. It is onto by definition.

It remains to be proven that $UE_1^+(\Delta')U^\dagger = E_2^+(\Delta')$ for every $\Delta' \in \mathcal{B}(X)$.We have

$$\begin{aligned}[UE_1^+(\Delta')U^\dagger](E_2^+(\Delta)\psi) &= UE_1^+(\Delta')E_1^+(\Delta)\psi \\ &= UE_1^+(\Delta' \cap \Delta)\psi \\ &= E_2^+(\Delta' \cap \Delta)\psi \\ &= [E_2^+(\Delta')](E_2^+(\Delta)\psi),\end{aligned}$$

which, by the linearity and the continuity of the inner product, can be extended to $\mathcal{H}_2^+$. □

As a consequence of proposition 2.20 the $\sqsubseteq^*$-minimality coincides with the minimality in the sense of Naimark (see Theorem 2.1). That puts the concept of minimality of the Naimark dilation in its natural framework: the theory of posets.

3. Conclusions

In the present note we reviewed some results about Naimark's dilation theorem. First we focused on its relationships with the integral characterization of a commutative POVM F. Here the sharp version A of F can be thought of as a kind of 'mirror image' of the Naimark operator A^+ in the extended Hilbert space $\mathcal{H}^+$. Next, we remarked on the role of Naimark's theorem in characterizing the joint measurability of generalized observables and showed that when we move from the Hilbert space $\mathcal{H}$ to its extension $\mathcal{H}^+$, the compatibility of two POVMs (which is in general different from their commutativity) becomes the commutativity of their Naimark dilations. The Naimark dilation theorem establishes a connection between a given POVM and a PVM (its Naimark dilation). The connection can be used to prove some of the properties of POVMs (subsection 2.2). Finally we provided a characterization of minimality of the Naimark dilation in the framework of partially ordered sets. Here minimality is connected to the order structure and it can be shown that minimal Naimark dilations always exist and coincide with the ones defined by the Naimark theorem.

4. Acknowledgement

The present work has been realized in the framework of the activities of the INDAM (Istituto Nazionale di Alta Matematica).

References

[1] Ludwig G 1983 *Foundations of Quantum Mechanics I* (New York: Springer)
[2] Davies E B 1986 *Quantum Mechanics of Open Systems* (London: Academic Press)
[3] Holevo A S 1982 *Probabilistic and Statistical Aspects of Quantum Theory* (Amsterdam: North Holland)
[4] Prugovečki E 1984 *Stochastic Quantum Mechanics and Quantum Spacetime* (Dordrecht: Reidel)
[5] Ali S T and Emch G G 1974 *J. Math. Phys.* **15** 176–82
[6] Ali S T 1985 *Riv. del Nuovo Cim.* **8** 1–127
[7] Busch P, Grabowski M and Lahti P 1995 *Operational Quantum Physics* (Berlin: Springer)
[8] Schroeck Jr F E 1996 *Quantum Mechanics on Phase Space* (Dordrecht: Kluwer)
[9] Busch P, Lahti P, Pellonpää J-P and Ylinen K 2016 *Quantum Measurement* (Berlin: Springer)
[10] Guz W 1984 **23** 157–84
[11] Stulpe W 1997 *Classical Representations of Quantum Mechanics Related to Statistically Complete Observables* (Berlin: Wissenschaft und Technik) *Preprint* quant-ph/0610122
[12] Busch P, Lahti P, and Werner R F 2013 *Phys. Rev. Lett.* **111** 160405
[13] Busch P, Lahti P and Werner R F 2014 *Rev. Mod. Phys.* **86** 1261–81
[14] Kraus K 1977 *The Uncertainty Principle and Foundations of Quantum Mechanics* ed W C Price and S S Chissick (London: Wiley) pp 293–320
[15] Hegerfeldt G C 1974 *Phys. Rev.* D **10** 3320–1
[16] Jauch J M and Piron C 1967 *Helv. Phys. Acta* **40** 559–70

[17] Castrigiano D P I 1981 *Lett. Math. Phys.* **5** 303–9
[18] Brooke J A and Schroeck Jr F E 1996 *J. Math. Phys.* **37** 5958–86
[19] Beneduci R and Schroeck F 2019 *Found. Phys.* **49** 561–76
[20] Beneduci R and Nisticó G 2003 *J. Math. Phys.* **44** 5461–73
[21] Cattaneo G and Nisticó G 2000 *J. Math. Phys.* **41** 4365–78
[22] Beneduci R 2006 *J. Math. Phys.* **47** 062104
[23] Jenčová A and Pulmannová S 2009 *Found. Phys.* **39** 613–24
[24] Beneduci R 2016 *J. Math. Anal. and App.* **442** 50–71
[25] Holevo A S 1972 *Trans. Mosc. Math. Soc.* **26** 133–47
[26] Ali S T 1982 *Differential Geometric Methods in Mathematical Physics* (Lecture Notes in Mathematics vol 905) ed H D Doebner *et al* (Berlin: Springer) pp 207–28
[27] Beneduci R 2018 *Russ. J. Math. Phys.* **25** 158–82
[28] Naimark M A 1940 *Izv. Akad. Nauk SSSR Ser. Mat.* **4** 277–318
[29] Riesz F and Nagy B S 1990 *Functional Analysis* (New York: Dover)
[30] Akhiezer N I and Glazman I M 1963 *Theory of Linear Operators in Hilbert Space* (New York: Ungar)
[31] Paulsen V 2020 *Completely Bounded Maps and Operator Algebras* (Cambridge: Cambridge University Press)
[32] Ozawa M 2014 *J. Phys.: Conf. Series* **504** 1–13
[33] Berberian S K 1966 *Notes on Spectral Theory* (Princeton: Van Nostrand)
[34] Reed M and Simon B 1980 *Methods of Modern Mathematical Physics* (New York: Academic Press)
[35] Lahti P, Pelloppää J-P and Ylinen K 1999 *J. Math. Phys.* **40** 2181–9
[36] Beneduci R 2010 *Bull. Lond. Math. Soc.* **42** 441–51
[37] Beneduci R 2010 *Lin. Alg. App.* **43** 1224–39
[38] Beneduci R 2006 *Int. J. Geom. Meth. Mod. Phys.* **3** 1559-71
[39] Beneduci R 2007 *J. Math. Phys.* **48** 022102
[40] Beneduci R 2010 *Int. J. Theor. Phys.* **49** 3030–8
[41] Beneduci R 2011 *Int. J. Theor. Phys.* **50** 3724–36
[42] Lahti P 2003 *Int. J. Theor. Phys.* **42** 893–906
[43] Beneduci R 2017 *Rep. Math. Phys.* **79** 197–214

Journal of Physics: Conference Series **1638** (2020) 012007 doi:10.1088/1742-6596/1638/1/012007

Sequential measurements and entropy

Heinz-Jürgen Schmidt and Jochen Gemmer
Universität Osnabrück, Fachbereich Physik, D-49069 Osnabrück, Germany
E-mail: hschmidt@uos.de, jgemmer@uos.de

Abstract. We sketch applications of the so-called J-equation to quantum information theory concerning fundamental properties of the von Neumann entropy. The J-equation has recently be proposed as a sort of progenitor of the various versions of the Jarzynski equation. It has been derived within a general framework of sequential measurements that is slightly generalised here.

1. Introduction

Entropy and its increase in closed systems, the so-called 2nd law, historically arose in classical thermodynamics and statistical physics of many-body systems. The status of these concepts in quantum theory is not completely clear despite the vast amount of literature on this subject. The time-honoured definition of the von Neumann entropy $S(\rho) = -\text{Tr}(\rho \log \rho)$ has the following properties: It remains constant under unitary time evolution and increases (always understood in the sense of including the case of remaining constant) during projective measurement [1]:

$$S(\rho) \leq S\left(\sum_n \underset{\sim}{P}_n \, \rho \, \underset{\sim}{P}_n\right) , \tag{1}$$

with self-explaining notation. These properties suggest to relate the 2nd law and, more generally, the basic concepts of quantum thermodynamics to sequential measurements, real or hypothetical ones.

A successful approach following these lines of thought has lead to the Jarzynski equation

$$\left\langle e^{-\beta\, w} \right\rangle = e^{-\beta \Delta F} , \tag{2}$$

that is the most famous representative of a class of similar fluctuation theorems [2–14]. The Jarzynski equation is an exact statement about the expectation value of a non-linear function of the work w, viewed as the energy difference of two sequential energy measurements. Between the two measurements an arbitrary unitary time evolution takes place. Although "work" is not an observable in the traditional sense of a self-adjoint operator [15] it can be understood as an example of the generalised observable concept described by a positive operator valued measure, see [16–18]. We have performed an analysis [19] of (2) and more general variants of the Jarzynski equation with the result that it can be derived from an equation, called J-equation, that concerns the statistics of sequential measurements and is initially independent of any realization in quantum theory. From the J-equation one derives in the usual way, via

Journal of Physics: Conference Series **1638** (2020) 012007 doi:10.1088/1742-6596/1638/1/012007

Jensen's inequality, an inequality that resembles the 2nd law. However, some caution is advised: The J-equation contains an undetermined probability distribution that can be chosen in such a way as leading to the Jarzynski equations or, alternatively, to an approach initially considered by W. Pauli [20] in the context of time-dependent perturbation theory ("Golden Rule"). Only the second choice gives a proper account of the 2nd law. A more general version of this approach has been published three years later by O. Klein [21] and is since known as "Klein's inequality", see [22], although detached from the thermodynamic context. We will see in Section 3.2 that it can also be derived from the J-equation.

Thus we come across the finding that the 2nd law has some aspects that can be viewed as statements about sequential measurements and are independent of many-body physics. One could object to this viewpoint that measurements are only possible by interactions with a macroscopic measuring device which in turn brings many-particle aspects into play. Without conclusively clarifying these issues, we note that applications of the J-equation result in a domain that could be seen as a pre-theory of quantum information theory and is, e.g., covered by [22, Chapter 11]. These applications are the subject of the present work.

However, we have to slightly extend the mathematical framework presented in [19], in order to include, for example, also statistical operators with the eigenvalue 0. This is done in Section 2. The next Section 3 deals with realisations of the statistical model of sequential measurements in quantum theory, first in general, see Section 3.1, and then in a special form tailored for current purposes, see Section 3.2. In addition to the mentioned Klein's inequality we also prove the statement (1) within our framework. One may ask: What is the purpose of proving familiar propositions anew? The obvious rationale is to uncover unexpected relationships between seemingly disjoint domains as non-equilibrium quantum statistics and quantum information theory. These relationships could, hopefully, also be used to obtain new results or simplified proofs of known ones, which is, however, beyond the scope of the present article. In the summary and outlook Section 4 we will shortly hint at these possibilities.

2. Statistical model of sequential measurements

We consider two sequential measurements at the same physical system at times $t_0 < t_1$ with respective outcome sets $\mathcal{I}$ and $\mathcal{J}$. These sets are assumed to be finite or countably infinite. Hence the joint outcome of the two measurements can be represented by the pair $(i, j) \in \mathcal{I} \times \mathcal{J}$. We define

$$\mathsf{E} \equiv \mathcal{I} \times \mathcal{J} \tag{3}$$

as the set of "elementary events". The probability of elementary events will be obtained by means of some auxiliary functions $\Pi, x, \widetilde{x}$ that have no direct statistical meaning. We assume the existence of the functions

$$\Pi \;:\; \mathsf{E} \to \mathbb{R}_{\geq}\,, \tag{4}$$

$$x \;:\; \mathcal{I} \to \mathbb{R}_{\geq}\,, \tag{5}$$

$$\widetilde{x} \;:\; \mathcal{J} \to \mathbb{R}_{\geq}\,, \tag{6}$$

where $\mathbb{R}_{\geq}$ denotes the set of non-negative reals. Π will be called the "conditional matrix" and its entries are written as $\Pi(j|i)$. Further, the $x(i)$ and $\widetilde{x}(j)$ will be called "abstract eigenvalues" of the first and second kind for reasons that will become clear in the next Section.

The marginal sums of Π will be denoted by

$$d(i) \equiv \sum_{j\in\mathcal{J}} \Pi(j|i) \leq \infty\,, \quad i \in \mathcal{I}\,, \tag{7}$$

$$\widetilde{d}(j) \equiv \sum_{i\in\mathcal{I}} \Pi(j|i) \leq \infty\,, \quad j \in \mathcal{J}\,, \tag{8}$$

and may assume values in $\mathbb{R}_{\geq} \cup \{\infty\}$.

Further, we define for all $(i,j) \in \mathsf{E}$

$$
\begin{aligned}
P(i,j) &\equiv \Pi(j|i)\, x(i) \,, && (9)\\
\widetilde{P}(j,i) &\equiv \Pi(j|i)\, \widetilde{x}(j) \,, && (10)
\end{aligned}
$$

and postulate our central axiom as

Assumption 1.

$$\sum_{(i,j)\in\mathsf{E}} P(i,j) = 1 \,, \tag{11}$$

and

$$\sum_{(i,j)\in\mathsf{E}} \widetilde{P}(j,i) = 1 \,. \tag{12}$$

Eq. (11) especially means that $x(i) = 0$ if $d(i) = \infty$, analogously Eq. (12) has to be understood as $\widetilde{x}(j) = 0$ if $\widetilde{d}(j) = \infty$. Both functions $P(i,j)$ and $\widetilde{P}(i,j)$ can be used to describe probabilities of elementary events. Correspondingly, we obtain the following four marginal probabilities

$$
\begin{aligned}
p(i) &\equiv \sum_{j\in\mathcal{J}} P(i,j) \overset{(9)}{=} \sum_{j\in\mathcal{J}} \Pi(j|i)\, x(i) \overset{(7)}{=} d(i)\, x(i) \,, && (13)\\
q(j) &\equiv \sum_{i\in\mathcal{I}} P(i,j) \overset{(9)}{=} \sum_{i\in\mathcal{I}} \Pi(j|i)\, x(i) \,, && (14)\\
\widetilde{p}(j) &\equiv \sum_{i\in\mathcal{I}} \widetilde{P}(j,i) \overset{(10)}{=} \sum_{i\in\mathcal{I}} \Pi(j|i)\, \widetilde{x}(j) \overset{(8)}{=} \widetilde{d}(j)\, \widetilde{x}(j) \,, && (15)\\
\widetilde{q}(i) &\equiv \sum_{j\in\mathcal{J}} \widetilde{P}(j,i) \overset{(8)}{=} \sum_{j\in\mathcal{J}} \Pi(j|i)\, \widetilde{x}(i) \,, && (16)
\end{aligned}
$$

where $p(i) = d(i)\,x(i)$ has to be set to 0 if $d(i) = \infty$, analogously for $\widetilde{p}(j) = \widetilde{d}(j)\,\widetilde{x}(j)$. According to Assumption 1 all four marginal probabilities sum to unity.

It may be instructive to calculate the conditional probability belonging to $P(i,j)$, where we preliminarily restrict ourselves to the case $p(i) > 0$ for all $i \in \mathcal{I}$:

$$\pi(j|i) \equiv \frac{1}{p(i)} P(i,j) \overset{(9)}{=} \frac{1}{p(i)} \Pi(j|i)\, x(i) \overset{(13)}{=} \frac{1}{d(i)} \Pi(j|i) \,, \tag{17}$$

for all $j \in \mathcal{J}$. It satisfies a kind of modified double stochasticity, namely

$$\sum_{i\in\mathcal{I}} \pi(j|i)\, d(i) \overset{(17)}{=} \sum_{i\in\mathcal{I}} \Pi(j|i) \overset{(8)}{=} \widetilde{d}(j) \,. \tag{18}$$

In accordance with the usual nomenclature of probability theory, functions $X : \mathsf{E} \to \mathbb{R}$ are also called "random variables". Their expectation value is defined as

$$\langle X \rangle \equiv \sum_{(i,j)\in\mathsf{E}} X(i,j)\, P(i,j) \,, \tag{19}$$

if the series converges. Using a sloppy notation the expectation value will be sometimes also written as $\langle X(i,j) \rangle$ if no misunderstanding is likely to occur.

For the applications we have in mind it is necessary to calculate the expectation value $\langle X(i,j)\rangle$ also if for some points $X(i,j)$ diverges and the probability $P(i,j)$ vanishes. It is not sufficient to simply exclude these points from the calculation of $\langle X(i,j)\rangle$. It seems that these mathematical difficulties are connected with the rare events sampling problem discussed in the literature, see, e.g., [23, 24]. For our purposes we need only to consider the points $(i,j) \in \mathsf{E}$ where $x(i) = 0$ and the $X(i,j)$ are of the form $X(i,j) = \frac{c(i,j)}{x(i)}$ with some finite numbers $c(i,j)$. Since $P(i,j) = \Pi(j|i)\,x(i)$ according to (9) the obvious regularisation of the otherwise undefined expectation value will be to cancel $x(i)$ and to set the contribution of $(i,j) \in \mathsf{E}$ to the expectation value to $P(i,j)\,X(i,j) = c(i,j)\,\Pi(j|i)$. Also the above considerations on the conditional probability would have to be reformulated by using this regularisation. This lends additional meaning to the auxiliary concept of the conditional matrix $\Pi(j|i)$ that has already been introduced in [19] in order to obtain a more symmetric formulation of the framework for sequential measurements.

Taking into account this regularisation procedure we have the following result:

Proposition 1. *Under the preceding conditions the following holds*

$$\left\langle \frac{\widetilde{x}(j)}{x(i)} \right\rangle = 1 \,. \tag{20}$$

The proof is elementary, see Appendix A.

We will call Eq. (20) the "J-equation" since we think that it contains the probabilistic core of the Jarzynski equation but should be distinguished from the latter for the sake of clarity. This claim has been further explained in [19]. Due to the symmetry of our assumptions a reciprocal J-equation could be proven using the second probability distribution $\widetilde{P}(i,j)$, but this will not be needed in what follows.

The probability distributions $q(j)$ and $\widetilde{p}(j)$ defined in (14) and (15) are completely independent. A possible specialization of the model for sequential measurements is given by the choice of $\widetilde{x}(j)$ that results in $\widetilde{p}(j) = q(j)$ for all $j \in \mathcal{J}$, namely

$$\widetilde{x}(j) = \frac{q(j)}{\widetilde{d}(j)} \,, \tag{21}$$

for $\widetilde{d}(j) < \infty$ and $\widetilde{x}(j) = 0$ else. This will be called the "minimal case" for reasons to be explained below.

In contrast to [22] we will always denote by "log" the natural logarithm. Since it is a concave function, Jensen's inequality yields

$$\langle \log X \rangle \le \log \langle X \rangle \tag{22}$$

for any random variable $X : \mathsf{E} \to \mathbb{R}_{\ge}$. We will define the "modified Shannon entropy", see [25], by

$$H(p) \equiv -\sum_i p_i \log \frac{p(i)}{d(i)} \,, \tag{23}$$

and obtain:

Proposition 2.

$$H(p) \le H(q) \le -\sum_{j \in \mathcal{J}} q(j) \log \frac{\widetilde{p}(j)}{\widetilde{d}(j)} \,. \tag{24}$$

The proof can be found in Appendix A. Obviously, $H(q)$ minimises the right hand side of (24), thereby justifying the denotation of the choice (21) resulting in $\widetilde{p}(j) = q(j)$ as the minimal case.

3. Applications to quantum theory

3.1. Sequential measurements in quantum theory

We consider a quantum system with a Hilbert space $\mathcal{H}$ and a finite number of mutually commuting self-adjoint operators $\underset{\sim}{E}_1, \ldots, \underset{\sim}{E}_L$ defined on (suitable domains of) $\mathcal{H}$. They are assumed to have a pure point spectrum and hence a family of common eigenprojections $(\underset{\sim}{P}_i)_{i\in\mathcal{I}}$ such that

$$\underset{\sim}{E}_\lambda = \sum_{i\in\mathcal{I}} E_i^{(\lambda)} \underset{\sim}{P}_i, \quad \lambda = 1, \ldots, L \,. \tag{25}$$

Here $\mathcal{I}$ is a finite or countable infinite index set to be identified with the outcome set of the first measurement according to Section 2. In general, the $\underset{\sim}{P}_i$ may be of infinite degeneracy; hence we define

$$\mathcal{I}' \equiv \{ i \in \mathcal{I} \,|\, d(i) \equiv \mathrm{Tr}\left(\underset{\sim}{P}_i\right) < \infty \} \,. \tag{26}$$

Further, the $\underset{\sim}{P}_i$ are chosen as maximal projections in the sense that $i \neq j$ implies $E_i^{(\lambda)} \neq E_j^{(\lambda)}$ for at least one $\lambda = 1, \ldots, L$. Note the completeness relation

$$\sum_{i\in\mathcal{I}} \underset{\sim}{P}_i = \mathbb{1} \,. \tag{27}$$

Physically, the $\underset{\sim}{E}_1, \ldots, \underset{\sim}{E}_L$ correspond to observables that can be jointly measured. We assume a (mixed) state of the system before the time $t = t_0$ described by a density operator ρ_0 and perform a joint Lüders measurement, cf. [16, Eq. (10.22)], of $\underset{\sim}{E}_1, \ldots, \underset{\sim}{E}_L$ at the time $t = t_0$. The probability of the outcome $i \in \mathcal{I}$ will be

$$p(i) = \mathrm{Tr}\left(\rho_0\, \underset{\sim}{P}_i\right) \,, \tag{28}$$

satisfying

$$\sum_{i\in\mathcal{I}} p(i) = 1 \,. \tag{29}$$

In accordance with the remarks after (16) we will make the following

Assumption 2.

$$p(i) > 0 \;\; \text{for all} \;\; i \in \mathcal{I}' \,. \tag{30}$$

After the first measurement of the $\underset{\sim}{E}_1, \ldots, \underset{\sim}{E}_L$ the system is subject to a further time evolution and a second measurement of (possibly) other observables. Thus the primary preparation together with the first measurement may be considered as another preparation of a certain state ρ, in general different from the initial state ρ_0. If a selection according to a particular outcome $i \in \mathcal{I}'$ is involved this state will be, according to the assumption of a Lüders measurement, cf. [16, Eq. (10.22)],

$$\rho_i = \frac{\underset{\sim}{P}_i\, \rho_0\, \underset{\sim}{P}_i}{\mathrm{Tr}\left(\rho_0\, \underset{\sim}{P}_i\right)} = \frac{\underset{\sim}{P}_i\, \rho_0\, \underset{\sim}{P}_i}{p(i)} \,. \tag{31}$$

If no selection according to a particular outcome is involved the state resulting after the first measurement will rather be the mixed state

$$\rho = \sum_{i\in\mathcal{I}'} p(i)\, \rho_i \overset{(31)}{=} \sum_{i\in\mathcal{I}'} \underset{\sim}{P}_i\, \rho_0\, \underset{\sim}{P}_i \,. \tag{32}$$

In order to apply the results of the preceding section we will make the following crucial assumption

Assumption 3.

$$\rho_i = \frac{1}{d(i)} \underset{\sim}{P}_i \quad \text{for all} \quad i \in \mathcal{I}' \,. \tag{33}$$

If $\underset{\sim}{P}_i$ is a one-dimensional projection, i.e., if $d(i) = 1$, the assumption (33) will be automatically satisfied. In the case of $d(i) > 1$ this assumption means that ρ_0 is diagonal with respect to any common eigenbasis of the $\underset{\sim}{E}_1, \ldots, \underset{\sim}{E}_L$. An important case where (33) holds is given if ρ_0 is a function of the operators $\underset{\sim}{E}_1, \ldots, \underset{\sim}{E}_L$, say,

$$\rho_0 = \mathcal{G}\left(\underset{\sim}{E}_1, \ldots, \underset{\sim}{E}_L\right) \,. \tag{34}$$

For example, the choice of $\mathcal{G}$ as the Boltzmann distribution leads to a Jarzynski equation of the form (2) for $L = 1$.

Next we consider a second set of observables described by the mutually commuting self-adjoint operators $\underset{\sim}{F}_1, \ldots, \underset{\sim}{F}_L$ subject to analogous assumptions. Hence the following holds:

$$\underset{\sim}{F}_\lambda = \sum_{j \in \mathcal{J}} F_j^{(\lambda)} \underset{\sim}{Q}_j, \quad \lambda = 1, \ldots, L \,, \tag{35}$$

$$\sum_{j \in \mathcal{J}} \underset{\sim}{Q}_j = \mathbb{1} \,, \tag{36}$$

$$\mathcal{J}' \equiv \left\{ j \in \mathcal{J} \,\middle|\, \widetilde{d}(j) \equiv \mathrm{Tr}\left(\underset{\sim}{Q}_j\right) < \infty \right\} , \tag{37}$$

and

Assumption 4.

$$\widetilde{p}(j) > 0 \quad \text{for all} \quad j \in \mathcal{J}' \,. \tag{38}$$

Here $\widetilde{p}(j)$ denotes an arbitrary probability distribution.

We have chosen another index set $\mathcal{J}$ for the second set of observables in order to stress that no natural identification between both index sets is required in what follows. Obviously, $\mathcal{J}$ has to be identified with the second outcome set introduced in Section 2. In general the $\underset{\sim}{E}_\lambda$ will not commute with the $\underset{\sim}{F}_\mu$. We assume that a second measurement of the $\underset{\sim}{F}_1, \ldots, \underset{\sim}{F}_L$ will be performed at the time $t = t_1 > t_0$, not necessarily of Lüders type. Between the two measurements in the time interval (t_0, t_1) the evolution of the system can be quite arbitrary and will be described by a unitary evolution operator $U = U(t_1, t_0)$.

In order to apply the results of the last section we will define the quantities $\Pi, x, \widetilde{x}$ and show that Assumption 1 will be satisfied in the quantum case. Moreover, we will show that the probability function $P(i, j)$ has its usual meaning here.

We set

$$\Pi(j|i) = \mathrm{Tr}\left(\underset{\sim}{Q}_j \, U \, \underset{\sim}{P}_i \, U^*\right) \quad \text{for all} \quad i \in \mathcal{I}, j \in \mathcal{J}, \tag{39}$$

$$x(i) = \frac{p(i)}{d(i)} \quad \text{for all} \quad i \in \mathcal{I}', \quad \text{and} \tag{40}$$

$$\widetilde{x}(j) = \frac{\widetilde{p}(j)}{\widetilde{d}(j)} \quad \text{for all} \quad j \in \mathcal{J}' \,. \tag{41}$$

It follows that the marginal sums of $\Pi(j|i)$ agree with the degeneracies $d(i)$ and $\widetilde{d}(j)$ defined above.

Moreover,

$$P(i,j) = \Pi(j|i)\,x(i) \tag{42}$$
$$\stackrel{(39,40)}{=} \text{Tr}\left(\underset{\sim}{Q}_j\,U\,\underset{\sim}{P}_i\,U^*\right)\frac{p(i)}{d(i)} \tag{43}$$
$$\stackrel{(33)}{=} \text{Tr}\left(\underset{\sim}{Q}_j\,U\,\rho_i\,U^*\right)p(i) \tag{44}$$
$$\stackrel{(31)}{=} \text{Tr}\left(\underset{\sim}{Q}_j\,U\,\underset{\sim}{P}_i\,\rho_0\,\underset{\sim}{P}_i\,U^*\right) \tag{45}$$

is the correct probability of the outcome (i,j) according to the rules of quantum theory. Moreover, the following holds:

Proposition 3. *If the above Assumptions 2, 3, and 4 are satisfied then the quintuple* $(\mathcal{I},\mathcal{J},\Pi,x,\widetilde{x})$ *defined in (39–41) also satisfies Assumption 1 and hence represents a model of sequential measurements.*

The proof can be found in Appendix A. Especially, the J-equation (20) holds in quantum theory as well as the 2nd law-like inequality (24). An essentially equivalent result has been published recently, see [26, Lemma IV.4].

As an application we consider a two-level system subject to an arbitrary time-dependent radiation field such that $|\mathcal{I}| = |\mathcal{J}| = 2$ without degeneracy. It follows that the matrix of conditional probabilities $\pi(j|i)$ according to (17) is doubly stochastic and hence symmetric. This is equivalent to the symmetry

$$B_{12} = B_{21} \tag{46}$$

of the corresponding *Einstein coefficients*. Another derivation of (46) based on the Jarzynski equation has been given in [27]. In the case of degeneracies d_1, d_2 of the two-level system Eq. (46) generalises to

$$B_{12}\,d_1 = B_{21}\,d_2\ , \tag{47}$$

see, e.g., [28, Eq.(10)]. This in turn can be easily derived from the modified double stochasticity (18) and the simple stochasticity of $\pi(j|i)$.

3.2. Results on the von Neumann entropy

Next we will prove some well-known results connected with the von Neumann entropy using the framework of sequential measurement sketched in Section 2. Recall the definition of the von Neumann entropy

$$S(\rho) \equiv -\text{Tr}\,(\rho\log\rho)\ , \tag{48}$$

for arbitrary statistical operators ρ. As usual, the limit $\lim_{x\downarrow 0} x\,\log x = 0$ is tacitly understood for vanishing eigenvalues of ρ.

For this subsection we will slightly specialise the definitions of the preceding subsection 3.1. We note that the eigenvalues of the operators $\underset{\sim}{E}_\lambda$ and $\underset{\sim}{F}_\lambda$ to be measured do not enter into the scheme of sequential measurement but only the corresponding eigenprojections. We use this freedom of choosing the eigenvalues in the following way. Let ρ and σ be two statistical operators with respective spectral decompositions and traces

$$\rho = \sum_{i\in\mathcal{I}} r_i \underset{\sim}{P}_i \,, \tag{49}$$

$$1 = \text{Tr}\;\rho = \sum_{i\in\mathcal{I}} r_i\,\text{Tr}\underset{\sim}{P}_i \equiv \sum_{i\in\mathcal{I}} r_i\, d(i) \equiv \sum_{i\in\mathcal{I}} p(i)\,, \tag{50}$$

$$\sigma = \sum_{j\in\mathcal{J}} s_j\, \underset{\sim}{Q}_j \,, \tag{51}$$

$$1 = \text{Tr}\;\sigma = \sum_{j\in\mathcal{J}} s_j\,\text{Tr}\underset{\sim}{Q}_j \equiv \sum_{j\in\mathcal{J}} s_j\, \widetilde{d}(j) \equiv \sum_{j\in\mathcal{J}} \widetilde{p}(j)\,. \tag{52}$$

Since ρ and σ are Hermitean operators they can also be viewed as observables. Thus we choose the initial state $\rho_0 = \rho$ and perform a first Lüders measurement of ρ at time $t = t_0$ with outcome $i \in \mathcal{I}$. The state after this measurement without selection is obviously again ρ. It follows that the condition (33) will be automatically satisfied. Between $t = t_0$ and $t = t_1 > t_0$ no interaction takes place, i.e., $U(t_1, t_0) = \mathbb{1}$. Then at time $t = t_1$ a second measurement of σ is performed with outcome $j \in \mathcal{J}$. The assumption $U(t_1, t_0) = \mathbb{1}$ does not imply any loss of generality since σ is completely arbitrary. The conditional matrix (39) of the sequential measurement will assume the simplified form

$$\Pi(j|i) = \text{Tr}\left(\underset{\sim}{P}_i\, \underset{\sim}{Q}_j\right)\,. \tag{53}$$

Moreover, the abstract eigenvalues can be identified with the actual eigenvalues of ρ and σ, i.e.,

$$x(i) = r_i \;\text{ and }\; \widetilde{x}(j) = s_j \;\text{ for all }\; i \in \mathcal{I} \text{ and } j \in \mathcal{J}\,. \tag{54}$$

For arbitrary statistical operators ρ, σ the "relative entropy" is defined as

$$S(\rho||\sigma) \equiv \text{Tr}\;(\rho \log \rho) - \text{Tr}\;(\rho \log \sigma)\,, \tag{55}$$

cf. [22, Eq. (11.50)]. The relative entropy may diverge for certain choices of ρ and σ (see [29]). It is never negative according to

Proposition 4. (Klein's inequality)

$$S(\rho||\sigma) \geq 0\,. \tag{56}$$

This is proven in [21] and, more recently, [22, Theorem 11.7]. Our alternative proof using sequential measurements can be found in Appendix A.

For the remainder of this subsection we will concentrate on the case where the statistical operator σ is chosen in such a way that the "minimal case" according to (21) is obtained. More precisely, we define

Definition 1. The pair (ρ, σ) of statistical operators will be called "minimal" iff

$$\widetilde{p}(j) = \text{Tr}\left(\sigma\, \underset{\sim}{Q}_j\right) = \text{Tr}\left(\rho\, \underset{\sim}{Q}_j\right) = q(j), \tag{57}$$

for all $j \in \mathcal{J}$, where the $\underset{\sim}{Q}_j$ are the eigenprojections of σ according to (51).

In this case Eq. (24) reduces to $H(p) \leq H(q)$ and the right hand side of this inequality can be identified with $S(\sigma)$:

Proposition 5. *If* (ρ, σ) *is minimal then* $S(\rho||\sigma) = S(\sigma) - S(\rho)$ *and hence, by Proposition 4,* $S(\rho) \leq S(\sigma)$.

The proof can be found in Appendix A.

It is worthwhile noting that the converse of this Proposition does not hold. We will present a counter-example where $S(\rho||\sigma) = S(\sigma) - S(\rho)$ without (ρ, σ) being minimal. The Hilbert space of this counter-example will be $\mathcal{H} = \mathbb{C}^4 \cong \mathbb{C}^2 \otimes \mathbb{C}^2$ and ρ will be the projector onto an entangled state ϕ:

$$\rho = P_\phi, \quad \text{where } \phi = \frac{1}{2}\left(\uparrow\downarrow + \sqrt{3}\downarrow\uparrow\right), \tag{58}$$

$$\sigma = \rho_1 \otimes \rho_2 , \tag{59}$$

and the ρ_i, $i = 1, 2$ denoting the partial traces of ρ with respect to the tensor factors of $\mathbb{C}^4 \cong \mathbb{C}^2 \otimes \mathbb{C}^2$. Note that the ρ_i, $i = 1, 2$ are isospectral since ρ is a pure state admitting a Schmidt decomposition. From this it follows that two eigenvalues of σ will be degenerate. Moreover, for the above choice of σ the equation $S(\rho||\sigma) = S(\sigma) - S(\rho)$ holds in general, see the proof of the subadditivity of the von Neumann entropy in [22, Sec. 11.3.4]. In our case σ is diagonal in the standard basis of $\mathbb{C}^4$:

$$\sigma = \begin{pmatrix} \frac{3}{16} & 0 & 0 & 0 \\ 0 & \frac{1}{16} & 0 & 0 \\ 0 & 0 & \frac{9}{16} & 0 \\ 0 & 0 & 0 & \frac{3}{16} \end{pmatrix}, \tag{60}$$

whereas

$$\rho = \begin{pmatrix} 0 & 0 & 0 & 0 \\ 0 & \frac{1}{4} & \frac{\sqrt{3}}{4} & 0 \\ 0 & \frac{\sqrt{3}}{4} & \frac{3}{4} & 0 \\ 0 & 0 & 0 & 0 \end{pmatrix}. \tag{61}$$

Obviously, the $q(j)$, the diagonal entries of ρ, are different from the $\widetilde{p}(j)$, the diagonal entries of σ, and hence the pair (ρ, σ) is not minimal.

Next we turn to the problem how the von Neumann entropy of a state changes during a quantum measurement. Obviously, this problem depends on the theoretical description of state changes during measurements and hence leads to the notions of operations and instruments, see [16]. The simplest case is that of a Lüders measurement $\mathfrak{I}$, see [16, Eq. (10.22)],

$$\rho \mapsto \mathfrak{I}(n)(\rho) = \underset{\sim}{P}_n\, \rho\, \underset{\sim}{P}_n , \tag{62}$$

where $\left(\underset{\sim}{P}_n\right)_{n\in\mathbb{N}}$ is a complete system of orthogonal projections, not necessarily finite-dimensional ones. The state change without any selection will be the trace preserving map

$$\rho \mapsto \sigma \equiv \sum_{n\in\mathbb{N}} \underset{\sim}{P}_n\, \rho\, \underset{\sim}{P}_n . \tag{63}$$

It is well-known, see [1] and [22, Theorem 11.9], that Lüders measurements increase entropy:

Proposition 6. *With the preceding definitions the following holds:*

$$S(\rho) \leq S(\sigma) \,. \tag{64}$$

A proof using the framework of sequential measurement can be found in Appendix A. We stress that the hypothetical sequential measurement used in this proof and the original Lüders measurement (62) are different, although related. This will be underscored by the following remarks:

- The first measurement of ρ used in the proof with outcome $i \in \mathcal{I}$ is not part of the Lüders measurement.
- The projections $\underset{\sim}{Q}_j$ of the second hypothetical measurement used in the proof are finite-dimensional in contrast to the $\underset{\sim}{P}_n$, they rather represent a refinement of the family $\left(\underset{\sim}{P}_n\right)_{n\in\mathbb{N}}$.
- Moreover, the second hypothetical measurement used in the proof need not be of Lüders type.

In the case of a more general instrument than that of Lüders type it is well-known that a statement analogous to (64) may fail, i.e., a generalised measurement can decrease entropy, see [22, Exercise 11.15]. This may sound paradoxical at first sight but can be understood by considering the "measurement dilation" of a general instrument, see [16, Chap. 7.7]. This means that the object system with Hilbert space $\mathcal{H}_1$ is coupled to a second system ("measuring device") with Hilbert space $\mathcal{H}_2$ and, after some interaction of the total system described by a unitary evolution operator U, a Lüders measurement with projectors $\underset{\sim}{P}_n$ is performed at the measuring device. The final step of the state change consists of a partial trace Tr_2 that yields a mixed state σ of the object system. We thus obtain for the state change without selection the following expression

$$\rho \mapsto \sigma = \mathrm{Tr}_2 \sum_{n\in\mathbb{N}} \left(\mathbb{1} \otimes \underset{\sim}{P}_n\right) U \left(\rho \otimes P_\phi\right) U^* \left(\mathbb{1} \otimes \underset{\sim}{P}_n\right) \,, \tag{65}$$

where we have used that the initial state of the measuring device can be chosen as a pure state P_ϕ, see [16, Chap. 7.7]. Without the partial trace Tr_2 this can be understood as a Lüders measurement of the total system with projectors

$$\underset{\sim}{Q}_n \equiv U^* \left(\mathbb{1} \otimes \underset{\sim}{P}_n\right) U \tag{66}$$

and the initial state $\rho \otimes P_\phi$. It follows from Proposition 6 that the total entropy does not decrease, i.e.,

$$S_1 \equiv S(\rho) = S(\rho \otimes P_\phi) \leq S\left(\rho'\right) \equiv S_2 \,, \tag{67}$$

where

$$\rho' \equiv \sum_{n\in\mathbb{N}} \underset{\sim}{Q}_n \left(\rho \otimes P_\phi\right) \underset{\sim}{Q}_n \,, \tag{68}$$

and we have used the fact that the entropy of a tensor product is additive and the entropy of a pure state vanishes. By forming partial traces the total entropy further increases, see [22, Sec. 11.3.4], and thus

$$S_1 \leq S_2 \leq S\left(\mathrm{Tr}_1(\rho')\right) + S\left(\mathrm{Tr}_2(\rho')\right) \equiv S_{32} + S_{31} \equiv S_3 \,. \tag{69}$$

Hence the total entropy during a generalised measurement does not decrease if the possible entropy increase of the measuring device is taken into account. The mentioned counter examples of a decreasing entropy of the object system occur if $S_{31} < S_1$ which is well possible in spite of $S_1 \leq S_{31} + S_{32}$. This means that the decrease of the entropy of the object system must be (over)compensated by an increase of the measurement device's entropy, see the corresponding discussion in [29] and [30, Chap. III.5].

4. Summary and Outlook

A general framework for sequential measurement including the J-equation has been recently formulated and shown to be realised in quantum theory [19]. The J-equation comprises the various variants of the famous Jarzynski equation. In the present paper we have slightly generalised this framework in order to cope with the problem of vanishing probabilities. A standard application of the J-equation results from Jensen's inequality and the fact that the logarithm is a concave function. The resulting inequality has been shown to be essentially equivalent to Klein's inequality already derived in 1931 in the context of quantum thermodynamics. This opens an unexpected connection between non-equilibrium quantum statistical mechanics and general entropy theory mainly used in context with quantum information, see [22]. The concept of sequential measurements is proving fruitful not only in its direct application to sequential measurements but also in the sense of a mathematical tool. Thus the new proofs of well-known laws like Proposition 6 opens up a new perspective in this field, insofar as the various 2nd law-like statements can be viewed as consequences of an underlying J-equation. It would be desirable to use these tools to simplify the involved proofs of, say, the strong subadditivity of the von Neumann entropy, see [22, Chap. 11.4], but this is definitely outside the scope of the present paper.

Acknowledgments

This work has been funded by the Deutsche Forschungsgemeinschaft (DFG), Grant No. 397107022 (GE 1657/3-1) within the DFG Research Unit FOR 2692. We thank all members of this research unit for stimulating and insightful discussions and hints to relevant literature. One of us (H.-J. S.) gratefully acknowledges the fruitful discussions with Paul Busch during the last years before his deplorable and untimely passing away.

Appendix A. Proofs

Proof of Proposition 1:

$$\left\langle \frac{\widetilde{x}(j)}{x(i)} \right\rangle \overset{(19)}{=} \sum_{(i,j)\in \mathsf{E}} P(i,j)\, \frac{\widetilde{x}(j)}{x(i)} \tag{A.1}$$

$$\overset{(9)}{=} \sum_{(i,j)\in \mathsf{E}} \Pi(j|i)\, \widetilde{x}(j) \tag{A.2}$$

$$\overset{(10)}{=} \sum_{(i,j)\in \mathsf{E}} \widetilde{P}(j,i) \overset{(12)}{=} 1\,. \tag{A.3}$$

We would like to point out that the summations in (A.1)–(A.3) have to be extended over the whole domain E including those points where $x(i)$ vanishes. The contribution to the expectation value from these points has to be calculated according to the regularisation procedure explained in the text following (19). □

Journal of Physics: Conference Series **1638** (2020) 012007 doi:10.1088/1742-6596/1638/1/012007

Proof of Proposition 2:

$$
\begin{aligned}
0 &= \log 1 \overset{(20)}{=} \log \left\langle \frac{\widetilde{x}(j)}{x(i)} \right\rangle && \text{(A.4)}\\
&\overset{(21)}{\geq} \left\langle \log \left(\frac{\widetilde{x}(j)}{x(i)} \right) \right\rangle && \text{(A.5)}\\
&= \langle \log \widetilde{x}(j) \rangle - \langle \log x(i) \rangle && \text{(A.6)}\\
&= \sum_{(i,j)\in \mathsf{E}} P(i,j) \log \widetilde{x}(j) - \sum_{(i,j)\in \mathsf{E}} P(i,j) \log x(i) && \text{(A.7)}\\
&\overset{(13,14)}{=} \sum_{j\in\mathcal{J}} q(j) \log \widetilde{x}(j) - \sum_{i\in\mathcal{I}} p(i) \log x(i) && \text{(A.8)}\\
&\overset{(15,23)}{=} \sum_{j\in\mathcal{J}} q(j) \log \frac{\widetilde{p}(j)}{\widetilde{d}(j)} + H(p)\ . && \text{(A.9)}
\end{aligned}
$$

Hence

$$H(p) \leq - \sum_{j\in\mathcal{J}} q(j) \log \frac{\widetilde{p}(j)}{\widetilde{d}(j)}\ , \tag{A.10}$$

and especially for the minimal case

$$H(p) \leq H(q)\ . \tag{A.11}$$

It remains to show the second inequality in (24). The log function is bounded by its tangent at $x = 1$:

$$\log x \leq x - 1 \quad \text{for } x > 0\ , \tag{A.12}$$

and thus

$$
\begin{aligned}
& - \sum_{j\in\mathcal{J}} q(j) \log \left(\frac{\widetilde{p}(j)}{\widetilde{d}(j)} \frac{\widetilde{d}(j)}{q(j)} \right) && \text{(A.13)}\\
\geq\ & - \sum_{j\in\mathcal{J}} q(j) \left(\frac{\widetilde{p}(j)}{\widetilde{d}(j)} \frac{\widetilde{d}(j)}{q(j)} - 1 \right) && \text{(A.14)}\\
=\ & - \sum_{j\in\mathcal{J}} \widetilde{p}(j) + \sum_{j\in\mathcal{J}} q(j) = -1 + 1 = 0\ . && \text{(A.15)}
\end{aligned}
$$

This entails

$$H(q) = - \sum_{j\in\mathcal{J}} q(j) \log \frac{q(j)}{\widetilde{d}(j)} \leq - \sum_{j\in\mathcal{J}} q(j) \log \frac{\widetilde{p}(j)}{\widetilde{d}(j)}\ , \tag{A.16}$$

thereby completing the proof of Proposition 2. □

Proof of Proposition 3: We have to prove (11) and (12). To this end we consider

$$\sum_{(i,j)\in \mathsf{E}} P(i,j) = \sum_{i\in\mathcal{I}'} \mathrm{Tr}\left(\left(\sum_{j\in\mathcal{J}} \underset{\sim}{Q}_j\right) U \underset{\sim}{P}_i U^*\right) \frac{p(i)}{d(i)}$$

$$\stackrel{(36)}{=} \sum_{i\in\mathcal{I}'} \mathrm{Tr}\left(U \underset{\sim}{P}_i U^*\right) \frac{p(i)}{d(i)} \tag{A.17}$$

$$= \sum_{i\in\mathcal{I}'} \frac{\mathrm{Tr}\left(\underset{\sim}{P}_i\right)}{d(i)} p(i) \stackrel{(26)}{=} \sum_{i\in\mathcal{I}'} p(i) = 1\,.$$

$$\tag{A.18}$$

Further,

$$\sum_{(i,j)\in \mathsf{E}} \widetilde{P}(i,j) = \sum_{(i,j)\in \mathsf{E}} \Pi(i,j)\,\widetilde{x}(j) \tag{A.19}$$

$$\stackrel{(41)}{=} \sum_{j\in\mathcal{J}'} \mathrm{Tr}\left(\underset{\sim}{Q}_j U \left(\sum_{i\in\mathcal{I}} \underset{\sim}{P}_i\right) U^*\right) \frac{\widetilde{p}(j)}{\widetilde{d}(j)}$$

$$\stackrel{(27)}{=} \sum_{j\in\mathcal{J}'} \frac{\mathrm{Tr}\left(\underset{\sim}{Q}_j\right)}{\widetilde{d}(j)} \widetilde{p}(j) \stackrel{(37)}{=} \sum_{j\in\mathcal{J}'} \widetilde{p}(j) = 1\,,$$

$$\tag{A.20}$$

which completes the proof of Proposition 3. □

Proof of Proposition 4: With the definitions of Section 3.2 we obtain

$$S(\rho) = -\mathrm{Tr}\left(\rho \log \rho\right) \stackrel{(49)}{=} -\sum_{i\in\mathcal{I}'} r_i \log(r_i)\,\mathrm{Tr}\,\underset{\sim}{P}_i$$

$$\tag{A.21}$$

$$\stackrel{(50)}{=} -\sum_{i\in\mathcal{I}'} p(i) \log\frac{p(i)}{d(i)} = H(p)\,, \tag{A.22}$$

and further

$$\mathrm{Tr}\left(\rho \log \sigma\right) \stackrel{(49,51)}{=} \sum_{(i,j)\in \mathsf{E}} r_i \log(s_j)\,\mathrm{Tr}\left(\underset{\sim}{P}_i \underset{\sim}{Q}_j\right) \tag{A.23}$$

$$\stackrel{(53,54)}{=} \sum_{(i,j)\in \mathsf{E}} x(i) \log(\widetilde{x}(j))\,\Pi(j|i) \tag{A.24}$$

$$\stackrel{(14)}{=} \sum_{j\in\mathcal{J}} q(j)\,\log(\widetilde{x}(j)) \tag{A.25}$$

$$\stackrel{(15)}{=} \sum_{j\in\mathcal{J}} q(j)\,\log\left(\frac{\widetilde{p}(j)}{\widetilde{d}(j)}\right)\,. \tag{A.26}$$

This implies

$$S(\rho||\sigma) = -S(\rho) - \mathrm{Tr}(\rho \log \sigma) \tag{A.27}$$

$$\overset{(A.22,A.26)}{=} -H(p) - \sum_{j \in \mathcal{J}} q(j) \log\left(\frac{\widetilde{p}(j)}{\widetilde{d}(j)}\right) \overset{(24)}{\geq} 0\,, \tag{A.28}$$

and the proof of Proposition 4 is complete. □

Proof of Proposition 5: It suffices to show that $-S(\sigma) = \mathrm{Tr}\,(\rho\, \log \sigma)$:

$$-S(\sigma) \overset{(51)}{=} \sum_{j \in \mathcal{J}} s_j \log s_j \,\mathrm{Tr}\, \underset{\sim}{Q}_j \tag{A.29}$$

$$\overset{(52)}{=} \sum_{j \in \mathcal{J}} \frac{\widetilde{p}(j)}{\widetilde{d}(j)} \log s_j\, \widetilde{d}(j) \tag{A.30}$$

$$= \sum_{j \in \mathcal{J}} \widetilde{p}(j) \log s_j \tag{A.31}$$

$$\overset{(57)}{=} \sum_{j \in \mathcal{J}} q(j) \log s_j \tag{A.32}$$

$$\overset{(14)}{=} \sum_{j \in \mathcal{J}} \left(\sum_{i \in \mathcal{I}} P(i,j) \right) \log s_j \tag{A.33}$$

$$\overset{(9)}{=} \sum_{j \in \mathcal{J}} \left(\sum_{i \in \mathcal{I}} \Pi(j|i) x(i) \right) \log s_j \tag{A.34}$$

$$\overset{(53),(54)}{=} \sum_{(i,j) \in \mathsf{E}} r_i \log s_j \,\mathrm{Tr}\left(\underset{\sim}{P}_i \underset{\sim}{Q}_j \right) . \tag{A.35}$$

$$\overset{(49),(51)}{=} \mathrm{Tr}\,(\rho\, \log\, \sigma)\ . \tag{A.36}$$

This completes the proof of Proposition 5. □

Proof of Proposition 6: The claim (64) follows from Proposition 5 if we can show that the pair (ρ, σ) is minimal in the sense of Definition 1. If

$$\sigma = \sum_{n \in \mathbb{N}} \underset{\sim}{P}_n\, \rho \underset{\sim}{P}_n = \sum_{j \in \mathcal{J}} s_j\, \underset{\sim}{Q}_j \tag{A.37}$$

is the spectral decomposition of σ it follows that all $\underset{\sim}{Q}_j$ commute with all $\underset{\sim}{P}_n$ for $j \in \mathcal{J}$ and

$n \in \mathbb{N}$. It follows that

$$\widetilde{p}(j) = \mathrm{Tr}\left(\sigma\, \underset{\sim}{Q}_j\right) \tag{A.38}$$
$$= \sum_{n\in\mathbb{N}} \mathrm{Tr}\left(\underset{\sim}{P}_n\, \rho \underset{\sim}{P}_n\, \underset{\sim}{Q}_j\right) \tag{A.39}$$
$$= \sum_{n\in\mathbb{N}} \mathrm{Tr}\left(\underset{\sim}{P}_n\, \rho\, \underset{\sim}{Q}_j\, \underset{\sim}{P}_n\right) \tag{A.40}$$
$$= \mathrm{Tr}\left(\rho\, \underset{\sim}{Q}_j \left(\sum_{n\in\mathbb{N}} \underset{\sim}{P}_n\right)\right) \tag{A.41}$$
$$= \mathrm{Tr}\left(\rho\, \underset{\sim}{Q}_j\right) = q(j)\,, \tag{A.42}$$

where we have used $[\underset{\sim}{Q}_j, \underset{\sim}{P}_n] = 0$ in (A.40) and $\sum_{n\in\mathbb{N}} \underset{\sim}{P}_n = \mathbb{1}$ in (A.42). Hence the Definition 1 of a minimal pair (ρ, σ) is satisfied and the proof of Proposition 6 is complete. □

References

[1] von Neumann J 1932 *Mathematische Grundlagen der Quantenmechanik* (Berlin: Springer) English translation 1955 *Mathematical Foundations of Quantum Mechanics* (Princeton: Princeton University Press)
[2] Jarzynski C 1997 Nonequilibrium equality for free energy differences *Phys. Rev. Lett.* **78** 2690–3
[3] Kurchan J 2000 A quantum fluctuation theorem *Preprint* arXiv:cond-mat/0007360v2
[4] Tasaki H 2000 Jarzynski relations for quantum systems and some applications *Preprint* arXiv:cond-mat/0000244v2
[5] Mukamel S 2003 Quantum extension of the Jarzynski relation: analogy with stochastic dephasing *Phys. Rev. Lett.* **90** 170604
[6] Schmiedl T and Seifert U 2007 Stochastic thermodynamics of chemical reaction networks *J. Chem. Phys.* **126** 044101
[7] Saito K and Utsumi Y 2008 Symmetry in full counting statistics, fluctuation theorem, and relations among nonlinear transport coefficients in the presence of a magnetic field *Phys. Rev.* B **78** 115429
[8] Andrieux D, Gaspard P, Monnai T and Tasaki S 2009 The fluctuation theorem for currents in open quantum systems *New J. Phys.* **11** 043014 Erratum 2009 *New J. Phys.* **11** 109802
[9] Yi J, Talkner P and Campisi M 2011 Nonequilibrium work statistics of an Aharonov-Bohm flux *Phys. Rev.* E **84** 011138
[10] Morikuni Y and Tasaki H 2011 Quantum Jarzynski-Sagawa-Ueda Relations *J. Stat. Phys.* **143** 1–10
[11] Esposito M 2012 Stochastic thermodynamics under coarse graining *Phys. Rev. E* **85** 041125
[12] Yi J, Kim Y W and Talkner P 2012 Work fluctuations for Bose particles in grand canonical initial states *Phys. Rev. E* **85** 051107
[13] Campisi M, Hänggi P and Talkner P 2011 Quantum fluctuation relations: foundations and applications *Rev. Mod. Phys.* **83** 771–91 Erratum 2011 *Rev. Mod. Phys.* **83** 1653
[14] Talkner P, Morillo M, Yi J and Hänggi P 2013 Statistics of work and fluctuation theorems for microcanonical initial states *New J. Phys.* **15** 095001
[15] Talkner P, Lutz E and Hänggi P 2007 Fluctuation theorems: work is not an observable *Phys. Rev.* E **75** 050102
[16] Busch P, Lahti P J, Pellonpää J-P and Ylinen K 2016 *Quantum Measurement* (Berlin: Springer)
[17] Roncaglia A J, Cerisola F and Paz J P 2014 Work measurement as a generalized quantum measurement *Phys. Rev. Lett.* **113** 250601
[18] De Chiara G, Roncaglia A J, Cerisola F and Paz J P 2015 Measuring work and heat in ultracold quantum gases *New J. Phys.* **17** 035004
[19] Schmidt H-J and Gemmer J 2020 A framework for sequential measurements and general Jarzynski equations *Z. Naturforsch. A* **75** 265–84
[20] Pauli W 1928 Über das H-Theorem vom Anwachsen der Entropie vom Standpunkt der neuen Quantenmechanik *Probleme der modernen Physik. Arnold Sommerfeld zum 60. Geburtstage gewidmet von seinen Schülern* ed P Debye (Leipzig: Hirzel); 1964 *W. Pauli, Collected Scientific Papers vol 1* ed R Kronig and V Weisskopf (New York: Interscience) pp 549–64

[21] Klein O 1931 Zur quantenmechanischen Begündung des zweiten Hauptsatzes der Wärmelehre *Z. Phys.* **72** 767–75
[22] Nielsen M A and Chuang I L 2000 *Quantum Computation and Quantum Information* (Cambridge: Cambridge University Press)
[23] Jarzynski C 2006 Rare events and the convergence of exponentially averaged work values *Phys. Rev.* E **73** 046105
[24] Berryman J T and Schilling T 2010 Sampling rare events in nonequilibrium and nonstationary systems *J. Chem. Phys.* **133** 244101
[25] Shannon C E 1948 A mathematical theory of communication *Bell Syst. Tech. J.* **27** 379–423
[26] Strasberg P and Winter A 2020 Dissipation in quantum systems: a unifying picture *Preprint* arXiv:2002.08817v2 [quant-ph]
[27] Gittes F 2018 Two famous results of Einstein derived from the Jarzynski equality *Am. J. Phys.* **86** 31–5
[28] Hilborn R C 1982 Einstein coefficients, cross sections, f-values, dipole moments, and all that *Am. J. Phys.* **50** 982–6
[29] Lindblad G 1973 Entropy, information and quantum measurements *Commun. Math. Phys.* **33** 305–22
[30] Busch P, Lahti P J and Mittelstaedt P 1996 *The Quantum Theory of Measurement* (Berlin: Springer) 2nd revised edition

A quantum reference frame size-accuracy trade-off for quantum channels

Takayuki Miyadera[1] and Leon Loveridge[2]

[1] Department of Nuclear Engineering, Kyoto University, Nishikyo-ku, Kyoto 615-8540, Japan
[2] Quantum Technology Group, Department of Science and Industry Systems, University of South-Eastern Norway, 3616 Kongsberg, Norway

E-mail: `miyadera@nucleng.kyoto-u.ac.jp`, `leon.d.loveridge@usn.no`

Abstract. The imposition of symmetry upon the nature and structure of quantum observables has recently been extensively studied, with quantum reference frames playing a crucial role. In this paper, we extend this work to quantum transformations, giving quantitative results showing, in direct analogy to the case of observables, that a "large" reference frame is required for non-covariant channels to be well approximated by covariant ones. We apply our findings to the concrete setting of $SU(2)$ symmetry.

Dedicated to the memory of our friend Paul Busch

1. Introduction

The role played by symmetry in the understanding and development of physics can hardly be overstated. It is an important part of the process of building mathematical models of the physical world, and is often crucial for their solubility in both classical and quantum settings. Less widely recognised, however, is the effect that symmetry has in limiting what is measurable. In fact, as is common in gauge theories, e.g., [1], and postulated in, e.g., [2,3] in the context of quantum reference frames, theoretical quantities which do not commute with a symmetry action are unobservable, even in principle. The tension between this apparent unobservability and the use of such quantities in the description of real physical systems is relieved by noting that any system of interest is a part of a larger whole - there is another system (called the environment, ancilla, reservoir, reference frame, or apparatus) whose presence is often assumed only implicitly and which does not appear in the formulation. It is then possible that unobservable quantities of the system of interest may be re-interpreted as representatives of observable *relative* quantities of system-plus-environment.

In this paper we study the extent to which the same kind of restrictions (due to symmetry) hold in a dynamical context. To make the problem concrete, we must identify, in analogy to the invariance requirement for observables, the right notion of restriction for channels, since it is not unique. This will be discussed in detail in a future publication. One natural example arises in the presence of conserved quantities. The Wigner-Araki-Yanase (WAY) theorem [4–9] shows that conservation laws restrict the accuracy and repeatability properties of a class of quantum measurements. Also in the presence of an additive conserved quantity, Åberg [10] has shown that any unitary dynamics can be approximately realized by preparing "coherent" (thus asymmetric) states of the environment. We here emphasize the necessity of the highly asymmetric (coherent) state (cf. [2, 3, 11]), which requires a large environment in a sense to be described. If the environment is not large, states cannot have enough coherence and the possible dynamics is restricted. In addition, if the symmetry is not Abelian, there may be some restriction due to the uncertainty relation, because large coherence with respect to some observable implies small coherence with respect to its conjugate.

Another possible symmetry constraint is covariance of the dynamics, whose relevance to the reference frame context will be discussed briefly in the next section. As will be shown later, this constraint is weaker than the one given by the presence of conserved quantities. In this paper, we study the possible dynamics of the system under the symmetry constraint on the whole system (object system plus reference). We consider a quantum channel on the system and study how well this (target) channel is approximately realized by a covariant channel on the whole system, contingent upon a choice of state of the environment. We derive a quantitative relation which shows that for the covariant channel to be well approximated by the target channel, high asymmetry/coherence is required for the state of the environment. As an example we apply our relation to a qubit system under $SU(2)$ symmetry.

Such size-versus-inaccuracy trade-offs are already present in the literature in various different contexts (see, e.g., [11–13]), and our findings are broadly in line with other findings - that good accuracy needs large size. Specific mention must be given to [14–16], which has already investigated the dynamical setting and some elegant bounds have been provided, particularly in the unitary case. However, we provide a novel quantitative bound in the dynamical context.

From a technical point of view, the present paper may be regarded as a descendant of [11], in which we derived a quantitative bound in approximating an arbitrary effect by a globally invariant effect. There we reinterpreted the issue as an approximate joint measurement problem of observables and employed a quantitative uncertainty relation [17]. In this paper, we show that a similar technique, which we may call uncertainty relation based method, can be applied also to the approximating channel problem.

2. Symmetry constraints on channels

The principle of symmetry limits the possible observables to invariant ones (see, e.g., [3]). There are some different generalizations of this constraint to channels.

Suppose that we have a system described by a Hilbert space $\mathcal{H}$. By $\mathbf{B}(\mathcal{H})$ we denote the algebra of bounded operators on $\mathcal{H}$. Throughout this paper every Hilbert space we encounter will be finite dimensional. We assume that a finite dimensional connected Lie group G is acting on $\mathcal{H}$. G is assumed to define a true smooth unitary representation on each system, denoted by $U(g)$.

One of the possible constraints on dynamics is given by a conservation law. Suppose that there exists a conserved charge N, which generates a $U(1)$ action. In the situation that the system is closed/isolated, the possible dynamics Λ must satisfy $\Lambda(N) = N$ (in the Heisenberg picture). We thus arrive at the following definition.

Definition 1. Invariant channels *are defined as those* $\Phi : \mathbf{B}(\mathcal{H}) \to \mathbf{B}(\mathcal{H})$ *for which* $\Phi(U(g)) = U(g)$ *for all* $g \in G$, *where* U *is an* $\mathcal{H}$*-representation of* G.

On the other hand, suppose that we have a system and are asked to rotate the state around the z-axis by some angle θ. The desired channel is $\Phi(X) = e^{iS_z\theta} X e^{-iS_z\theta}$, where S_z is the z-component of angular momentum. (We work in units in which $\hbar = 1$.) This channel implicitly assumes the existence of a reference frame/system which specifies the z-axis. If we employ another reference frame, the z-component of angular momentum is represented as $U(R)S_zU(R)^*$ with some $R \in SO(3)$ and the corresponding channel becomes $\Phi_R(X) = U(R)\Phi(U(R)^*XU(R))U(R)^*$. If we are not informed which reference frame is to be used, we may choose one randomly, in which case the channel is described as

$$\overline{\Phi}(X) = \int \mu(dR)\Phi_R(X), \tag{1}$$

where $\mu(\cdot)$ is (the) Haar measure on $SO(3)$. $\overline{\Phi}$ is an example of a *covariant channel* which we now define.

Definition 2. *A channel* $\Lambda : \mathbf{B}(\mathcal{H}) \to \mathbf{B}(\mathcal{H})$ *is called covariant if and only if*

$$\Lambda(U(g)^* X U(g)) = U(g)^* \Lambda(X) U(g) \tag{2}$$

holds for all $X \in \mathbf{B}(\mathcal{H})$ *and for all* $g \in G$.

As the next proposition shows, invariant channels form an important subclass of covariant channels.

Proposition 1. *Invariant channels are covariant.*

Proof. We begin by using a channel Λ to define an "operator-valued inner product" $\langle\langle A|B\rangle\rangle := \Lambda(A^*B) - \Lambda(A)^*\Lambda(B)$, which satisfies a Cauchy-Schwarz type inequality (see [18] and Lemma 3 in [11]):

$$\|\langle\langle A|B\rangle\rangle\|^2 \le \|\langle\langle A|A\rangle\rangle\|\|\langle\langle B|B\rangle\rangle\|, \tag{3}$$

where $||\cdot||$ denotes the standard operator norm in $\mathbf{B}(\mathcal{H})$. Suppose that a unitary U is a fixed point, i.e., $\Lambda(U) = U$. Then it holds that

$$\langle\langle U|U\rangle\rangle = \mathbb{1} - \Lambda(U)^*\Lambda(U) = 0.$$

Thus for such a U and arbitrary A we find

$$\langle\langle A|U\rangle\rangle = \Lambda(A^*U) - \Lambda(A^*)\Lambda(U) = \Lambda(A^*U) - \Lambda(A)^*U = 0, \text{ by (3)}.$$

Now Let $\Lambda : \mathbf{B}(\mathcal{H}) \to \mathbf{B}(\mathcal{H})$ be an invariant channel. Then for all $g \in G$,

$$\Lambda(AU(g)) = \Lambda(A)\Lambda(U(g)) = \Lambda(A)U(g).$$

Similarly $\Lambda(U(g)^*B) = U(g)^*\Lambda(B)$, and thus $\Lambda(U(g)^*AU(g)) = U(g)^*\Lambda(A)U(g)$. □

We note that there exist covariant channels which are not invariant; for instance, for any invariant state ω_0, the channel $\Lambda(X) = \omega_0(X)\mathbb{1}$ is covariant but not invariant. However, covariance and invariance are equivalent for unitary channels:

Proposition 2. *Unitary covariant channels are invariant.*

Proof. Let us consider a unitary (and thus automorphic), covariant channel $\Lambda : \mathbf{B}(\mathcal{H}) \to \mathbf{B}(\mathcal{H})$, i.e., $\Lambda(X) = V^*XV$ and $U(g)^*\Lambda(X)U(g) = \Lambda(U(g)^*XU(g))$. Then it holds that for all X

$$\Lambda(U(g)^*XU(g)) = V^*U(g)^*VV^*XVV^*U(g)V = U(g)^*(V^*XV)U(g).$$

Now putting $V^*XV = Z$, we apply $U(g) \cdot U(g)^*$ to the above equation to obtain

$$(U(g)V^*U(g)^*V)Z(V^*U(g)VU(g)^*) = Z,$$

which implies $V^*U(g)VU(g)^* = c(g)\mathbb{1}$ with $|c(g)| = 1$ and $V^*U(g)V = c(g)U(g)$. As the left-hand side satisfies $V^*U(g)VV^*U(g')V = V^*U(gg')V$, $c(g)c(g') = c(gg')$ holds for all $g, g \in G$. Now in a neighborhood of $e \in G$, for each element l of Lie algebra the corresponding generator L exists satisfying $U(e^{ls}) = e^{iLs}$ for sufficiently small $|s|$. If we put the generator of $V^*U(e^{ls})V$ as L', it satisfies $V^*LV = L'$. It in addition satisfies $L' = L + k\mathbb{1}$ for some $k \in \mathbf{R}$ as $V^*U(e^{ls})V = c(e^{ls})U(e^{ls})$ must hold. But as L is bounded (as $\mathcal{H}$ is finite dimensional) and $\|L\| = \|L'\|$ holds, $k = 0$ is the only possible choice. Thus we have shown that for a neighbourhood N_e of $e \in G$ $V^*U(g)V = U(g)$ is satisfied. As G is connected, it is generated by N_e. It implies that $c(g) = 1$ for all $g \in G$. □

Journal of Physics: Conference Series **1638** (2020) 012008 doi:10.1088/1742-6596/1638/1/012008

3. The setting and results

As we have seen in the last section, we cannot implement (for instance) the rotation around the z-axis without using a "correct" reference frame. More precisely, we may implement the right rotation but this occurs only by chance. The averaged channel is a covariant $\overline{\Phi}$ which is different from the desired rotation. In the worst case, the discrepancy is larger than the averaged case. Thus we must have a reference frame. Since a reference frame is also a physical system, there should be a quantum description. Our question is then to ask what is the condition on the quantum reference frame so that it works well to implement the desired channel. In the following we formulate the problem in a general setting.

Let G be a connected Lie group. We have a system and a reference frame described by (as always, finite dimensional) Hilbert spaces $\mathcal{H}_{\mathcal{S}}$ and $\mathcal{H}_{\mathcal{R}}$, and on each space, G has a smooth true unitary representation $U_{\mathcal{S}}(g)$ and $U_{\mathcal{R}}(g)$. Their composition is written as $U(g) = U_{\mathcal{S}}(g) \otimes U_{\mathcal{R}}(g)$ which acts on $\mathcal{H} = \mathcal{H}_{\mathcal{S}} \otimes \mathcal{H}_{\mathcal{R}}$. Our purpose is to study how well a general channel $\Lambda : \mathbf{B}(\mathcal{H}_{\mathcal{S}}) \to \mathbf{B}(\mathcal{H}_{\mathcal{S}})$ is approximately realized by the restriction of a covariant channel $\Phi : \mathbf{B}(\mathcal{H}_{\mathcal{S}} \otimes \mathcal{H}_{\mathcal{R}}) \to \mathbf{B}(\mathcal{H}_{\mathcal{S}} \otimes \mathcal{H}_{\mathcal{R}})$. We view Φ as representing the "true" transformation, with its restriction representing the transformation with the additional system suppressed. Therefore, Φ satisfies

$$\Phi(U(g)^* X U(g)) = U(g)\Phi(X)U(g)$$

for all $g \in G$ and $X \in \mathbf{B}(\mathcal{H}_{\mathcal{S}} \otimes \mathcal{H}_{\mathcal{R}})$. On the level of observables, the restriction to the system $\Gamma_{\rho_{\mathcal{R}}} : \mathbf{B}(\mathcal{H}_{\mathcal{S}} \otimes \mathcal{H}_{\mathcal{R}}) \to \mathbf{B}(\mathcal{H}_{\mathcal{S}})$ is determined by a state $\rho_{\mathcal{R}}$ on $\mathbf{B}(\mathcal{H}_{\mathcal{R}})$ and is defined by the completely positive conditional expectation

$$\mathrm{tr}[\rho_{\mathcal{S}} \Gamma_{\rho_{\mathcal{R}}}(X)] = \mathrm{tr}[(\rho_{\mathcal{S}} \otimes \rho_{\mathcal{R}})X],$$

which holds for all states $\rho_{\mathcal{S}}$ of the system and $X \in \mathbf{B}(\mathcal{H}_{\mathcal{S}} \otimes \mathcal{H}_{\mathcal{R}})$. In order to define the restriction for channels, we use the natural inclusion $\iota : \mathbf{B}(\mathcal{H}_{\mathcal{S}}) \to \mathbf{B}(\mathcal{H}_{\mathcal{S}} \otimes \mathcal{H}_{\mathcal{R}})$, given as

$$\iota(A) = A \otimes \mathbb{1}_{\mathcal{R}}.$$

Then the realized channel is written as $\Phi_{\rho_{\mathcal{R}}} := \Gamma_{\rho_{\mathcal{R}}} \circ \Phi \circ \iota : \mathbf{B}(\mathcal{H}_{\mathcal{S}}) \to \mathbf{B}(\mathcal{H}_{\mathcal{S}})$, and we wish to quantify the discrepancy between $\Phi_{\rho_{\mathcal{R}}}$ and Λ. As a quantity to characterize the discrepancy, one may employ the norm difference between two channels defined by

$$\|\Phi_{\rho_{\mathcal{R}}} - \Lambda\|_{\mathrm{channel}} := \sup_{X \in \mathbf{B}(\mathcal{H}_{\mathcal{S}}), \|X\|=1} \|\Phi_{\rho_{\mathcal{R}}}(X) - \Lambda(X)\|.$$

For each element of the Lie algebra $\mathfrak{g}$ of a Lie group G there exists a corresponding self-adjoint operator (the generator) acting in $\mathcal{H}_{\mathcal{S}}$. For each $l \in \mathfrak{g}$, there exist operators $L_{\mathcal{S}}$ and $L_{\mathcal{R}}$ satisfying $U_{\mathcal{S}}(e^{ls}) = e^{iL_{\mathcal{S}}s}$ and $U_{\mathcal{R}}(e^{ls}) = e^{iL_{\mathcal{R}}s}$ and therefore $U(e^{ls}) = e^{i(L_{\mathcal{S}} \otimes \mathbb{1}_{\mathcal{R}} + \mathbb{1}_{\mathcal{S}} \otimes L_{\mathcal{R}})}$. As unitary operators have norm 1, we obtain an inequality for each $U_{\mathcal{S}}(e^{ls_0})$,

$$\|\epsilon(l : s_0)\| := \| \left(\Gamma_{\rho_{\mathcal{R}}} \circ \Phi \circ \iota\right) \left(U_{\mathcal{S}}(e^{ls_0})\right) - \Lambda(U_{\mathcal{S}}(e^{ls_0}))\| \leq \|\Phi_{\rho_{\mathcal{R}}} - \Lambda\|_{\mathrm{channel}}.$$

$F(\rho_0, \rho_1)$ represents the fidelity between two states ρ_0 and ρ_1 defined by $F(\rho_0, \rho_1) := \mathrm{tr}[\sqrt{\rho_0^{1/2} \rho_1 \rho_0^{1/2}}]$. This quantity is positive and equals 1 if and only if $\rho_0 = \rho_1$ holds.

Theorem 1. *Let $L_{\mathcal{S}}$ and $L_{\mathcal{R}}$ be generators of unitary representations of e^{ls} ($s \in \mathbb{R}$) on $\mathcal{H}_{\mathcal{S}}$ and $\mathcal{H}_{\mathcal{R}}$ for $l \in \mathfrak{G}$. Define $U_{\mathcal{S}}(l : s) := e^{iL_{\mathcal{S}}s}$ and $U_{\mathcal{R}}(l : s) = e^{iL_{\mathcal{R}}s}$. Then, for any $s_0 \in \mathbb{R}$, $\epsilon(l : s_0) := \left(\Gamma_{\rho_{\mathcal{R}}} \circ \Phi \circ \iota\right)(U_{\mathcal{S}}(l : s_0)) - \Lambda(U_{\mathcal{S}}(l : s_0))$ is bounded for all $s \in \mathbb{R}$ by:*

$$\begin{aligned}
&\|[\Lambda(U_{\mathcal{S}}(l : s_0)), U_{\mathcal{S}}(l : s)]\| \leq 2\|U_{\mathcal{S}}(l : s) - \mathbb{1}\|\|\epsilon(l : s_0)\| \\
&+ \left(\frac{1}{F(\rho_{\mathcal{R}}, U_{\mathcal{R}}(l : s)\rho_{\mathcal{R}} U_{\mathcal{R}}(l : s)^*)^2} - 1\right)^{1/2} \left(\left(\|\mathbb{1}_{\mathcal{S}} - \Lambda(U_{\mathcal{S}}(l : s_0))^* \Lambda(U_{\mathcal{S}}(l : s_0))\| + 2\|\epsilon(l : s_0)\|\right)^{1/2} \right. \\
&\left. + \left(\|\mathbb{1}_{\mathcal{S}} - \Lambda(U_{\mathcal{S}}(l : s_0))\Lambda(U_{\mathcal{S}}(l : s_0))^*\| + 2\|\epsilon(l : s_0)\|\right)^{1/2} \right).
\end{aligned}$$

Before proving Theorem 1, we present some immediate implications. We first observe that the left hand side of the above inequality vanishes for covariant Λ, since

$$\begin{aligned}\|[\Lambda(U_{\mathcal{S}}(l:s_0)), U_{\mathcal{S}}(l:s)]\| &= \|U_{\mathcal{S}}(l:s)^*[\Lambda(U_{\mathcal{S}}(l:s_0)), U_{\mathcal{S}}(l:s)]\| \\ &= \|U_{\mathcal{S}}(l:s)^*\Lambda(U_{\mathcal{S}}(l:s_0))U_{\mathcal{S}}(l:s) - \Lambda(U_{\mathcal{S}}(l:s_0))\|, \qquad (4)\end{aligned}$$

and

$$U_{\mathcal{S}}(l:s)^*\Lambda(U_{\mathcal{S}}(l:s_0))U_{\mathcal{S}}(l:s) = \Lambda(U_{\mathcal{S}}(l:s)^*U_{\mathcal{S}}(l:s_0)U_{\mathcal{S}}(l:s)) = \Lambda(U_{\mathcal{S}}(l:s_0)).$$

Therefore, there is no bound for approximating covariant channels $\mathbf{B}(\mathcal{H}_{\mathcal{S}}) \to \mathbf{B}(\mathcal{H}_{\mathcal{S}})$ by restrictions of covariant channels $\mathbf{B}(\mathcal{H}) \to \mathbf{B}(\mathcal{H})$. Indeed, any covariant Λ can trivially be written as the restriction of a covariant channel Φ on $\mathbf{B}(\mathcal{H})$, i.e., as $\Phi_{\rho_{\mathcal{R}}}$ for all $\rho_{\mathcal{R}}$, by setting $\Phi = \Lambda \otimes \mathrm{id}$. If Λ is a unitary channel, Theorem 1 takes a much simpler form.

Corollary 1. *Under the same assumptions as Theorem 1, but for unitary Λ, it holds that*

$$\begin{aligned}\|[\Lambda(U_{\mathcal{S}}(l:s_0)), U_{\mathcal{S}}(l:s)]\| \leq& 2\|U_{\mathcal{S}}(l:s) - \mathbb{1}\|\|\epsilon(l:s_0)\| \\ &+2\left(\frac{1}{F(\rho_{\mathcal{R}}, U_{\mathcal{R}}(l:s)\rho_{\mathcal{R}}U_{\mathcal{R}}(l:s)^*)^2} - 1\right)^{1/2}\|\epsilon(l:s_0)\|^{1/2}.\end{aligned}$$

The proof follows from the observation that if Λ is a unitary channel (or indeed, multiplicative), then for any unitary operator $U \in \mathbf{B}(\mathcal{H}_{\mathcal{S}})$, we have $\Lambda(U)^*\Lambda(U) = \mathbb{1}$.

Therefore, we see that in order to make possible good agreement between Λ and $\Phi_{\rho_{\mathcal{R}}}$, a highly "asymmetric" reference state $\rho_{\mathcal{R}}$ is necessary, since $F(\rho_{\mathcal{R}}, U_{\mathcal{R}}(l:s)\rho_{\mathcal{R}}U_{\mathcal{R}}(l:s)^*)$ must decrease rapidly with respect to $|s|$ as otherwise the left-hand side of the inequality can be large for non-covariant Λ.

Furthermore, this asymmetry, or *coherence* factor, can be bounded by the "spread" of the (symmetry) generator $L_{\mathcal{R}}$:

Corollary 2. *In the same scenario as Theorem 1, it holds that*

$$\begin{aligned}\|[\Lambda(U_{\mathcal{S}}(l:s_0)), L_{\mathcal{S}}]\| \leq& 2\|L_{\mathcal{S}}\|\|\epsilon(l:s_0)\| \\ &+(\Delta_{\rho_{\mathcal{R}}}L_{\mathcal{R}})\Big((\|\mathbb{1}_{\mathcal{S}} - \Lambda(U_{\mathcal{S}}(l:s_0))^*\Lambda(U_{\mathcal{S}}(l:s_0))\| + 2\|\epsilon(l:s_0)\|)^{1/2} \\ &+(\|\mathbb{1}_{\mathcal{S}} - \Lambda(U_{\mathcal{S}}(l:s_0))\Lambda(U_{\mathcal{S}}(l:s_0))^*\| + 2\|\epsilon(l:s_0)\|)^{1/2}\Big),\end{aligned}$$

where $\Delta_{\rho_{\mathcal{R}}}L_{\mathcal{R}} := \sqrt{\mathrm{tr}[\rho_{\mathcal{R}}L_{\mathcal{R}}^2] - \mathrm{tr}[\rho_{\mathcal{R}}L_{\mathcal{R}}]^2}$ represents the standard deviation of $L_{\mathcal{R}}$ in the state $\rho_{\mathcal{R}}$.

Corollary 3. *Under the same assumptions as Theorem 1, but for unitary Λ, it holds that*

$$\|[\Lambda(U_{\mathcal{S}}(l:s_0)), L_{\mathcal{S}}]\| \leq 2\|L_{\mathcal{S}}\|\|\epsilon(l:s_0)\| + 2\sqrt{2}(\Delta_{\rho_{\mathcal{R}}}L_{\mathcal{R}})\|\epsilon(l:s_0)\|^{1/2}.$$

This immediately follows from Corollary 2. The inequality is easy to interpret. For non-covariant Λ which yields non-vanishing left-hand side, $\Delta_{\rho_{\mathcal{R}}}L_{\mathcal{R}}$ must be large to attain small $\|\epsilon(l:s_0)\|$. Thus it implies that the reference system $\mathcal{R}$ must be large (macroscopic). This result has some qualitative similarity to the bounds obtained in [14, 15], where large size/coherence/energy fluctuation of the reference is shown to be necessary for implementing unitary dynamics.

We now present proofs of Theorem 1 and Corollary 2. To prove Theorem 1, we need the following lemma [11, 18]:

Lemma 1. *Consider a channel $\Gamma : \mathbf{B}(\mathcal{H}) \to \mathbf{B}(\mathcal{K})$ for Hilbert spaces $\mathcal{H}$ and $\mathcal{K}$. If $A, B \in \mathbf{B}(\mathcal{H})$ satisfy $[A, B] = 0$, then*

$$\|[\Gamma(A), \Gamma(B)]\| \leq \|\Gamma(A^*A) - \Gamma(A)^*\Gamma(A)\|^{1/2}\|\Gamma(BB^*) - \Gamma(B)\Gamma(B)^*\|^{1/2} \tag{5}$$
$$+\|\Gamma(AA^*) - \Gamma(A)\Gamma(A)^*\|^{1/2}\|\Gamma(B^*B) - \Gamma(B)^*\Gamma(B)\|^{1/2}. \tag{6}$$

We now present the proof of Theorem 1.

Proof. If the state $\rho_\mathcal{R}$ and $s \in \mathbb{R}$ satisfy $F(\rho_\mathcal{R}, U_\mathcal{R}(l : s)\rho_\mathcal{R} U_\mathcal{R}(l : s)^*) = 0$, the claim follows trivially, and thus we assume otherwise. For notational simplicity, we omit the dependence on l and write $U_\mathcal{S}(s)$ for $U_\mathcal{S}(l : s)$, $U_\mathcal{R}(s)$ for $U_\mathcal{R}(l : s)$ and $\epsilon(s_0)$ for $\epsilon(l : s_0)$. We first write

$$[\Lambda(U_\mathcal{S}(s_0)), U_\mathcal{S}(s)] = [U_\mathcal{S}(s), \epsilon(s_0)] + [\Gamma_{\rho_\mathcal{R}}\Phi(U_\mathcal{S}(s_0) \otimes \mathbb{1}_\mathcal{R}), U_\mathcal{S}(s)]. \tag{7}$$

The first term on the right hand side is bounded as

$$\|[U_\mathcal{S}(s), \epsilon(s_0)]\| = \|[U_\mathcal{S}(s) - \mathbb{1}, \epsilon(s_0)]\| \leq 2\|U_\mathcal{S}(s) - \mathbb{1}\|\|\epsilon(s_0)\|.$$

To estimate the second term on the right hand side of (7), we introduce a purification of $\rho_\mathcal{R}$ to $|\phi_{RZ}\rangle \in \mathcal{H}_\mathcal{R} \otimes \mathcal{H}_Z$, where we choose the purification space $\mathcal{H}_Z$ to be minimal, i.e., its dimension coincides with the rank of $\rho_\mathcal{R}$. We denote $\Gamma_{|\phi_{RZ}\rangle\langle\phi_{RZ}|} : \mathbf{B}(\mathcal{H}_\mathcal{S} \otimes \mathcal{H}_\mathcal{R} \otimes \mathcal{H}_Z) \to \mathbf{B}(\mathcal{H}_\mathcal{S})$ by Γ for simplicity. Now, for an arbitrary operator W_Z on $\mathcal{H}_Z$, we have

$$\Gamma(U_\mathcal{S}(s) \otimes U_\mathcal{R}(s) \otimes W_Z) = U_\mathcal{S}(s)\langle\phi_{RZ}|U_\mathcal{R}(s) \otimes W_Z|\phi_{RZ}\rangle.$$

In the following we denote $\Phi \otimes \mathrm{id}_Z$ by $\hat{\Phi}$, and in order to simplify some long expressions we will make the abbreviations $A_0 = U_\mathcal{S}(s_0) \otimes \mathbb{1}_\mathcal{R} \otimes \mathbb{1}_Z$ and $A_s = U_\mathcal{S}(s) \otimes U_\mathcal{R}(s) \otimes W_Z$ when convenient. Thus we have, for W_Z with $\langle\phi_{RZ}|U_\mathcal{R}(s) \otimes W_Z|\phi_{RZ}\rangle \neq 0$,

$$[\Gamma(\hat{\Phi}(A_0)), U_\mathcal{S}(s)] = \frac{[\Gamma(\hat{\Phi}(A_0)), \Gamma(A_s)]}{\langle\phi_{RZ}|U_\mathcal{R}(s) \otimes W_Z|\phi_{RZ}\rangle}.$$

Since Φ is a covariant channel, it holds that

$$(U_\mathcal{S}(s)^* \otimes U_\mathcal{R}(s)^*)\Phi(U_\mathcal{S}(s_0) \otimes \mathbb{1}_\mathcal{R})(U_\mathcal{S}(s) \otimes U_\mathcal{R}(s)) = \Phi(U_\mathcal{S}(s_0) \otimes \mathbb{1}_\mathcal{R}).$$

Therefore we find

$$[\Phi(U_\mathcal{S}(s_0) \otimes \mathbb{1}_\mathcal{R}) \otimes \mathbb{1}_Z, U_\mathcal{S}(s) \otimes U_\mathcal{R}(s) \otimes W_Z] = 0,$$

which enables us to apply Lemma 1. In the following, W_Z is chosen to be unitary. Now we bound

$$\|[\Gamma(\hat{\Phi}(A_0)), U_\mathcal{S}(s)]\| = \frac{\|[\Gamma(\hat{\Phi}(A_0)), \Gamma(A_s)]\|}{|\langle\phi_{RZ}|U_\mathcal{R}(s) \otimes W_Z|\phi_{RZ}\rangle|}.$$

Then Lemma 1 yields the numerator of the above equation to be bounded as

$$\|[\Gamma(\hat{\Phi}(A_0)), \Gamma(A_s)]\| \leq \|\Gamma(\hat{\Phi}(A_0)^*\hat{\Phi}(A_0)) - \Gamma(\hat{\Phi}(A_0))^*\Gamma(\hat{\Phi}(A_0))\|^{1/2}\|\mathbb{1} - \Gamma(A_s)\Gamma(A_s)^*\|^{1/2}$$
$$+\|\Gamma(\hat{\Phi}(A_0)\hat{\Phi}(A_0)^*) - \Gamma(\hat{\Phi}(A_0))\Gamma(\hat{\Phi}(A_0))^*\|^{1/2}\|\mathbb{1} - \Gamma(A_s)^*\Gamma(A_s)\|^{1/2}.$$

We first estimate the norm of

$$A := \Gamma(\hat{\Phi}(A_0)^*\hat{\Phi}(A_0)) - \Gamma(\hat{\Phi}(A_0)^*\Gamma(\hat{\Phi}(A_0)).$$

Due to the two-positivity of Γ (i.e., $\Gamma(X^*X) \geq \Gamma(X)^*\Gamma(X)$ for all X) the operator A is positive. Furthermore applying the two-positivity of $\hat{\Phi}$, we obtain

$$\hat{\Phi}(A_0)^*\hat{\Phi}(A_0) \leq \hat{\Phi}(A_0^*A_0) = \mathbb{1}.$$

Since Γ is a positive map we find

$$\mathbf{0} \leq A \leq \mathbb{1} - \Gamma(\hat{\Phi}(U_{\mathcal{S}}(s_0) \otimes \mathbb{1}_{\mathcal{R}} \otimes \mathbb{1}_Z))^*\Gamma(\hat{\Phi}(U_{\mathcal{S}}(s_0) \otimes \mathbb{1}_{\mathcal{R}} \otimes \mathbb{1}_Z)),$$

from which we conclude

$$\|A\| \leq \|\mathbb{1} - \Gamma(\hat{\Phi}(U_{\mathcal{S}}(s_0) \otimes \mathbb{1}_{\mathcal{R}} \otimes \mathbb{1}_Z))^*\Gamma(\hat{\Phi}(U_{\mathcal{S}}(s_0) \otimes \mathbb{1}_{\mathcal{R}} \otimes \mathbb{1}_Z))\|.$$

The term

$$\|\Gamma(\hat{\Phi}(A_0)\hat{\Phi}(A_0)^*) - \Gamma(\hat{\Phi}(A_0))\Gamma(\hat{\Phi}(A_0))^*\|^{1/2}$$

can be treated similarly. Writing $c_{RZ} \equiv \langle\phi_{RZ}|U_{\mathcal{R}}(s) \otimes W_Z|\phi_{RZ}\rangle$, we thus obtain

$$\begin{aligned}\|[\Gamma(\hat{\Phi}(A_0)), U_{\mathcal{S}}(s)]\| \leq \frac{1}{|c_{RZ}|}\Big(&\|\mathbb{1} - \Gamma(\hat{\Phi}(A_0))^*\Gamma(\hat{\Phi}(A_0))\|^{1/2}\|\mathbb{1} - \Gamma(A_s)\Gamma(A_s)^*\|^{1/2} \\ &+ \|\mathbb{1} - \Gamma(\hat{\Phi}(A_0))\Gamma(\hat{\Phi}(A_0))^*\|^{1/2}\|\mathbb{1} - \Gamma(A_s)^*\Gamma(A_s)\|^{1/2}\Big),\end{aligned}$$

which is bounded above by

$$\begin{aligned}\frac{1}{|c_{RZ}|}(1 - |c_{RZ}|^2)^{1/2}\Big(&\|\mathbb{1} - \Gamma_{\rho_{\mathcal{R}}}(\Phi(U_{\mathcal{S}}(s_0) \otimes \mathbb{1}_{\mathcal{R}}))^*\Gamma_{\rho_{\mathcal{R}}}(\Phi(U_{\mathcal{S}}(s_0) \otimes \mathbb{1}_{\mathcal{R}}))\|^{1/2} \\ &+ \|\mathbb{1} - \Gamma_{\rho_{\mathcal{R}}}(\Phi(U_{\mathcal{S}}(s_0) \otimes \mathbb{1}_{\mathcal{R}}))\Gamma_{\rho_{\mathcal{R}}}(\Phi(U_{\mathcal{S}}(s_0) \otimes \mathbb{1}_{\mathcal{R}}))^*\|^{1/2}\Big).\end{aligned}$$

We estimate

$$\begin{aligned}&\|\mathbb{1}-\Gamma_{\rho_{\mathcal{R}}}(\Phi(U_{\mathcal{S}}(s_0) \otimes \mathbb{1}_{\mathcal{R}}))^*\Gamma_{\rho_{\mathcal{R}}}(\Phi(U_{\mathcal{S}}(s_0) \otimes \mathbb{1}_{\mathcal{R}}))\| \\ &= \|\mathbb{1} - \Lambda(U_{\mathcal{S}}(s_0))^*\Lambda(U_{\mathcal{S}}(s_0)) - \epsilon(s_0)^*\epsilon(s_0) - \epsilon(s_0)^*\Lambda(U_{\mathcal{S}}(s_0)) - \Lambda(U_{\mathcal{S}}(s_0))^*\epsilon(s_0)\| \\ &\leq \|\mathbb{1} - \Lambda(U_{\mathcal{S}}(s_0))^*\Lambda(U_{\mathcal{S}}(s_0))\| + 2\|\epsilon(s_0)\|.\end{aligned}$$

Similarly we obtain

$$\begin{aligned}&\|\mathbb{1}-\Gamma_{\rho_{\mathcal{R}}}(\Phi(U_{\mathcal{S}}(s_0) \otimes \mathbb{1}_{\mathcal{R}}))\Gamma_{\rho_{\mathcal{R}}}(\Phi(U_{\mathcal{S}}(s_0) \otimes \mathbb{1}_{\mathcal{R}}))^*\| \\ &\leq \|\mathbb{1} - \Lambda(U_{\mathcal{S}}(s_0))\Lambda(U_{\mathcal{S}}(s_0))^*\| + 2\|\epsilon(s_0)\|.\end{aligned}$$

Finally, one can choose W_Z so as to maximize $|\langle\phi_{RZ}|U_{\mathcal{R}}(s) \otimes W_Z|\phi_{RZ}\rangle|$, which coincides with $F(\rho_{\mathcal{R}}, U_{\mathcal{R}}(s)\rho_{\mathcal{R}}U_{\mathcal{R}}(s)^*)$ due to Uhlmann's theorem [23], thereby completing the proof. □

We now provide a proof of Corollary 2.

Proof. Adopting the shorthand $F \equiv F(\rho_{\mathcal{R}}, U_{\mathcal{R}}(l:s)\rho_{\mathcal{R}}U_{\mathcal{R}}(l:s)^*)$, the equality (4) replaces the inequality of Theorem 1 by,

$$\begin{aligned}\|U_{\mathcal{S}}(l:s)^*\Lambda(U_{\mathcal{S}}(l:s_0))U_{\mathcal{S}}(l:s) - \Lambda(U_{\mathcal{S}}(l:s_0))\| &\leq 2\|U_{\mathcal{S}}(l:s) - \mathbb{1}\|\|\epsilon(l:s_0)\| \\ &+ (\frac{1}{F^2} - 1)^{1/2}\Big((\|\mathbb{1}_{\mathcal{S}} - \Lambda(U_{\mathcal{S}}(l:s_0))^*\Lambda(U_{\mathcal{S}}(l:s_0))\| + 2\|\epsilon(l:s_0)\|)^{1/2} \\ &+ (\|\mathbb{1}_{\mathcal{S}} - \Lambda(U_{\mathcal{S}}(l:s_0))\Lambda(U_{\mathcal{S}}(l:s_0))^*\| + 2\|\epsilon(l:s_0)\|)^{1/2}\Big).\end{aligned}$$

To bound the first term on the right hand side we write

$$U_{\mathcal{S}}(s) = \mathbb{1}_{\mathcal{S}} + i\int_0^s dt U_{\mathcal{S}}(t)L_{\mathcal{S}},$$

and therefore

$$\|U_{\mathcal{S}}(s) - \mathbb{1}_{\mathcal{S}}\| \le |s|\|L_{\mathcal{S}}\|.$$

For the second term, we bound $F(\rho_{\mathcal{R}}, U_{\mathcal{R}}(l:s)\rho_{\mathcal{R}}U_{\mathcal{R}}(l:s)^*)$ by choosing a purification of $\rho_{\mathcal{R}}$ as $|\phi\rangle \in \mathcal{H}_{\mathcal{R}} \otimes \mathcal{H}_Z$. Then Uhlmann's theorem states that the fidelity is written as

$$F(\rho_{\mathcal{R}}, U_{\mathcal{R}}(l:s)\rho_{\mathcal{R}}U_{\mathcal{R}}(l:s)^*) = \sup_{|\phi\rangle} |\langle\phi|e^{iL_{\mathcal{R}}s} \otimes \mathbb{1}_Z|\phi\rangle|.$$

For each purification $|\phi\rangle$, the Mandelstam-Tamm uncertainty relation [19, 20] provides a bound for $0 \le \Delta_{\rho_{\mathcal{R}}} L_{\mathcal{R}} \cdot s \le \pi/2$,

$$|\langle\phi|e^{iL_{\mathcal{R}}s} \otimes \mathbb{1}_Z|\phi\rangle| \ge \cos(\Delta_{\rho_{\mathcal{R}}} L_{\mathcal{R}} \cdot s).$$

Thus we obtain

$$\left(\frac{1}{F(\rho_{\mathcal{R}}, U_{\mathcal{R}}(l:s)\rho_{\mathcal{R}}U_{\mathcal{R}}(l:s)^*)^2} - 1\right)^{1/2} \le \tan(\Delta_{\rho_{\mathcal{R}}} L_{\mathcal{R}} \cdot s).$$

We divide the both terms by $|s|$ and take $|s| \to 0$ to obtain,

$$\|[\Lambda(U_{\mathcal{S}}(s_0)), L_{\mathcal{S}}]\| \le 2\|L_{\mathcal{S}}\|\|\epsilon(s_0)\| + \Delta_{\rho_{\mathcal{R}}} L_{\mathcal{R}}\Big((\|\mathbb{1} - \Lambda(U_{\mathcal{S}}(s_0))^*\Lambda(U_{\mathcal{S}}(s_0))\| + 2\|\epsilon(s_0)\|)^{1/2} + (\|\mathbb{1} - \Lambda(U_{\mathcal{S}}(s_0))\Lambda(U_{\mathcal{S}}(s_0))^*\| + 2\|\epsilon(s_0)\|)^{1/2}\Big).$$

□

4. Rotational symmetry

As an example of the general behaviour we have investigated, we consider the possible dynamics of a qubit with Hilbert space $\mathcal{H}_{\mathcal{S}} = \mathbb{C}^2$ under $SO(3)$ symmetry, realized by a true irreducible unitary representation of its universal covering group $SU(2)$. Since only a trivial unitary operator proportional to $\mathbb{1}$ commutes with all $SU(2)$ generators (angular momenta), one cannot change the state of the qubit in isolation (i.e., unitarily). The environment $\mathcal{H}_{\mathcal{R}}$ also has $SU(2)$ as a symmetry. We denote the angular momenta of the system and the reference frame by s_j and S_j ($j = x, y, z$) respectively. We consider an $SU(2)$-covariant channel $\Phi : \mathbf{B}(\mathcal{H}_{\mathcal{S}} \otimes \mathcal{H}_{\mathcal{R}}) \to \mathbf{B}(\mathcal{H}_{\mathcal{S}} \otimes \mathcal{H}_{\mathcal{R}})$. The following corollary is immediately obtained from Corollary 2.

Corollary 4. *Let G be a Lie group. For a covariant channel $\Phi : \mathbf{B}(\mathcal{H}_{\mathcal{S}} \otimes \mathcal{H}_{\mathcal{R}}) \to \mathbf{B}(\mathcal{H}_{\mathcal{S}} \otimes \mathcal{H}_{\mathcal{R}})$, its restriction $R \equiv \Gamma_{\rho_{\mathcal{R}}} \circ \Phi \circ \iota : \mathbf{B}(\mathcal{H}_{\mathcal{S}}) \to \mathbf{B}(\mathcal{H}_{\mathcal{S}})$, satisfies*

$$\|[R(U_{\mathcal{S}}(s_0)), L_{\mathcal{S}}]\| \le (\Delta_{\rho_{\mathcal{R}}} L_{\mathcal{R}})\Big(\|\mathbb{1} - R(U_{\mathcal{S}}(s_0))^*R(U_{\mathcal{S}}(s_0))\|^{1/2} + \|\mathbb{1} - R(U_{\mathcal{S}}(s_0))R(U_{\mathcal{S}}(s_0))^*\|^{1/2}\Big).$$

We apply this to the case $G = SU(2)$ and $\mathcal{H}_{\mathcal{S}} = \mathbb{C}^2$. For $l = s_x$, $U_{\mathcal{S}}(s)$ is written as $U_{\mathcal{S}}(s) = e^{is_x s} = e^{i\frac{\sigma_x}{2}s}$. We set $s_0 = \pi$ to obtain $U_{\mathcal{S}}(s_0) = i\sigma_x$. Then the restriction to the system $\Lambda := \Gamma_{\rho_{\mathcal{R}}} \circ \Phi \circ \iota : \mathbf{B}(\mathcal{H}_{\mathcal{S}}) \to \mathbf{B}(\mathcal{H}_{\mathcal{S}})$ satisfies the following three inequalities:

$$\|[\Lambda(\sigma_x), s_x]\| \le 2(\Delta_{\rho_{\mathcal{R}}} S_x)\|\mathbb{1} - \Lambda(\sigma_x)^2\|^{1/2}$$
$$\|[\Lambda(\sigma_y), s_y]\| \le 2(\Delta_{\rho_{\mathcal{R}}} S_y)\|\mathbb{1} - \Lambda(\sigma_y)^2\|^{1/2}$$
$$\|[\Lambda(\sigma_z), s_z]\| \le 2(\Delta_{\rho_{\mathcal{R}}} S_z)\|\mathbb{1} - \Lambda(\sigma_z)^2\|^{1/2},$$

where $s_x = \frac{1}{2}\sigma_x$, *etc.* The uncertainty relations for angular momenta gives a non-trivial bound on sums of their fluctuations. We consider

$$\begin{aligned}(\Delta S_x)^2 + (\Delta S_y)^2 + (\Delta S_z)^2 &= \langle S_x^2 + S_y^2 + S_z^2\rangle - (\langle S_x\rangle^2 + \langle S_y\rangle^2 + \langle S_z\rangle^2) \\ &\leq l(l+1) - (\langle S_x\rangle^2 + \langle S_y\rangle^2 + \langle S_z\rangle^2),\end{aligned}$$

where l is the magnitude of the largest spin of the environment. (Note that $\mathcal{H}_{\mathcal{R}}$ is written as a direct sum of irreducible representations of $SU(2)$ as $\mathcal{H}_{\mathcal{R}} = \oplus_s \mathbb{C}^{2s+1}$. l is the largest value of s in the summation.) It is easy to show that $\langle S_x\rangle^2 + \langle S_y\rangle^2 + \langle S_z\rangle^2$ is rotationally invariant. We consider the quantity $\langle \mathbf{S}\cdot\mathbf{n}\rangle$ for $|\mathbf{n}| = 1$. This is a smooth function over the sphere and therefore has a maximum value at a certain point. To estimate the value of $\langle S_x\rangle^2+\langle S_y\rangle^2+\langle S_z\rangle^2$, we assume that the maximum of $\langle \mathbf{S}\cdot\mathbf{n}\rangle$ is attained at $\mathbf{n} = \mathbf{e}_z$. By differentiating in polar coordinates, one can conclude that this state shows $\langle S_x\rangle = \langle S_y\rangle = 0$. Thus we have $\langle S_x\rangle^2+\langle S_y\rangle^2+\langle S_z\rangle^2 = \langle S_z\rangle^2$. Using $0 \leq \langle S_z\rangle^2 \leq l^2$, we conclude that

$$l \leq (\Delta S_x)^2 + (\Delta S_y)^2 + (\Delta S_z)^2 \leq l(l+1).$$

Thus we obtain the bound

$$\begin{aligned}&\|[\Lambda(\sigma_x), \sigma_x]\| + \|[\Lambda(\sigma_y), \sigma_y]\| + \|[\Lambda(\sigma_z), \sigma_z]\| \\ &\leq 2\sqrt{l(l+1)}(\|\mathbb{1} - \Lambda(\sigma_x)^2\| + \|\mathbb{1} - \Lambda(\sigma_y)^2\| + \|\mathbb{1} - \Lambda(\sigma_z)^2\|)^{1/2},\end{aligned}$$

where we used the Cauchy-Schwarz inequality. One can confirm, as expected, that any realizable non-covariant channel is inevitably dissipative, as non-dissipative (=unitary) dynamics satisfies $\mathbb{1} = \Lambda(\sigma_x)^2 = \Lambda(\sigma_y)^2 = \Lambda(\sigma_z)^2$. The right-hand side can be regarded as a quantity measuring the "dissipativity" of Λ, while the left-hand side represents the "magnitude" of dynamics. If the environment consists of N qubits, as $l = \frac{N}{2}$ holds the term $\sqrt{l(l+1)}$ in the right-hand side of the above inequality is proportional to N. Thus for Λ whose magnitude of dynamics is $O(1)$, its dissipativity cannot be smaller than $O\left(\frac{1}{N}\right)$ in the presence of N environment qubits.

We employ Stokes parameterization [24] to illustrate possible channels. Any qubit state is written as $\rho = \frac{1}{2}(\mathbb{1}_{\mathcal{S}} + \mathbf{x}\cdot\sigma)$ with $|\mathbf{x}| \leq 1$. Λ^*, the dual of Λ, maps ρ to another state $\rho' = \frac{1}{2}(\mathbb{1}_{\mathcal{S}} + \mathbf{y}\cdot\sigma)$. This map $(1, \mathbf{x}) \mapsto (1, \mathbf{y})$ is a linear map on $\mathbf{R}^4$ since Λ is self-adjoint. We denote this map by ${}^t\tilde{T}_\Lambda$ with a parameterization,

$${}^t\tilde{T}_\Lambda = \begin{pmatrix} 1 & 0 & 0 & 0 \\ t_1 & t_{11} & t_{12} & t_{13} \\ t_2 & t_{21} & t_{22} & t_{23} \\ t_3 & t_{31} & t_{32} & t_{33} \end{pmatrix} = \begin{pmatrix} 1 & \mathbf{0} \\ \mathbf{t} & T \end{pmatrix},$$

where T is a 3×3 matrix. Back in the Heisenberg picture, we obtain

$$\Lambda(a_0\mathbb{1}_{\mathcal{S}} + \mathbf{a}\cdot\sigma) = (a_0 + \mathbf{t}\cdot\mathbf{a})\mathbb{1}_{\mathcal{S}} + (T\mathbf{a})\cdot\sigma.$$

T can be written as

$$T = R_1 D R_2,$$

where R_1 and R_2 are elements of $SO(3)$ and D is a diagonal matrix as,

$$D = \begin{pmatrix} \lambda_1 & 0 & 0 \\ 0 & \lambda_2 & 0 \\ 0 & 0 & \lambda_3 \end{pmatrix}.$$

One can choose the coordinate system so that $R_2 = \mathbb{1}$ is satisfied. Thus we will consider T with form $T = RD$. Then we obtain, for redefined $\mathbf{t}$,

$$\Lambda(a_0 \mathbb{1}_{\mathcal{S}} + \mathbf{a} \cdot \sigma) = (a_0 + \mathbf{t} \cdot \mathbf{a})\mathbb{1}_{\mathcal{S}} + \sum_{ij=1}^{3} R_{ij}\lambda_j a_j \sigma_i,$$

where $R_{ij} \in SO(3)$. Assume that R is written as a rotation around the z-axis, with the vector $\mathbf{t} = \mathbf{0}$, as,

$$R = \begin{pmatrix} \cos\theta & \sin\theta & 0 \\ -\sin\theta & \cos\theta & 0 \\ 0 & 0 & 1 \end{pmatrix}.$$

Then we have

$$\lambda_x |\sin\theta| \leq (\Delta_{\rho_{\mathcal{R}}} S_x)\sqrt{1-\lambda_x^2}$$
$$\lambda_y |\sin\theta| \leq (\Delta_{\rho_{\mathcal{R}}} S_y)\sqrt{1-\lambda_y^2}.$$

That is, we have a relation between the dissipative and symmetry breaking natures.

$$\lambda_x^2 \leq \frac{(\Delta_{\rho_{\mathcal{R}}} S_x)^2}{(\Delta_{\rho_{\mathcal{R}}} S_x)^2 + (\sin\theta)^2};$$
$$\lambda_y^2 \leq \frac{(\Delta_{\rho_{\mathcal{R}}} S_y)^2}{(\Delta_{\rho_{\mathcal{R}}} S_y)^2 + (\sin\theta)^2}.$$

5. Concluding remarks

We have seen that there is a positive lower bound on the difference between an arbitrary quantum channel and the restriction of a covariant channel, and moreover, that in order to reduce this discrepancy a large spread in the generator of the symmetry is needed in the reference system. This result bears similarities with the WAY theorem, and is in line with the relational view of quantum mechanics, wherein we interpret non-symmetric channels as representatives of their symmetric counterparts of system and reference taken together. The large spread required for good approximation of relative (symmetric) by non-relative (asymmetric) can be understood as a condition on the quality of the reference frame, in the sense of the findings of [3] and [11]. As a final remark, we mention that there is yet another symmetry condition on channels that differs from the one employed in this paper and arises naturally in the context of quantum reference frames. We will return to this issue elsewhere.

Acknowledgments

TM acknowledges financial support from JSPS (KAKENHI Grant Number 20K03732).

References

[1] Haag R 1996 *Local Quantum Physics* (Berlin: Springer)
[2] Loveridge L, Busch P and Miyadera T 2017 *EPL* **117** 40004
[3] Loveridge L, Miyadera T and Busch P 2018 *Found. Phys.* **48** 135–98
[4] Wigner E 1952 *Z. Phys.* **133** 101–8
[5] Araki H and Yanase M M 1960 *Phys. Rev.* **120** 622
[6] Loveridge L and Busch P 2011 *Eur. Phys. Jour.* D **62** 297–307
[7] Busch P and Loveridge L 2011 *Phys. Rev. Lett.* **106** 110406
[8] Loveridge L 2020 A relational perspective on the Wigner-Araki-Yanase theorem *Preprint* arXiv:2006.07047 [quant-ph]

[9] Miyadera T and Imai H 2006 *Phys. Rev.* A **74** 024101
[10] Åberg J 2014 *Phys. Rev. Lett.* **113** 150402
[11] Miyadera T, Loveridge L and Busch P 2016 *J. Phys. A: Math. Theor.* **49** 185301
[12] Ahmadi M, Jennings D and Rudolph T 2013 *New J. Phys* **15** 013057
[13] Bartlett S D, Rudolph T, Spekkens R W and Turner P S 2009 *New J. Phys.* **11** 063013
[14] Tajima H, Shiraishi N and Saito K 2018 *Phys. Rev. Lett.* **121** 110403
[15] Tajima H, Shiraishi N and Saito K 2019 Coherence cost for violating conservation laws *Preprint* arXiv:1906.04076 [quant-ph]
[16] Takagi R and Tajima H 2020 *Phys. Rev.* A **101** 022315
[17] Miyadera T and Imai H 2008 *Phys. Rev.* A **78** 052119
[18] Janssens B 2017 *Lett. Math. Phys.* **107** 1557–79
[19] Mandelstam L I and Tamm I E 1945 *J. Phys. USSR* **9** 249–54
[20] Busch P 2008 *Time in Quantum Mechanics* (Lecture Notes in Physics vol 734) ed G Muga *et al* (Berlin: Springer) pp 73–105
[21] Marvian I and Spekkens R W 2012 An information-theoretic account of the Wigner-Araki-Yanase theorem *Preprint* arXiv:1212.3378 [quant-ph]
[22] Piani M, Cianciaruso M, Bromley T R, Napoli C, Johnston N and Adesso G 2016 *Phys. Rev.* A **93** 042107
[23] Uhlmann A 1976 *Rep. Math. Phys.* **9** 273–9
[24] Heinosaari T and Ziman M 2011 *The Mathematical Language of Quantum Theory* (Cambridge: Cambridge University Press)

Journal of Physics: Conference Series **1638** (2020) 012009 doi:10.1088/1742-6596/1638/1/012009

A relational perspective on the Wigner-Araki-Yanase theorem

Leon Loveridge

Quantum Technology Group, Department of Science and Industry Systems, University of South-Eastern Norway, 3616 Kongsberg, Norway

E-mail: `leon.d.loveridge@usn.no`

Abstract. We present a novel interpretation of the Wigner-Araki-Yanase (WAY) theorem based on a relational view of quantum mechanics. Several models are analysed in detail, backed up by general considerations, which serve to illustrate that the moral of the WAY theorem may be that in the presence of symmetry, a measuring apparatus must fulfil the dual purpose of both reflecting the statistical behaviour of the system under investigation, and acting as a physical reference system serving to define those quantities which must be understood as relative.

In this way quantum theory reminds us, as Bohr has put it, of the old wisdom that when searching for harmony in life one must never forget that in the drama of existence we are ourselves both players and spectators. Werner Heisenberg [1]

Dedicated to the memory of Paul Busch

1. Introduction

In this paper we present an overview of some ideas developed in the decade 2008–2018, which form part of a large and ongoing programme aimed at providing a novel understanding of quantum theory based on what one might call a "symmetry-enforced relationalism". Some of this work was included in the author's PhD thesis, and was undertaken in collaboration with Takayuki Miyadera and the author's supervisor and friend Paul Busch, whose attention, guidance and encouragement made the work what it is, and is now so sorely missing it often feels impossible to proceed.

The work presented in this contribution to the volume dedicated to Paul remains unfinished, with both technical and interpretational work remaining. The idea is to combine the point of view that the world described by quantum mechanics is a deeply and irreducibly relational one, with a technical result in the quantum theory of measurement called the Wigner-Araki-Yanase (WAY) theorem, which imposes measurement limitations in the presence of conserved quantities, in the hope of both better understanding the message of this important theorem, and of gaining more general qualitative insight into the relationship between the formal aspects of quantum theory and the world that it is supposed to describe.

We will argue that the WAY theorem may be seen as an expression of the fact that the quantum theoretical formalism as it is traditionally used speaks not of quantum systems as they are, but of the relation between a quantum system and a reference system which possesses certain properties that we are accustomed to viewing as "classical". This point of view, which emerged from our mutual discussion and reflection over the years, was referred to by Paul with the acronym "RRR", to stand for *radically relational reality*, the consequences of which we had only seriously begun to consider. Of course broadly speaking such an attitude is not new: Bohr emphasised very early the epistemological view that quantum phenomena must be understood *only* in relation to the "whole experimental arrangement" [2] and therefore pertain to both system and measuring apparatus combined – one cannot speak of system without apparatus. There are many other forms of quantum "relationalism" which are morally in harmony with what we are advocating, for instance the relational quantum mechanics of Rovelli [3] and the perspectival approach of Bene and Dieks [4–6]. However, the specific form of relationalism presented here and in our other works (e.g., [7, 8]) is founded on symmetry in a fundamental way, and in that respect has more in common with the large body of work on *quantum frames of reference* [9–11] – a topic which also forms a major part of the current investigation.

After briefly introducing the rudiments of the quantum theory of measurement, we present (section 3) the WAY theorem as it appears in [12–15], with some additional comments on developments in the years after those papers were published. The message of this work is principally that sharp observables not commuting with a conserved quantity cannot be measured exactly, but that good approximation is possible provided that the apparatus is able to be, and is, prepared in a state with large fluctuation in the conserved quantity. In section 4 we present two arguments for the adoption of the view that "truly" observable quantities are invariant under symmetry transformations. The first arises from the basic observation that this is standard in gauge theories, and in a Galilean spacetime context (for example) the possibility of measuring "absolute" position or time (for instance) is at odds with Galilean relativity. The second is dynamical, and can be stated as a theorem in the quantum theory of measurement wherein the system under investigation possesses its own conserved quantity. In both cases, we find that reference frames/systems play a crucial role in resolving the tension between the unobservability of "absolute" (or non-invariant) quantities, and the routine use of such quantities in the accurate description of the physical world. We find that non-invariant quantities are theoretical shorthands for the true, relational/invariant quantities which relate system and reference. With this in hand, we will argue (section 5) that the WAY theorem is an expression of the relativity, or relationalism, of physical quantities under symmetry, where the measuring apparatus is also required to fulfil the role of reference. We conclude with a brief summary and discussion of further avenues of investigation for the RRR programme.

2. Preliminary material

2.1. Basic framework

To a quantum system $\mathcal{S}$ is assigned a complex, separable Hilbert space $\mathcal{H}$ with inner product $\langle \cdot | \cdot \rangle : \mathcal{H} \times \mathcal{H} \to \mathbb{C}$, linear in the second argument (and hence conjugate linear in the first). In case $\mathcal{S}$ consists of exactly two separate subsystems so that $\mathcal{S}$ can be viewed as a composite of $\mathcal{S}_1$ and $\mathcal{S}_2$, $\mathcal{H}$ takes the form of the tensor product $\mathcal{H}_1 \otimes \mathcal{H}_2$. States of $\mathcal{S}$ are represented by (and identified with) positive operators in the trace class $\mathcal{T}(\mathcal{H})$ of $\mathcal{H}$ which have unit trace. This is a convex subset of $\mathcal{T}(\mathcal{H})$, which we will write as $\mathcal{S}(\mathcal{H})$. Pure states are singled out as the one dimensional projections, to be written as P_φ, with φ a unit vector in $\mathcal{H}$, and we will occasionally indulge the traditional laissez-faire inaccuracy and refer to φ itself as a pure state of $\mathcal{S}$ (sometimes also to be called a *vector state*).

Observables are identified with positive operator-valued measures (POMs) $\mathsf{E} : \mathcal{F} \to \mathcal{L}(\mathcal{H})$, where $\mathcal{F}$ is a σ-algebra of subsets of some set Ω which represents the *value space* of E, i.e., the space of possible outcomes that may arise in a measurement of E, and $\mathcal{L}(\mathcal{H})$ denotes the (algebra of) bounded linear operators in $\mathcal{H}$. We recall that POMs are positive in the sense that all operators in the range of a POM are positive (we write $\mathcal{L}(\mathcal{H})^+$ for the positive elements of $\mathcal{L}(\mathcal{H})$),

Journal of Physics: Conference Series **1638** (2020) 012009 doi:10.1088/1742-6596/1638/1/012009

normalised ($\mathsf{E}(\Omega) = \mathbb{1}_{\mathcal{H}}$) and have the property of being countably additive. The operators in the range of a POM are called *effects*.

That we use $\mathcal{F}$ rather than Ω owes itself to the probabilistic nature of the basic framework, to which we will shortly turn. *Sharp* observables are singled out amongst all observables as the projection-valued measures (PVMs), characterised as those POMs E for which $\mathsf{E}(X)^2 = \mathsf{E}(X)$ for all $X \in \mathcal{F}$, or equivalently, $\mathsf{E}(X \cap Y) = \mathsf{E}(X)\mathsf{E}(Y)$ for all $X, Y \in \mathcal{F}$. In this case, and if $(\Omega, \mathcal{F}) = (\mathbb{R}, \mathcal{B}(\mathbb{R}))$ (the real numbers and the corresponding Borel sets) or subsets thereof, then $\mathsf{E} \equiv \mathsf{E}^A$ is the spectral measure of some unique self-adjoint operator $A \in \mathcal{L}(\mathcal{H})$, where A and E^A are connected via the spectral integral $A = \int x\mathsf{E}^A(dx)$. We also refer to A itself as an observable when it is more convenient.

The state-observable pairings (ρ, E) thus defined capture the empirical content of the theory via the "Born rule" probability measures $X \mapsto \mathrm{tr}[\rho\mathsf{E}(X)]$, understood as representing the probability that if the system is prepared in a state ρ and a measurement of E is performed, this measurement yields an outcome in the set X. Indeed, for an arbitrary mapping $\mathsf{E} : \mathcal{F} \to \mathcal{L}(\mathcal{H})^+$, the assignment $X \mapsto \mathrm{tr}[\rho\mathsf{E}(X)]$ being a probability measure for all states guarantees that E is a POM, highlighting that this description of observables saturates the full probabilistic content of the Hilbert space framework. For a detailed account of the conceptual and mathematical development of the use of POMs in the foundations of quantum mechanics and measurement we refer the reader to [16–18].

2.2. Measurement

Measurements of quantum objects inevitably involve an interaction of the object $\mathcal{S}$ under investigation with some second system, to be called the *measuring apparatus* $\mathcal{A}$ which fulfils the dual role of being "quantum enough" to interact properly with $\mathcal{S}$ (specifically, to become entangled with $\mathcal{S}$, which is indeed necessary for the measurement process—as information extraction—to take place – see [19]) and "classical enough" to produce measurement records. We will not discuss the problem of how such desiderata may be satisfied (a difficult question to answer, even in principle [17,20]), but rather proceed to model such a situation operationally [16].

Given some observable E of $\mathcal{S}$ the question therefore naturally arises as to the existence of a suitable apparatus (or *probe*) system $\mathcal{A}$ which may be used to faithfully reproduce the statistics of E, and therefore to function as a measurement of E. Put more precisely, one wishes to specify an observable Z of $\mathcal{A}$ (to be called a *pointer observable*), here assumed to be the spectral measure of a self-adjoint operator Z (an assumption which will soon be justified) acting in $\mathcal{H}_{\mathcal{A}}$, a state $\rho_{\mathcal{A}} \in \mathcal{S}(\mathcal{H}_{\mathcal{A}})$, and a suitable unitary mapping $U(\tau)$ acting in $\mathcal{H}_{\mathcal{S}} \otimes \mathcal{H}_{\mathcal{A}}$ which serves to correlate $\mathcal{S}$ and $\mathcal{A}$ in a way which we will shortly spell out. The τ in appearing in $U(\tau)$ is a parameter related to the interaction strength and duration, which we will sometimes suppress, and suitable here means that we may arrange for the pre-interaction statistics of E of $\mathcal{S}$, namely, the probability measures $X \mapsto \mathrm{tr}[\rho\mathsf{E}(X)]$ to be equal to the post-interaction probabilities for the apparatus observable, i.e., $X \mapsto \mathrm{tr}\,(\mathsf{Z}(X)\,\mathrm{tr}_{\mathcal{S}}(U(\tau)\rho \otimes \rho_{\mathcal{A}} U(\tau)^*))$. This amounts to the statement of the *probability reproducibility condition* (PRC) which serves as a minimal operational condition for the tuple $\mathcal{M} := \langle \mathcal{H}_{\mathcal{A}}, U(\tau), Z, \rho_{\mathcal{A}} \rangle$, called *measurement scheme* [16], to serve as a measurement of E. This we write as

$$\mathrm{tr}[\rho\mathsf{E}(X)] = \mathrm{tr}[U(\tau)\rho \otimes \rho_{\mathcal{A}} U(\tau)^* \mathbb{1} \otimes \mathsf{Z}(X)]. \tag{1}$$

The existence of such an apparatus which satisfies the above condition is established by two central mathematical results. The first is *Naimark's dilation theorem*, which states that given a POM $\mathsf{E} : \mathcal{F} \to \mathcal{L}(\mathcal{H})$ there exists a Hilbert space $\mathcal{K}$, a PVM $\mathsf{F} : \mathcal{F} \to \mathcal{K}$ and an isometry $V : \mathcal{H} \to \mathcal{K}$ for which

$$\mathsf{E}(X) = V^*\mathsf{F}(X)V. \tag{2}$$

The second, which builds upon this, is found in the remarkable contribution of Ozawa [21], who showed that E always admits a *measurement dilation*—a Naimark dilation of a specific form—comprising a Hilbert space $\mathcal{H}_{\mathcal{A}}$ for which $\mathcal{K} = \mathcal{H} \otimes \mathcal{H}_{\mathcal{A}}$, a fixed unit vector $\phi \in \mathcal{H}_{\mathcal{A}}$, and V which can be chosen to be of the form $V = UV_{\phi}$ with U a unitary mapping $\mathcal{K} \to \mathcal{K}$ and $V_{\phi} : \mathcal{H} \to \mathcal{K}$ the isometric embedding $V_{\phi}(\varphi) = \varphi \otimes \phi$. Moreover, it can be arranged so that $\mathsf{F}(X) = \mathbb{1} \otimes \mathsf{Z}(X)$ where Z is the spectral measure of a self-adjoint operator Z. Thus we may write the probability reproducibility condition in this case (dropping the τ for now) and for pure states as

$$\langle \varphi \,|\, \mathsf{E}(X)\varphi \rangle = \langle U(\varphi \otimes \phi) \,|\, \mathbb{1} \otimes \mathsf{Z}(X) U(\varphi \otimes \phi) \rangle \,, \tag{3}$$

stipulated to hold for all unit vectors $\varphi \in \mathcal{H}$ and $X \in \mathcal{F}$. This states that the probabilities for the pointer observable Z in the vector state $U\varphi \otimes \phi$ perfectly reproduce those of E in the vector state φ, and that therefore the measuring process for Z (and thus for Z) in the final state can be understood as a measuring process for E in the initial state. There are two conceptually distinct readings of the PRC (3). One is that the right hand side (i.e., U, Z and ϕ, and thus implicitly also $\mathcal{H}_{\mathcal{A}}$) is chosen so as to be a measurement (scheme) for some fixed E of interest to an experimenter. The other is to first fix $\mathcal{M}$, which reflects the common situation in empirical practice, wherein only certain apparatus preparations, dynamics, *etc.*, are available. Then the right hand side of (3) uniquely identifies the observable E actually measured by this scheme, which may or may not function as a good measurement of the observable of interest.

Note that whilst the measurement outcome probabilities are fully characterised by the measured observable, the measurement scheme provides a far more comprehensive account of the entire measurement process and the information transfer between system and probe. The measurement scheme is not uniquely determined by the measured observable, and therefore it is often useful to consider the equivalence class of operationally indistinguishable schemes.

Occasionally it is necessary to include a measurable function $f : \Omega_{\mathcal{A}} \to \Omega$ in the definition of a measurement scheme in order to account for the measured observable and the pointer having different value spaces or scales; in this case equation (3) becomes

$$\left\langle U\varphi \otimes \phi \,|\, \mathbb{1} \otimes \mathsf{Z}(f^{-1}(X)) U\varphi \otimes \phi \right\rangle = \langle \varphi \,|\, \mathsf{E}(X)\varphi \rangle \,. \tag{4}$$

The second point of view mentioned above is perhaps more clearly expressed at the operator level by the equation

$$\mathsf{E}(X) = V_{\phi}^{*} U^{*} (\mathbb{1} \otimes \mathsf{Z}(X)) U V_{\phi}; \tag{5}$$

the mapping $\Gamma_{\phi}(\cdot) \equiv V_{\phi}^{*}(\cdot)V_{\phi} : \mathcal{L}(\mathcal{K}) \to \mathcal{L}(\mathcal{H})$ is a (normal) conditional expectation and can be viewed as a "restriction" mapping on observables, arising from fixing a state of $\mathcal{H}_{\mathcal{A}}$. The observable defined through restriction may not be equal to the "target" observable which we actually wish to measure; if not, then statistical techniques are required in order to quantify the discrepancy between the observable we wish to measure (which will often be sharp) and the actually measured observable $\Gamma_{\phi}(U^{*}\mathbb{1} \otimes \mathsf{Z}(X))U)$, which may be unsharp and function as an unsharp approximator for E. Writing $Z(\tau) = U^{*}\mathbb{1} \otimes \mathsf{Z}(X))U$ highlights the fact that the measured observable E is then just the restriction of the time-evolved pointer observable.

The restriction map is not limited to describing measurement processes, can be naturally extended to mixtures, and represents a natural means by which to describe a subsystem "as though" it were being considered in isolation. There are various equivalent definitions which will be employed. It is the dual of the isometric embedding $\mathcal{S}(\mathcal{H}) \to \mathcal{S}(\mathcal{H} \otimes \mathcal{H}_{\mathcal{A}})$ defined by $\rho \mapsto \rho \otimes \sigma$. It is the continuous linear extension of the map $A \otimes B \mapsto A\,\mathrm{tr}[B\sigma]$, and finally, it may be defined through the formula

$$\mathrm{tr}[\rho\Gamma_{\sigma}(\Lambda)] = \mathrm{tr}[\rho \otimes \sigma\Lambda], \text{ for all } \rho \in \mathcal{S}(\mathcal{H}) \text{ and } \Lambda \in \mathcal{L}(\mathcal{H} \otimes \mathcal{H}_{\mathcal{A}}). \tag{6}$$

Journal of Physics: Conference Series **1638** (2020) 012009 doi:10.1088/1742-6596/1638/1/012009

For a vector state ϕ we continue to write Γ_ϕ rather than Γ_{P_ϕ}. In this work, we view the restriction as an *externalisation* of the second system, i.e., a way to describe the subsystem contingent upon some preparation of the other system, which is being suppressed in the theoretical description but nevertheless functions crucially in the ascription of properties of $\mathcal{S}$ which pertain also to the state of $\mathcal{A}$.

We conclude this section by noting that the *repeatability* of a measurement, which is often assumed to comprise part of the definition (of measurement) in standard textbook treatments, is not logically implied by the probability reproducibility condition and must therefore be viewed as an additional stipulation which a measurement may or may not satisfy. We recall that repeatable measurements are defined as those for which the same outcome is guaranteed (with probability 1) to be realised upon immediate repetition, i.e.,

$$\langle U(\varphi \otimes \phi) \,|\, \mathsf{E}(X) \otimes \mathsf{Z}(X) U(\varphi \otimes \phi) \rangle = \langle \varphi \,|\, \mathsf{E}(X)\varphi \rangle , \tag{7}$$

holding for all φ and X. Indeed, with the broader definition of observable than is traditional (i.e., the representation by POMs), it is possible that no repeatable measurements exist at all. This is the case if, for example, some of the effects in the range of an observable do not have 1 as an eigenvalue. Moreover, as shown by Ozawa in his 1984 contribution [21], observables with continuous spectrum do not admit *any* repeatable measurements, and therefore the extension of the notion of measurement to include those which are not repeatable is a crucial ingredient in the description of physical experiments.

Historically, the question of whether all observables can, at least theoretically, be measured, perhaps also repeatably, was raised for the first time some twenty years after the landmark text of von Neumann [22], which takes as a working hypothesis that all self-adjoint operators correspond to observable quantities. The Wigner-Araki-Yanase theorem rules this out, and is the subject of the next section.

3. The Wigner-Araki-Yanase theorem

3.1. Background

The framework presented above demonstrates that, in principle, any observable can be measured by specifying an appropriate measurement scheme. However, it may not be practicable to arrange a suitable scheme for measuring arbitrary observables (for instance, it may be difficult to prepare the required ϕ in practice) and, *a fortiori*, there may be *fundamental* restrictions to the realizability of the elements of $\mathcal{M}$ required for a measurement (scheme) for E, arising for instance from their incompatibility with other basic principles (see, e.g., [23] for a recent example in quantum field theory).

Two such proposed fundamental restrictions were raised in 1952, with Wigner involved in both. One is the limitation imposed by a *superselection rule* [24], which reduces the notion of observable to include only those self-adjoint operators which sit in the commutant of some particular superselection "charge" or, equivalently, prevents the preparation of coherent superpositions of eigenstates of differing superselection charge. This subject has its own long and controversial history; see [8] and references therein for a recent review and analysis of this story, to aspects of which we will also return in the present manuscript.

The other 1952 paper on fundamental obstructions to measurement was authored solely by Wigner (see [25] for the original article, and [26] for a facsimile-style translation into English), and investigated the possibility that conservation laws may interfere with the existence of measurement schemes satisfying the PRC.

Specifically, Wigner observed that a simple *repeatable* measurement, given as a specified state evolution resulting in a transformation of von Neumann/Lüders type, of the x-component of spin of a spin-$\frac{1}{2}$ particle conflicts with angular momentum conservation. We note four key observations made by Wigner in [25]: 1) That the specified model of an S_x measurement violates the conservation of $S_z \otimes \mathbb{1} + \mathbb{1} \otimes L_z$, the latter denoting the (z-component of) the angular momentum of the measuring apparatus, 2) That a modification of the dynamics so as to incorporate an "error" state of the apparatus allows the conservation law to be satisfied, but the accuracy and repeatability of the measurement are lost, 3) Good accuracy and repeatability properties can be recovered to an arbitrarily high degree if the initial state of the apparatus has a large uncertainty with respect to L_z, which can be viewed as a "size" requirement on the apparatus and 4) That if the repeatability condition is dropped completely, then perfectly accurate measurements are possible, even under the constraint of the conservation law, without any size condition on the apparatus.

Regarding the fourth point, Wigner made the following observation (see also [27]): the reason for the measurement limitation is the noncommutativity between the spin components of the system under investigation, *viz* $[S_x, S_z] \neq 0$. Dropping the repeatability condition in his model inevitably leads to a state evolution in which the final pointer states are eigenstates of a quantity not commuting with L_z. Hence the question arises of how to measure the pointer.

The topic of conservation-law-induced limitations to quantum measurements has been revisited in the literature until the present day. The main advancement after Wigner's paper was the 1960 contribution of Araki and Yanase [28], in which Wigner's example was shown to be an instance of a theorem about quantum measurements in general, wherein points 1) to 4) above are all seen to generalise in a natural way. See also [12] for a detailed discussion and reconstruction of Wigner's early work, a generalisation of the main result of [28] which will play an important role in this manuscript, and a critical examination of the evolution of the WAY theorem up to 2011.

We now present the Wigner-Araki-Yanase theorem. The setting is that there is a system $\mathcal{S}$, an apparatus $\mathcal{A}$, and a quantity $L = L_{\mathcal{S}} \otimes \mathbb{1} + \mathbb{1} \otimes L_{\mathcal{A}} \equiv L_{\mathcal{S}} + L_{\mathcal{A}}$ which is *conserved* in the sense that $[U, L] = 0$. Following Ozawa [29], we shall call the condition that $[Z, L_{\mathcal{A}}] = 0$ (which may or may not be satisfied) the *Yanase condition.*

3.2. WAY theorem for discrete observables

The WAY theorem may be stated in the quantum theory of measurement as follows.

Theorem. *Let* $\mathcal{M} := \langle \mathcal{H}_{\mathcal{A}}, U, \phi, Z, \rangle$ *be a measurement scheme for a discrete-spectrum self-adjoint operator* A *on* $\mathcal{H}_{\mathcal{S}}$*, and let* $L_{\mathcal{S}}$ *and* $L_{\mathcal{A}}$ *be bounded self-adjoint operators on* $\mathcal{H}_{\mathcal{S}}$ *and* $\mathcal{H}_{\mathcal{A}}$*, respectively, such that* $[U, L_{\mathcal{S}} + L_{\mathcal{A}}] = 0$*. Assume that* $\mathcal{M}$ *is repeatable or satisfies the Yanase condition. Then* $[A, L_{\mathcal{S}}] = 0$.

See [12] for a proof and discussion of some generalisations. The most important one for us is that the unboundedness of $L_{\mathcal{A}}$ can be easily accommodated by taking the exponential of $L_{\mathcal{A}}$ (which gives a bounded operator) and running the argument as before – see footnote 4 in [28]. The theorem entails that if $[A, L_{\mathcal{S}}] \neq 0$, then any measurement of A is necessarily non-repeatable, and must violate the Yanase condition. The *SWAP map*, defined when $\mathcal{H}_{\mathcal{S}} \cong \mathcal{H}_{\mathcal{A}}$ by $\varphi \otimes \phi \mapsto \phi \otimes \varphi$, is perhaps the simplest example of such a realization [12].

The Yanase condition was not known to be a condition in the statement of the WAY theorem until fairly recently (its first appearance in that capacity is in [30]; see also [12]) and is worth dwelling on further, as it will play an important role in the sequel. From a physical perspective, such a requirement appears to be natural: applying the WAY theorem to the pointer (imagine measuring the pointer) prohibits repeatable measurements of Z, which surely nullifies its purpose as a stable pointer that can be repeatedly checked by an experimenter.

The statement of the WAY theorem does not rule out approximate measurements of A, understood in the sense that there may be some unsharp observable E whose statistics are close to those of (the spectral measure of) A and which may nevertheless *not* commute with $L_{\mathcal{S}}$.

Moreover, the WAY theorem as stated also does not prohibit such approximate measurements from having approximate repeatability properties. Such a possibility can indeed be realised, and can be viewed as a positive counterpart to the strict impossibility verdict, in line with Wigner's earlier observations. Even if $[A, L_S] \neq 0$, approximate measurements of A, with approximate repeatability properties, are feasible, with increasingly good approximation properties becoming possible with a large "spread" of the apparatus part of the conserved quantity in the initial state of the apparatus. Large spread in those early works corresponded to a large number of components of the given state when written as a linear combination of eigenstates of the conserved quantity with different eigenvalue. Other measures are also possible—variance for instance—and as argued in [12], such a spread, appropriately quantified, is *necessary* for good measurements.

The case of position measurements obeying momentum conservation is not covered by the WAY theorem as presented above, yet is of clear physical importance. This case also carries an interesting message regarding the relational perspective on the WAY theorem, with which we will eventually make contact, and therefore we will discuss the position-momentum case in some detail next.

3.3. Position and momentum

WAY-type constraints persist also in the continuous-variable and unbounded case, not covered by the original proofs, as reported for the first time in [13]; we will be interested here in position (measurements) and momentum (conservation). We present a model due to Ozawa (see [31] and [13, 14] for further analysis of this model) which demonstrates the presence of WAY-type limitations for position measurements, and contrast this with a second model [16] which, although also has momentum conservation built in, shows no such WAY-type behaviour. We then present a heuristic argument in favour of WAY-type constraints persisting in general for Q and P.

3.3.1. Two models Ozawa's measurement scheme is characterised by the following unitary coupling acting in $\mathcal{H}_S \otimes \mathcal{H}_\mathcal{R} \otimes \mathcal{H}_\mathcal{A} \otimes \mathcal{H}_\mathcal{B}$ (we omit the identity factors and tensor product symbols for notational clarity):

$$U = e^{i\frac{\lambda}{2}(Q-Q_\mathcal{R})(Q_\mathcal{A}-Q_\mathcal{B})}. \tag{8}$$

All Hilbert spaces are taken to be isomorphic to $L^2(\mathbb{R})$, understood as position representation spaces for $\mathcal{H}_S$ and $\mathcal{H}_\mathcal{R}$, and momentum representation spaces for $\mathcal{H}_\mathcal{A}$ and $\mathcal{H}_\mathcal{B}$. We may therefore write $\mathcal{H}_\mathcal{A} \otimes \mathcal{H}_\mathcal{B} \cong L^2(\mathbb{R}^2)$ with $\mathbb{R}^2$ parametrised by $(p_\mathcal{A}, p_\mathcal{B})$. Then under the orthogonal transformation $(p_\mathcal{A}, p_\mathcal{B}) \mapsto (u, v) = \frac{1}{\sqrt{2}}(p_\mathcal{B}-p_\mathcal{A}, p_\mathcal{B}+p_\mathcal{A})$, we have $\mathcal{H}_\mathcal{A}\otimes\mathcal{H}_\mathcal{B} \cong \tilde{\mathcal{H}}_\mathcal{A}\otimes\tilde{\mathcal{H}}_\mathcal{B} \cong L^2(\mathbb{R}^2)$ with coordinates (u, v), which are spectral values of $P_\mathcal{B} - P_\mathcal{A}$ and $P_\mathcal{B} + P_\mathcal{A}$ respectively. Finally, the pointer observable is chosen to be $Z = P_\mathcal{B} - P_\mathcal{A}$, which satisfies the Yanase condition (we ignore subtleties arising from domain questions). We continue to denote by Z the spectral measure of Z. Crucially, U conserves the total momentum $P + P_\mathcal{R} + P_\mathcal{A} + P_\mathcal{B}$.

The scheme is constructed so as to measure (an unsharp approximation of) Q—the position observable of $\mathcal{S}$—and therefore the remaining systems $\mathcal{R}$, $\mathcal{A}$ and $\mathcal{B}$ are all treated as part of the apparatus (though the $\mathcal{R}$ system does not comprise part of the pointer and plays a role which we will investigate in more detail in due course). The apparatus is initially prepared in the vector state $\phi \otimes \xi_a \otimes \xi_b \in \mathcal{H}_\mathcal{R} \otimes \tilde{\mathcal{H}}_\mathcal{A} \otimes \tilde{\mathcal{H}}_\mathcal{B}$, and the measured observable is extracted from the probability reproducibility condition (where we in this instance label identity factors to properly identify subsystems)

$$\langle \varphi \,|\, \mathsf{E}(X)\varphi \rangle = \Big\langle U(\varphi \otimes \phi \otimes \xi_a \otimes \xi_b) \,|\, \mathbb{1}_{\mathcal{H}_S} \otimes \mathbb{1}_{\mathcal{H}_\mathcal{R}} \otimes \mathsf{Z}(f^{-1}(X)) \otimes \mathbb{1}_{\mathcal{H}_\mathcal{B}} U(\varphi \otimes \phi \otimes \xi_a \otimes \xi_b) \Big\rangle, \tag{9}$$

Journal of Physics: Conference Series **1638** (2020) 012009 doi:10.1088/1742-6596/1638/1/012009

and can be shown [14] to be of the form

$$\mathsf{E}(X) = (\chi_X * e^{(\lambda)})(Q), \tag{10}$$

where χ_X denotes the characteristic set function, $*$ is convolution and the scaling function f (cf. eq. (4)) has been chosen as $f(x) = -(2/\lambda)x$. Note that the ξ_b plays no role at all. The POM E represents a *smeared position observable* [16, 18], with $e^{(\lambda)}$ a density (or confidence function), itself also a convolution given by

$$e^{(\lambda)}(x) = |\phi|^2 * |\xi_a^{(\lambda)}|^2(x), \tag{11}$$

where $\xi_a^{(\lambda)}(s) = \sqrt{\lambda}\xi_a(\lambda s)$. The smearing, or inaccuracy, is dictated by the (spread of the) density function $e^{(\lambda)}$; the more tightly peaked $e^{(\lambda)}$, as quantified for instance by the variance $\mathrm{Var}(e^{(\lambda)})$, the better the measurement of Q. This is understood in the sense of statistical proximity between Q (the spectral measure of Q) and the actually measured observable E.

In the above case we find that $\mathrm{Var}(e^{(\lambda)}) = \mathrm{Var}\,|\phi|^2 + \frac{4}{\lambda^2}\,|\xi_a|^2$, which is strictly positive, and therefore $\mathcal{M}$ always realised an approximate determination of position. Since λ can be made arbitrarily large by choosing a suitably large interaction strength, $\mathrm{Var}(e^{(\lambda)})$ is therefore bounded below by the spread (as variance) of the distribution $|\phi|^2$, which can be made small only by a high degree of localisation with respect to position $Q_{\mathcal{R}}$ or, by the uncertainty relation, a large uncertainty/delocalisation/spread with respect to $P_{\mathcal{R}}$, providing the continuous-variable analogue of the positive part of WAY.

A better measure of spread than variance (or standard deviation) of a probability distribution is the *overall width* $\mathcal{W}(e; 1-\epsilon)$ at confidence level $1-\epsilon$, defined as the smallest possible size of an interval J for which $\int_J e(x)dx \geq 1-\epsilon$ (see [18]). The overall width is finite for all $\epsilon > 0$, and the overall width of a convolution of probability measures is bounded below by the largest, i.e., $\mathcal{W}(e^{\lambda}; 1-\epsilon) \geq \max\{\mathcal{W}(|\phi|^2; 1-\epsilon), \mathcal{W}(\frac{4}{\lambda^2}\,|\xi_a|^2; 1-\epsilon)\}$. Hence, $\mathcal{W}(e^{\lambda}; 1-\epsilon) \geq \mathcal{W}(|\phi|^2; 1-\epsilon)$. In other words, the quality of the approximation of Q by E is dictated precisely by the $Q_{\mathcal{R}}$ localisation of ϕ and is bounded from below by this quantity. We see therefore that it is the second ($\mathcal{R}$) system which plays a vital role in the quantitative part of WAY here, and *not* the $\mathcal{A}$ and $\mathcal{B}$ systems pertaining to the pointer.

The Ozawa model can be contrasted with another model [16] (again we simplify the notation by omitting tensor product and identity symbols) described by (with $\lambda > 1$)

$$U = \exp\left[-i\frac{\lambda}{2}\left((Q - Q_{\mathcal{A}})P_{\mathcal{A}} + P_{\mathcal{A}}(Q - Q_{\mathcal{A}})\right)\right], \tag{12}$$

with pointer $Q_{\mathcal{A}}$. This model manifestly conserves momentum, and clearly violates the Yanase condition. This model may be viewed as a kind of "symmetrised version" of the position measurement of von Neumann as described in the final few pages of his classic text [22].[1]

The measured observable is again extracted from the PRC by noting that the initial state of the system and apparatus combined $\Psi_0 = \varphi \otimes \phi \equiv \varphi\phi \in L^2(\mathbb{R}) \otimes L^2(\mathbb{R})$ (with position-representation wavefunction $\Psi_0(x, y)$) evolves to $\Psi_t(x, y) = e^{\frac{\lambda t}{2}}\varphi(x)\phi\left((1 - e^{\lambda t})x + e^{\lambda t}y\right)$ and the measurement is completed at $t = 1/2$. The scaling function in this model is $f(x) = (1 - e^{-\lambda})^{-1}x$, and the observable measured is again a smeared position $\mathsf{E}(X) = (\chi_X * e^{(\lambda)})(Q)$,

[1] Von Neumann's model is described by the unitary coupling $U(\lambda) = e^{i\lambda Q \otimes P_{\mathcal{A}}}$. As pointed out in [32], von Neumann compared the particle position and the pointer reading after the cessation of the interaction period, thereby analysing the *repeatability* properties of the scheme. Had he also investigated the correlation of the position *before* the interaction to the pointer afterwards, i.e., performed a calculation associated with the PRC, he may well have discovered the representation of observables by POMS in 1932, around 40 years before their eventual introduction.

with $e^{(\lambda)}$ the confidence function $e^{(\lambda)}(x) = (e^{\lambda} - 1)|\phi(-x)(e^{\lambda} - 1))|^2$ (recall that ϕ represents the initial state of $\mathcal{A}$.)

This model is therefore arbitrarily accurate for arbitrarily large interaction strength λ (in the sense that the actually measured observable $\mathsf{E}(X) \equiv \chi_X * e^{(\lambda)}(Q)$ can be made as close as one likes to the sharp $\chi_X(Q) \equiv \mathsf{Q}(X)$), and as it turns out has arbitrarily good repeatability properties, both without any condition on ϕ.[2] Therefore, with the Yanase condition violated, arbitrarily accurate position measurements can be realised whilst respecting the conservation of momentum, and there is no WAY-type limitation in this model.[3] See [13, 14, 16, 18] for further discussion.

3.3.2. A general inequality Finally we give an argument using an inequality due to Ozawa [29], based the concept of a *noise operator* $N := Z(\tau) - Q$. As an unbounded operator there are of course sensitive issues here relating to domains, which we continue to surreptitiously ignore. The use of such a quantity has also been (correctly) criticised (see, for example, [35]) in general, the reason being that the observables being compared (in this case, $Z(\tau)$ and $Q \otimes \mathbb{1}$) do not necessarily commute, therefore requiring separate experimental procedures for the following estimate. Hence what comes next must be taken as heuristic and not operationally meaningful in general. Setting $\epsilon^2 := \sup_{\varphi} \epsilon(\varphi)^2 := \langle \varphi \otimes \phi \,|\, N^2 \varphi \otimes \phi \rangle \geq (\Delta N)^2$, the uncertainty relation then gives

$$\epsilon^2 \geq \epsilon(\varphi)^2 \geq \frac{1}{4} \frac{|\langle [Z(\tau) - Q, P + P_{\mathcal{A}}] \rangle|^2}{(\Delta P_T)^2}; \tag{13}$$

the supremum is taken over unit vectors and is assumed to be finite across all states, with ϵ then understood as a global measure of "error", and $(\Delta P_T)^2 = (\Delta_{\varphi}(P))^2 + (\Delta_{\phi}(P_{\mathcal{A}}))^2$. The measurement is deemed accurate exactly when $\epsilon = 0$. From here we observe that i) if the right hand side is non-zero for some system states, there is a measurement limitation, ii) that if the numerator is non-vanishing, the only way to make the lower bound to the error small, without reference to the properties of the system, is by making $(\Delta_{\phi}(P_{\mathcal{A}}))$ large, and (iii) that the numerator $[Z(\tau) - Q, P + P_{\mathcal{A}}]$ vanishes if $[Z, P_{\mathcal{A}}] = i$, which is a statement of the violation of the Yanase condition and is exactly the situation encountered in the second model above. Therefore, the Q, P case in many ways reflects the discrete/bounded scenario (subject to some caveats; see [12]).

We finally mention that Eq. (13) generalises, subject to the same objections, to arbitrary

[2] This model played the central role in providing an "approximate solution" of an old question of Shimony and Stein [33] regarding the (im-) possibility of a two-valued (coarse-grained) position measurement respecting the conservation of momentum. The problem was posed in the research problems section of *The American Mathematical Monthly*, which contained "easily stated research problems dealing with notions ordinarily encountered in undergraduate mathematics." It reappeared in 2009 [34] as the first of a list of six unsolved "bequest" problems of Shimony, based presumably on a 2006 talk of his at the Perimeter Institute. The question of whether the solution presented in [12] (which is not in perfect accordance with the stated problem but is operationally indistinguishable from a positive answer to it) was—to the enormous regret of the author—never put to Abner.

[3] That the inaccuracy scales with $e^{-\lambda}$ in this model, in comparison to von Neumann's model which scales with λ^{-1} (and does not conserve momentum) was pointed out to be surprising by Paul in a private communication in 2009 – surprising because one would have thought, on WAY-type grounds, that incorporating momentum conservation would make matters worse, not exponentially better. In fact, this observation led Paul to doubt his old result in [16] on the scaling behaviour in the momentum-conserving model, which in turn led to the author's first investigation into WAY-type limitations for continuous observables, from which a large body of work has followed. Of course, in 2009 we did not know about the role of the Yanase condition: Paul's old calculation was right, and by violating the Yanase condition good measurements are feasible without any constraint on the apparatus.

quantities, giving rise to an inequality of the following form:

$$\epsilon^2 \geq \epsilon(\varphi)^2 \geq \frac{1}{4}\frac{|\langle[Z(\tau) - A, L_{\mathcal{S}} + L_{\mathcal{A}}]\rangle|^2}{(\Delta L)^2}, \tag{14}$$

where again the measurement is accurate exactly when $\epsilon = 0$, with a broadly similar conclusion to above.

3.4. Further developments

From 2011 until the present, there have been several interesting developments in relation to the WAY theorem. Technical progress for instance includes the generalisation of the impossibility part due to Tukiainen [36], where additivity and even conservation are dropped but the theorem nevertheless retains its main flavour, and the generalisation due to Łuczak [37] based on von Neumann algebras and instruments. The WAY theorem has inspired further analysis of the case of energy conservation in [38], and has even found applications in the rapidly emerging field of quantum thermodynamics; see, e.g., [39, 40].

Perhaps the main lines of development have come via the general perspective that conservation laws are instances of symmetry constraints. The 2013 contribution of Ahmadi, Jennings and Rudolph [41] proposes that the WAY theorem may be understood in the context of *resource theory*—a subject which is by now a research programme in its own right (see [42] for a recent review)—in which there is collection of preparations, transformations, measurements which are "permissible" from the point of view of some constraint, and everything else is viewed as a *resource*. In this instance, permissible states and observables are symmetry-invariant, and the large spread in the conserved quantity of the initial state of the apparatus is then the asymmetric resource, which can be appropriately quantified.

There remains a large variety of views on the "true meaning" of the WAY theorem. These include the resource-theoretic perspective mentioned above, the more informational point of view as first proposed in [43], and the informal connection to measurement disturbance [38], to name a few. However, it has been intimated by many that the WAY theorem is not intuitive, or at least that the conceptual message is unclear. Here we put forward the idea that in certain cases, namely when the conserved quantity has a (covariant) conjugate quantity, the WAY theorem is a statement of the limitation on the possibility of "absolute" quantities being well approximated by invariant quantities of system-plus-reference, wherein the reference is also playing the role of measuring apparatus. We begin with a general analysis of observability under symmetry, putting WAY to one side for a moment.

4. Symmetry and observability

We investigate the relationship between three important notions: symmetry, the natural relativity of certain physical quantities, and the requirement of reference frames or, more properly, reference systems, for the relativity and symmetry to manifest.

4.1. Observables as invariants

There are strong reasons to adopt the view that in the presence of symmetry, *only* those quantities (represented as self-adjoint operators or POMs) which are invariant under relevant symmetry transformations should be considered to have the status of (being) observable. This is standard in the case of gauge symmetry (e.g., [44]), and can be argued more generally. For example, "absolute" position or angle or time are not observable classically, and are manifestly non-invariant quantities under the relevant transformations. On the other hand, *relative* position, angle and time (see [45] for an analysis of the case of time in quantum theory), are invariant and observable. The non-observability of "absolute" quantities is in accordance

Journal of Physics: Conference Series **1638** (2020) 012009 doi:10.1088/1742-6596/1638/1/012009

with fundamental (in this case, space-time) symmetries. Therefore we see that many standard theoretical notions such as position acquire their meaning *only* in relation to other physical objects. Classically, the observable relative position can be replaced with the non-observable absolute position by suppressing the (variables of the) second system which functions as a *reference* with no difficulty. The question of whether this is possible in quantum mechanics is one of the main questions addressed in [8], to which we will repeatedly refer.

The situation in quantum theory is complicated by the fact that the reference system should (must?) be treated as a quantum system, and is therefore subject to properties held by quantum systems, such as indeterminacy, unsharpness, and entanglement, among others [8]. On the other hand, physicists routinely use "absolute" quantities in the description of microscopic physical systems, seemingly without contradiction or even imprecision.

There is another reason for demanding the invariance of observables, which comes from conservation and from the quantum theory of measurement. Suppose first that $\mathcal{S}$ itself has a conserved quantity, which we continue to write as $L_{\mathcal{S}}$. Then the following theorem shows that *any* observable (sharp or unsharp) must commute with $L_{\mathcal{S}}$ in order to be measured (or, we argue, deemed truly (an) "observable") – a situation which was dubbed the "strong WAY theorem" in [8] due to the stronger form of the conservation law than in the original WAY theorem (where the conservation is stipulated only for system and apparatus combined).

Proposition. *Consider a measurement scheme $\mathcal{M}$ for E, and write $U_{\mathcal{S}}(\ell) = e^{i\ell L_{\mathcal{S}}}$ for some real parameter ℓ and suppose that $[U, U_{\mathcal{S}}(\ell)] = 0$ for all ℓ (this is equivalent to the conservation of $L_{\mathcal{S}}$). Then*

$$U_{\mathcal{S}}(\ell)^* \mathsf{E}(X) U_{\mathcal{S}}(\ell) = \mathsf{E}(X), \tag{15}$$

for all ℓ and X.

Proof. Without loss of generality suppose that $f = 1$. Then the PRC is

$$\langle U\varphi \otimes \phi \,|\, \mathbb{1} \otimes \mathsf{Z}(X) U\varphi \otimes \phi \rangle = \langle \varphi \,|\, \mathsf{E}(X)\varphi \rangle ; \tag{16}$$

substituting φ for $U_{\mathcal{S}}(\ell)\varphi$ leaves the left hand side of the above unchanged, and hence the right hand side is unchanged as well, proving the result. □

This holds for general observables and conserved quantities; we make no assumptions about their spectra or even their boundedness. The result imposes stronger constraints than the WAY theorem, ruling out any measurements of any (sharp or unsharp) quantities not commuting with a conserved quantity. Note that this is strongly reminiscent of the situation encountered when there is a superselection rule.

The two perspectives (on the unobservability of certain theoretical quantities) are of a different source and nature. However, in both cases there is a tension between the apparently defensible viewpoint that "absolute" quantities such as position do not represent observable quantities, and the routine use of such quantities in the (accurate) description of the physical world. And in both cases, we will argue, the resolution is found via the introduction of quantum frames of reference, which we now discuss.

4.2. Quantum reference systems and relationalism

That certain physical quantities obtain their meaning only in relation to a standard of reference is by now presumably uncontroversial. However, there are differing viewpoints on whether such references are to be understood as abstract "coordinate systems" or as physical objects. Our position is that the latter should be the starting point: that relations between physical entities is primary, and that abstract coordinatisation is an idealisation that must reflect actual physical possibility. For instance, in a spatial Newtonian setting, positions of objects are understood as

Journal of Physics: Conference Series **1638** (2020) 012009 doi:10.1088/1742-6596/1638/1/012009

relative to other objects, these objects being localised in space and any of which can serve as a reference. Once it is established that such physical references exist and function as they do, the process of abstraction to "mathematical" references can take place and these references are, in the setting just discussed, systems of Cartesian coordinates.

As Einstein has stressed in the classical relativistic setting, "Every description of events in space involves the use of a rigid body to which such events have to be referred. The resulting relationship takes for granted that the laws of Euclidean geometry hold for "distances", the "distance" being represented physically by means of the convention of two marks on a rigid body" [46]. That the step of abstraction is possible appears to be so natural in classical physics that it is seldom considered. Einstein's remark encapsulates the operational philosophy: the relative position is understood as an observable quantity and one must therefore provide a physical means by which to measure it. The theoretical use of the absolute position is only justified as a mathematical "shorthand", if indeed it is justified.

If quantum theory is understood to be universally applicable, the reference bodies must be quantum physical systems in the world, governed by the laws of quantum theory. The manoeuvre by which the reference system is externalised/abstracted must then be carefully examined and justified. We discuss this possibility now, referring again to [8] for the full treatment, and [7] for the condensed version.

The observation that symmetry (invariance) and the relativity of quantities seem to come together is interesting and worth investigating in its own right. We proceed under the point of view that the symmetry is somehow "prior", if only because one can formulate the question we wish to discuss in a clean way, and because it may be more general.

4.3. Relativisation

Suppose we have some group G (with some properties which we shall soon discuss) acting both in the value space Ω (with σ-algebra $\mathcal{F}$) of some observable F of $\mathcal{R}$ and via strongly continuous unitary representations $U_{\mathcal{S}}$ and $U_{\mathcal{R}}$ in the corresponding spaces, and suppose moreover that F is *covariant* under the action of G:

$$U_{\mathcal{R}}(g)\mathsf{F}(X)U_{\mathcal{R}}(g)^* = \mathsf{F}(g.X) \text{ for all } g \in G,\ X \in \mathcal{F}. \tag{17}$$

If we can identify Ω and G then the following map may be used to construct invariant quantities in $\mathcal{S}+\mathcal{R}$ from arbitrary ones of $\mathcal{S}$:[4]

$$¥ : \mathcal{L}(\mathcal{H}_{\mathcal{S}}) \to \mathcal{L}(\mathcal{H}_{\mathcal{S}} \otimes \mathcal{H}_{\mathcal{R}}); \quad A \mapsto \int_G U_{\mathcal{S}}(g)AU_{\mathcal{S}}(g)^* \otimes \mathsf{F}(dg). \tag{18}$$

This extends naturally to POMs by setting $(¥ \circ \mathsf{E})(X) := ¥(\mathsf{E}(X))$. Equation (18) represents the integral of an operator-valued function with respect to an operator measure; see [8] for the construction. We note that a sufficient condition for the integral to be well defined is that G is compact and metrisable if $\mathcal{H}_{\mathcal{R}}$ is finite [47], and that G is abelian and second countable in the general case [8]. The mapping ¥ can be seen to enjoy a host of agreeable qualities: it is (completely) positive, normal, unital, *-preserving, and moreover an algebraic *-homomorphism if F is projection-valued. Crucially, as is easily confirmed, $¥(A)$ is invariant under $¥(A) \mapsto U_{\mathcal{S}}(g) \otimes U_{\mathcal{R}}(g)¥(A)U_{\mathcal{S}}(g)^* \otimes U_{\mathcal{R}}(g)^* \equiv \alpha_g(A)$.

The ¥ map therefore ascribes to each POM of $\mathcal{S}$ a new observable of $\mathcal{S}+\mathcal{R}$ which is invariant under the given unitary representation of G. As will be exemplified below, we view this procedure as the explicit incorporation of a system of reference into the theoretical description.

[4] The unusual notation owes itself to a suggestion of Busch, circa 2011, that since ¥ generalises and makes rigorous the map $ appearing in [11], and since the map may be accredited to Miyadera who introduced it in a visit to York in 2011, a currency exchange may be appropriate.

We give some examples. Suppose $G = (\mathbb{R}, +)$ (the additive group on the real line; we henceforth denote this group by $\mathbb{R}$) with strongly continuous unitary representations $U_{\mathcal{S}}$ and $\mathcal{U}_{\mathcal{R}}$ acting in $\mathcal{H}_{\mathcal{S}}$ and $\mathcal{H}_{\mathcal{R}}$, respectively. We write $U_{\mathcal{S}}(x) = e^{iP_{\mathcal{S}}x}$ and $U_{\mathcal{R}}(x) = e^{iP_{\mathcal{R}}x}$, where $P_{\mathcal{S}}$ and $P_{\mathcal{R}}$ are shift-generators, interpreted physically as momenta. Choosing F to be the sharp position observable $\mathsf{F} = \mathsf{E}^{Q_{\mathcal{R}}}$, we find that

$$(\yen \circ \mathsf{E}^{Q_{\mathcal{S}}})(X) = \int_{\mathbb{R}} \mathsf{E}^{Q_{\mathcal{S}}}(X+x) \otimes \mathsf{E}^{Q_{\mathcal{R}}}(dx) = \int_{\mathbb{R}}\int_{\mathbb{R}} \chi_X(x'-x)\mathsf{E}^{Q_{\mathcal{S}}}(dx') \otimes \mathsf{E}^{Q_{\mathcal{R}}}(dx) = \mathsf{E}^{Q_{\mathcal{S}}-Q_{\mathcal{R}}}(X). \tag{19}$$

Since $\mathsf{E}^{Q_{\mathcal{S}}-Q_{\mathcal{R}}}$ is the spectral measure of the self-adjoint relative position observable $Q_{\mathcal{S}} - Q_{\mathcal{R}}$, we observe that $\yen$ in this case has the effect of "relativising" the absolute position $Q_{\mathcal{S}}$, yielding $Q_{\mathcal{S}} - Q_{\mathcal{R}}$.

We may also consider the case of the circle S^1, defined by the interval $(-\pi, \pi]$ with the end points identified and addition mod 2π. As a group this is then isomorphic to $U(1)$ and in the example we are about to give, $\mathcal{H}_{\mathcal{R}} \cong L^2(S^1)$. Choosing $\mathsf{F} = \mathsf{E}^{\Phi_{\mathcal{R}}}$ to be the spectral measure of the azimuthal angle operator Φ conjugate to (say) the z-component L_z of angular momentum, we have

$$(\yen \circ \mathsf{E}^{\Phi_{\mathcal{S}}})(X) = \mathsf{E}^{\Phi_{\mathcal{S}}-\Phi_{\mathcal{R}}}(X); \tag{20}$$

here, $\mathsf{E}^{\Phi_{\mathcal{S}}-\Phi_{\mathcal{R}}}$ is the spectral measure of the relative angle $\Phi_{\mathcal{S}} - \Phi_{\mathcal{R}}$, and therefore we may write $\yen(\Phi_{\mathcal{S}}) = \Phi_{\mathcal{S}} - \Phi_{\mathcal{R}}$.

We also mention the physically important quantity of phase conjugate to number, which generalises the angle/angular momentum pair; we write $\mathsf{F} : \mathcal{B}(S^1) \to \mathcal{L}(\mathcal{H}_{\mathcal{R}})$ for such a quantity. Conjugacy here means that phase shifts are generated by number operators, i.e., $e^{iN_{\mathcal{R}}\theta}\mathsf{F}(X)e^{-iN_{\mathcal{R}}\theta} = \mathsf{F}(X + \theta)$, where addition is understood modulo 2π. The relativisation $\yen$ gives rise to a relative phase observable [48, 49], as shown in [8]. If the spectrum of $N_{\mathcal{R}}$ is bounded from below, as in the harmonic oscillator for example, F cannot be sharp. Note that if $N_{\mathcal{R}}$ can be defined in a Hilbert space of low dimension, in which case F is highly unsharp – the spin phase [16] acting in $\mathbb{C}^2$ is such an example.

Therefore we observe that the effect of relativising "absolute" position and phase (with angle as the special case where $N_{\mathcal{R}}$ is two-sided-unbounded) is to produce relative position and phase, as one might expect from a relativising map. However, $\yen$ is not confined to acting only on POMs which have familiar relative counterparts as (the spectral measures of) difference operators, or already established relative versions such as relative phase, and is quite general. We therefore view any observable obtained under $\yen$ as relative, or *relational*, and, even more generally, *any* invariant quantity of $\mathcal{S} + \mathcal{R}$.[5] Again, we refer to [8, 14] for further discussion and examples.

4.4. Localisation and approximation

One of the main results of [8] is in regard to the possibility of making A and $\yen(A)$ statistically close (or more generally with A replaced by a POM). The interest in this question stems directly from the fact that in standard applications of quantum mechanics, "absolute" quantities such as position and phase are empirically adequate theoretical objects, yet, as we know, cannot be represented in physical reality because they are not invariant. The question therefore arises as to whether, and under what conditions, the unobservable "absolute" and the truly observable relative quantities can at least approximately agree.

Using the $\yen$ map we find that good approximation is possible *provided* the relativising quantity F possesses a technical localisability property called the norm-1 property (e.g., [51]), namely, that for any X for which $\mathsf{F}(X) \neq 0$, there exists a sequence of unit vectors $(\phi_n) \subset \mathcal{H}_{\mathcal{R}}$ for which $\lim_{n\to\infty} \langle \phi_n \,|\, \mathsf{F}(X)\phi_n \rangle = 1$. We note that this is satisfied for all PVMs. Then we have

[5] It is known that in general $\yen$ is not surjective on the invariant part of $\mathcal{L}(\mathcal{H}_{\mathcal{S}} \otimes \mathcal{H}_{\mathcal{R}})$ [50].

Theorem. *Let G be $\mathbb{R}$ or S^1 and Γ be defined as in equation (6). There is a sequence of unit vectors $(\phi_n) \subset \mathcal{H}_{\mathcal{R}}$ for which*

$$\lim_{n\to\infty} (\Gamma_{\phi_n} \circ \text{¥})(A) = A, \tag{21}$$

where the convergence is understood weakly, i.e., in the topology of pointwise convergence of expectation values.

The essence of the proof (see [8] for the details) lies in the observation that

$$(\Gamma_{\phi_n} \circ \text{¥})(A) = \int_G U(g)^* A U(g) \mu^{\mathsf{F}}_{\phi_n}(dg), \tag{22}$$

and that if F satisfies the norm-1 property then we can make $\mu^{\mathsf{F}}_{\phi_n}$ as concentrated as we like, i.e., approximately contained in any measurable set. Fixing this to be centred on 0 (the identity of G) then yields the result. The localisation of the state at the identity of G acts as a sort of "encoding" of this element of G in $\mathcal{H}_{\mathcal{R}}$ and serves as the zero-reference.

There are various further observations to be made. The noise operator, subject to the already-discussed substantial caveat of not being operationally meaningful in general, can be used to compare ¥(A) and A for representations of G generated by the additive quantity $L = L_{\mathcal{S}} + L_{\mathcal{R}}$ on $\mathcal{H}_{\mathcal{S}} \otimes \mathcal{H}_{\mathcal{R}}$, with state $\varphi \otimes \xi$ (see 3.3.2):

$$\epsilon^2 \geq \epsilon(\varphi)^2 \geq \frac{1}{4} \frac{|\langle [\text{¥}(A) - A, L_{\mathcal{S}} + L_{\mathcal{R}}] \rangle|^2}{(\Delta L)^2}; \tag{23}$$

with perfect agreement coming with $\epsilon = 0$. Now, $[\text{¥}(A), L] = 0$ and therefore, since $(\Delta L)^2 = (\Delta_\varphi L_{\mathcal{S}})^2 + (\Delta_\xi L_{\mathcal{R}})^2$ and φ is arbitrary, if $[A, L_{\mathcal{S}}] \neq 0$, there is a positive lower bound on the discrepancy between A and ¥(A) which can only be made smaller by increasing the spread of $L_{\mathcal{R}}$ in ξ [14]. This spread of course coincides with a highly localised conjugate quantity, if one exists, for instance position conjugate to momentum, angle conjugate to angular momentum, or even unsharp phase conjugate to number.

Actually, the above results can be substantially improved and can be shown to hold in operational terms – see [52]. There it is also shown that a "large" apparatus is needed for good agreement between A and ¥(A) (for general effects A), and that poor localisation gives poor approximation. Moreover, it is proven there that this type of behaviour persists even when relative observables are not obtained via the ¥ map, i.e., good agreement between an arbitrary quantity of $\mathcal{S}$ and an invariant one of $\mathcal{S}+\mathcal{R}$ requires large spread in the symmetry generator on $\mathcal{H}_{\mathcal{R}}$. This corresponds to high localisation in a conjugate quantity, should there be one.

This demonstrates that the reference frame, as a physical system, is fulfilling its role appropriately in serving to define the absolute quantities of $\mathcal{S}$. That badly localised states give bad approximation indicates that the approximation of relative by absolute is only as good as the reference system allows. This is most vividly born out in the case of position or angle, where the resulting absolute quantities $\Gamma_\phi(\cdot)$ after restriction are also not localisable.

If the reference is not sharply localised this is not a good reference for the "absolute" position, as one might well expect. In the extreme situation that the reference state is completely delocalised (possible only in an approximate sense for position), the restricted quantity of $\mathcal{S}$ is shift-invariant, and thus commutes with momentum, and therefore says nothing about "absolute" position at all. This situation is reminiscent of the apt metaphor in [53], where the concept of the "edge of a mountain" is presented as one that is inherently vague, in order to facilitate the exposition of the idea of *unsharp reality*, which may be seen as one of Paul Busch's major contributions to the philosophy of physics. The edge of a mountain, as intrinsically vague, does not function as a good reference for position. The position of a person relative to the edge of a mountain cannot be any more precisely specified than one can specify the edge of a mountain.

In quantum mechanics, as also elucidated in [53], the vagueness arises from the *nature* of the physical entities themselves, and not in deficiency of the concept, and therefore the question of the existence and properties of suitable references for quantum physical quantities is a deep issue.

The relative localisation of physical objects appears to be so ubiquitous in the classical world that it feels perhaps incommodious to view localisation itself as a relational property. However, this may be one of the essential differences between classical and quantum physics, and we see the reference localisation (and also the possibility of networks of systems which are localised relative to each other) as something of a classicality condition – a topic that we believe deserves further attention.

4.5. Measurement

We now return to the quantum theory of measurement, and the measurement of relational quantities, specifically those of the form $(\yen \circ \mathsf{E})(X)$ where $\yen$ is implicitly defined for some G and relativising quantity F. Measurements of $\yen \circ \mathsf{E}$ will then be compared to measurements of E, and the need of reference localisation for good approximation will again be highlighted.

We therefore make the shift from viewing the system to be measured as $\mathcal{S}$ to $\mathcal{S}+\mathcal{R}$, i.e., let $\mathcal{H}_\mathcal{S} \otimes \mathcal{H}_\mathcal{R}$ denote the Hilbert space of the system to be measured, and introduce the space $\mathcal{H}_\mathcal{A}$ to represent the apparatus. Therefore the total space is $\mathcal{H}_\mathcal{S} \otimes \mathcal{H}_\mathcal{R} \otimes \mathcal{H}_\mathcal{A}$. Let $\tilde{\mathsf{E}} = \yen \circ \mathsf{E} : X \to L(\mathcal{H}_\mathcal{S} \otimes \mathcal{H}_\mathcal{R})$, which as an invariant quantity can be measured without constraint by a suitable pointer Z on $\mathcal{H}_\mathcal{A}$. The probability reproducibility condition for the measurement of $\tilde{\mathsf{E}}$ can be written

$$\langle \psi \,|\, (\yen \circ \mathsf{E})(X)\psi \rangle = \left\langle \psi \,|\, \tilde{\mathsf{E}}(X) \quad \right\rangle = \langle U(\psi \otimes \phi) \,|\, \mathbb{1} \otimes \mathsf{Z}(X) U(\psi \otimes \phi) \rangle, \tag{24}$$

holding for all X and unit vectors $\psi \in \mathcal{H}_\mathcal{S} \otimes \mathcal{H}_\mathcal{R}$. Writing $\mathsf{Z}(X)(\tau) := U^* \mathbb{1} \otimes \mathsf{Z}(X) U$, this can be rewritten as

$$(\yen \circ \mathsf{E})(X) = \tilde{\mathsf{E}}(X) = \Gamma_\phi(\mathsf{Z}(X)(\tau)), \tag{25}$$

and $\tilde{\mathsf{E}}$ is the measured observable.

Now we want to see whether the same scheme can be re-purposed so as to function as a measurement scheme for E. To do this, we fix some unit vector η of the reference $\mathcal{R}$, which defines the observable $\mathsf{E}^{(\eta)}$ of $\mathcal{S}$ through the formula

$$\left\langle \varphi \,|\, \mathsf{E}^{(\eta)}(X)\varphi \right\rangle = \left\langle \varphi \otimes \eta \,|\, \tilde{\mathsf{E}}(X) \varphi \otimes \eta \right\rangle \tag{26}$$

to hold, as always, for all unit φ and X, which can also be written therefore at the operator level as $\mathsf{E}^{(\eta)}(X) = \Gamma_\eta(\tilde{\mathsf{E}}(X))$. Then we may write

$$\left\langle \varphi \,|\, \mathsf{E}^{(\eta)}(X)\varphi \right\rangle = \langle U(\varphi \otimes \eta \otimes \phi) \,|\, \mathbb{1} \otimes \mathsf{Z}(X) U(\varphi \otimes \eta \otimes \phi) \rangle, \tag{27}$$

where both ϕ and η are fixed and may be viewed as part of the measuring apparatus. This can be rewritten as $\mathsf{E}^{(\eta)}(X) = \Gamma_{\eta\otimes\phi}(\mathsf{Z}(X)(\tau) = \Gamma_\eta(\Gamma_\phi(\mathsf{Z}(\tau)(X)))$.

Finally, we let $(\eta)_n$ be a localising sequence for F at the origin (additive identity) of G, and take the high localisation limit, yielding

$$\lim_{n\to\infty} \Gamma_{\eta_n}(\Gamma_\phi(\mathsf{Z}(\tau)(X))) = \mathsf{E}(X). \tag{28}$$

This observation highlights the central point of this investigation. The observable $\tilde{\mathsf{E}}$ is a relational observable of $\mathcal{S}+\mathcal{R}$. This can be measured whilst respecting the strong WAY theorem, and qualifies as a "true observable" under the principle that only invariants represent genuine physical quantities. However, under suitable reference preparation, arbitrary POMs of $\mathcal{S}$, after the suppression of $\mathcal{R}$ (or the absorption of $\mathcal{R}$ into the apparatus) can function *as though* they are observable, even though they are not actually represented in physical reality. The localisation of the reference state is the *sine qua non* of a well-functioning reference: it provides the zero to which all else is then referred. What's more, high localisation with respect to one quantity corresponds to large spread in a conjugate quantity, should one exist. High phase localisation corresponds to large spread in number, high position localisation to large momentum spread, and so on.

The Ozawa position measurement scheme provides a good example of this behaviour. Instead of the absolute position, we instead use the same set-up to measure the relative position $Q \otimes \mathbb{1} - \mathbb{1} \otimes Q_{\mathcal{R}} \equiv Q - Q_{\mathcal{R}}$. Crucially, $P + P_{\mathcal{R}}$ is separately conserved and therefore falls under the remit of the strong WAY theorem. The unique measured POM in this scheme is $\widetilde{\mathsf{E}} : \mathcal{B}(\mathbb{R}) \to \mathcal{L}(\mathcal{H}_{\mathcal{S}} \otimes \mathcal{H}_{\mathcal{R}}) \equiv \mathcal{L}(L^2(\mathbb{R}^2))$, extracted from the condition (writing $\Psi_\tau = U(\tau)\Psi_0$)

$$\left\langle \Psi_\tau | \mathbb{1} \otimes \mathbb{1} \otimes \mathsf{Z}(f^{-1}(X)) \otimes \mathbb{1} \Psi_\tau \right\rangle = \left\langle \varphi \otimes \phi | \widetilde{\mathsf{E}}(X) \varphi \otimes \phi \right\rangle, \tag{29}$$

required to hold for all unit $\varphi \in \mathcal{H}_{\mathcal{S}}$ and $\phi \in \mathcal{H}_{\mathcal{R}}$. It then follows that

$$\widetilde{\mathsf{E}}(X) = \chi_X * \widetilde{e}^{(\lambda)}(Q - Q_{\mathcal{R}}), \tag{30}$$

where the right-hand side is the convolution of the set indicator function χ_X with the probability distribution $\widetilde{e}^{(\lambda)}(x) = \left|\xi_a^{(\lambda)}(x)\right|^2$ with $\xi_a^{(\lambda)}(s) = \sqrt{\lambda}\xi_a(\lambda s)$.

We see that $\widetilde{\mathsf{E}}$ is a smeared or unsharp version of the sharp $\mathsf{E}^{Q-Q_{\mathcal{R}}}$, and the quality of the approximation is again dictated by the spread of the density/confidence function $\widetilde{e}^{(\lambda)}$. The variance of $\widetilde{e}^{(\lambda)}$ is $\mathrm{Var}(\widetilde{e}^{(\lambda)}) = \frac{4}{\lambda^2}\mathrm{Var}\,|\xi_a|^2$. Therefore by tuning λ to be large, arbitrarily accurate measurements of $Q - Q_{\mathcal{R}}$ can be achieved, with no size/localisation requirements, and conclusion holds also for the overall width [14].

It is only when trying to rehabilitate Q as the observable-to-be-measured that the localisation requirement reappears. By fixing ϕ, the measurement scheme can be viewed as "measuring" a POM E for $\mathcal{S}$, in which we then return to the situation previously investigated:

$$\left\langle \varphi \otimes \phi \,|\, \widetilde{\mathsf{E}}(X) \varphi \otimes \phi \right\rangle =: \langle \varphi \,|\, \mathsf{E}(X) \varphi \rangle .$$

The probability distribution for the relative position has thereby been re-expressed in terms of a "smeared" distribution for the "absolute" position by considering a fixed reference state ϕ of $\mathcal{R}$. The approximation error of E relative to E^Q is given by $\mathrm{Var}(e^{(\lambda)}) = \mathrm{Var}\,|\phi|^2 + \frac{4}{\lambda^2}\mathrm{Var}\,|\xi_a|^2$ (or, again, the overall width measure). The probability distributions corresponding to the relative coordinate in the states $\varphi \otimes \phi$ become indistinguishable from those of the "absolute" coordinate Q in the limit that the localisation of the state ϕ with respect to $Q_{\mathcal{R}}$ is arbitrarily good (provided also that λ is tuned to be large).

Therefore we see that the limitation arises only when we attempt to "measure" the non-invariant Q, and not the invariant $Q - Q_{\mathcal{R}}$ or, from the more WAY-type reasoning, the measurement of Q, which does not commute with P, gives a limitation, but $Q - Q_{\mathcal{R}}$ which does commute with $P + P_{\mathcal{R}}$ does not, and this can be seen from within the measurement scheme of Ozawa.

The Ozawa model differs from the standard WAY set-up in that it includes the possibility of incorporating the reference as either part of the system or part of the apparatus. The standard WAY scenario, in our view, must be understood as pertaining to the situation wherein the single apparatus system must function also as the reference – a situation to which we now turn.

5. Relational view of the WAY theorem

We are now in position to outline the main conclusion of this work; first we briefly recap. In section 3 we discussed the WAY theorem in its usual incarnation as presented in [12–14], learning that observables (represented by self-adjoint operators) not commuting with the system part of an additive conserved quantity cannot be measured precisely in any measurement scheme satisfying the Yanase condition. However, we also saw that approximate measurements are possible, in the sense that the actually measured POM becomes statistically close to the "target" sharp observable that we wish to measure, and that this comes at the price of having to prepare an initial apparatus state with a large spread in the conserved quantity.

We then approached the question of characterising observability from a different perspective. We provided two arguments regarding the imposition of symmetry on observability. The first was that fundamental symmetries limit what is observable to those quantities which are invariant under the symmetry transformation. This may be, for example, invariance under the action of the Galilei group or certain subgroups, the invariance then being a statement of the lack of absolute space, time and angle. The case of internal symmetries such as phase are bound by the same restriction, though the interpretation seems to be more directly linked to the fact that phase is inherently relative and is more akin to gauge.

The second argument was derived from the quantum theory of measurement itself, and pertained to systems which are isolated in the sense that they possess conserved quantities. Given this, from both points of view the conclusion is the same: no POM not invariant under the symmetry action is (an) observable. We argued that the use of non-invariant quantities in the empirically accurate description of quantum systems is a reflection of the fundamental point that such quantities are defined only with respect to a second system—a reference system—which can be externalised *contingent* upon the preparation of a reference state highly localised in the appropriate variable.

Given that the compatibility condition $[A, L_{\mathcal{S}}] = 0$ in the context of the WAY theorem can be recast as the symmetry/invariance statement $A = e^{i\ell L_{\mathcal{S}}} A e^{-i\ell L_{\mathcal{S}}}$, we observe that the conclusion of WAY, though conveying a similar moral to the point of view espoused in the subsequent section, does not completely agree with it, because it does not rule out approximate measurements, i.e., unsharp approximators of A, described by a POM E which is not compatible with $L_{\mathcal{S}}$.[6] In our view, the situation encountered in the WAY theorem is an expression of the general perspective that we have been advocating, and the possibility of approximate "measurements" is actually an expression of the possibility of approximating relative observables by their unobservable/unmeasurable counterparts. We will now describe this in more detail.

5.1. *Measuring apparatus as reference frame*

We first recast the relevant compatibility conditions as invariance conditions, collecting together also some extra pieces of relevant information. Here the real parameter ℓ can be taken to lie in $\mathbb{R}$ or S^1. The physical setting is arranged for a measurement to take place, and thus we have systems $\mathcal{S}$ and $\mathcal{A}$, described together by the space $\mathcal{H}_{\mathcal{S}} \otimes \mathcal{H}_{\mathcal{A}}$, together with a unitary coupling U, and so on.

- Invariance of $\mathcal{S}$-observables: $[A, L_{\mathcal{S}}] = 0$, if and only if $A = e^{i\ell L_{\mathcal{S}}} A e^{-i\ell L_{\mathcal{S}}} \equiv \alpha_\ell^{\mathcal{S}}(A)$, if and only if $[\mathsf{E}^A(X), L_{\mathcal{S}}] = 0$, if and only if $\alpha_\ell^{\mathcal{S}}(\mathsf{E}^A(X)) = \mathsf{E}^A(X)$ for all X. All of this generalises to the case where E^A is replaced by a POM E.
- Conservation law for $\mathcal{S} + \mathcal{A}$: $[U, L_{\mathcal{S}} \otimes \mathbb{1} + \mathbb{1} \otimes L_{\mathcal{A}}] \equiv [U, L] = 0$.
- Yanase condition: $[Z, L_{\mathcal{A}}] = 0$, which is equivalent to $Z = e^{i\ell L_{\mathcal{A}}} Z e^{-i\ell L_{\mathcal{A}}} \equiv \alpha_\ell^{\mathcal{A}}(Z)$, which under the conservation law $[U, L] = 0$ is equivalent to $[U^*(\mathbb{1} \otimes Z)U, L] = 0$, or $[Z(\tau), L] = 0$.

[6] The compatibility of a POM E and a self-adjoint operator M is equivalent to the statement that $[\mathsf{E}(X), \mathsf{E}^M(Y)] = 0$ for all X and Y, where $\mathsf{E}^M(Y)$ is the spectral measure of M; see [16].

This latter condition says that the time evolved pointer $Z(\tau)$ must be invariant under $Z(\tau) \mapsto e^{i\ell L} Z(\tau) e^{-i\ell L}$.

- *Weak Yanase condition*: $[Z(\tau), L] = 0$, i.e., the time evolved pointer $Z(\tau)$ must be invariant under $Z(\tau) \mapsto e^{i\ell L} Z(\tau) e^{-i\ell L}$ (this can be stipulated without the conservation or additivity of L).[7]

The WAY theorem in its usual reading, with the Yanase condition in place of repeatability, is ostensibly about the compatibility of $L_{\mathcal{A}}$ and Z being "inherited" by A and $L_{\mathcal{S}}$. Refashioning the compatibility conditions as invariance conditions allows for the impossibility part of the WAY theorem to be recast in the following (slightly informal) form: For any Z satisfying the (weak) Yanase condition, there is no measurement scheme which (perfectly) realises a measurement of (the self-adjoint operator) A unless $A = \alpha_\ell^{\mathcal{S}}(A)$. Put yet another way, no non-invariant A can satisfy $A = \Gamma_\phi(Z(\tau))$. Given that $Z(\tau)$ is an invariant, the connection with the discussion on reference frames becomes apparent.

A fortiori, the "positive" part of WAY asserts that A and $\Gamma_\phi(Z(\tau))$ can become close, exactly when ϕ has a large $L_{\mathcal{A}}$–spread. In many instances of physical interest, for instance when $L_{\mathcal{A}}$ is the momentum or angular momentum, this corresponds exactly to localisation in the conjugate variable (here, position or angle), and therefore we see an extremely compelling link to the theory of quantum reference frames, where reference localisation was the key property for good approximation.

Now, we argue, that the WAY theorem as traditionally stated can be understood as pertaining to the case wherein the apparatus $\mathcal{A}$ is functioning *both* as a measuring apparatus in the traditional sense, *and* as a reference system serving to define the relational observables which are traditionally treated as absolute. Indeed, there is no other system present which could serve this purpose – in the standard set-up of WAY, there is no "$\mathcal{R}$" system in addition to system and apparatus, but $\mathcal{S} + \mathcal{A}$ is certainly isolated, and the conservation of $L_{\mathcal{S}} + L_{\mathcal{A}}$ entails by strong WAY that no quantity of $\mathcal{S} + \mathcal{A}$ which does not commute with $L = L_{\mathcal{S}} + L_{\mathcal{A}}$/is not invariant under the symmetry action generated by L can be measured "from the outside" of $\mathcal{S} + \mathcal{A}$. This is precisely the statement of the weak Yanase condition, i.e., that the time-evolved pointer $Z(\tau)$ is invariant.

The localisation condition on the preparation of the reference frame corresponds exactly to large spread in the conjugate quantity in the WAY setting, *in the initial state of the apparatus*. There is therefore a strong harmony, both with respect to the impossibility part and the positive part. What is still unclear is the reference frames interpretation of WAY in the setting that the conserved quantity does not admit a conjugate quantity, in which case the large spread in the conserved quantity does not in any natural way correspond to a localised state with respect to another quantity. Understanding this remains as work to be done, although we note that almost all physically interesting conserved quantities do admit conjugates.

6. Discussion, future directions and closing Remarks

Our conclusion, simply put, is that the Wigner-Araki-Yanase theorem, originally an expression of the limitations of the measurability of quantities not commuting with (the system part of) an additive conserved quantity, can be instead interpreted as a statement about the (im)possibilities of (unmeasurable) absolute quantities being good representations of observable relative quantities.

The work presented comprises part of the vision to understand all aspects of quantum theory from a relational viewpoint. There has already been much progress in this direction, and contemporary efforts are perhaps most strongly focused on understanding better the form

[7] That the WAY theorem admits a generalisation in terms of the weak Yanase condition has been pointed out by Tukiainen in [36].

of relationalism associated with quantum reference frames. By now the relational description of states and observables is fairly well established, and the situation for quantum channels is described in a concurrent publication in this volume. The rejuvenation of foundational interest in quantum reference frames has spurred further recent progress in manifold directions. One major example arises from the need to describe transformations between quantum frames in the spirit of relativistic physics; see for example [54–57]. *Inter alia*, other recent work has provided insights into the relative nature of coherence and superpositions [8, 58], entanglement [57], the 'Wigner's friend' thought experiment [54], and the nature of time in quantum mechanics [45, 59, 60]. This latter topic is related in deep ways to the "problem of time" in quantum gravity, and there are encouraging signs that the reference frames point of view may provide new insight there and in (quantum) gravitational physics more generally [61–63].

The relational nature of quantum mechanics bears of course on the deep questions in quantum mechanics and beyond – the distinction between subject and object, observer and observed or, to return to the elegant words of Heisenberg, between player and spectator. The most important question, however, takes us back to the beginning and to the WAY theorem, and is also the one that was closest to Paul's heart: what does all of this say about quantum measurement?

References

[1] Heisenberg W 2000 *Physics and Philosophy* (London: Penguin)
[2] Bohr N 1949 Discussion with Einstein on epistemological problems in atomic physics *Albert Einstein, Philosopher-Scientist* (The Library of Living Philosophers vol 7) ed A P Schilpp (Wisconsin: Banta) pp 199–241
[3] Rovelli C 1996 Relational quantum mechanics *Int. J. Theor. Phys.* **35** 1637–78
[4] Bene G 1997 Quantum reference systems: a new framework for quantum mechanics *Physica A* **242** 529–65
[5] Bene G and Dieks D 2002 A perspectival version of the modal interpretation of quantum mechanics and the origin of macroscopic behavior *Found. Phys.* **32** 645–71
[6] Dieks D 2019 Quantum reality, perspectivalism and covariance *Found. Phys.* **49** 629–46
[7] Loveridge L, Busch P and Miyadera T 2017 Relativity of quantum states and observables *EPL* **117** 40004
[8] Loveridge L, Miyadera T and Busch P 2018 Symmetry, reference frames and relational quantities in quantum mechanics *Found. Phys.* **48** 135–98
[9] Aharonov Y and Kaufherr T 1983 Quantum frames of reference *Proc. International Symposium on the Foundations of Quantum Mechanics in the Light of New Technology* ed S Kamefuchi and N B Gakkai (Tokyo: Physical Society of Japan) pp 190–4
[10] Aharonov Y and Kaufherr T 1984 Quantum frames of reference *Phys. Rev.* D **30** 368–85
[11] Bartlett S D, Rudolph T and Spekkens R W 2007 Reference frames, superselection rules, and quantum information *Rev. Mod. Phys.* **79** 555–609
[12] Loveridge L D and Busch P 2011 'Measurement of quantum mechanical operators' revisited *Eur. Phys. J.* D **62** 297–307
[13] Busch P and Loveridge L 2011 Position measurements obeying momentum conservation *Phys. Rev. Lett.* **106** 110406
[14] Loveridge L D 2012 *Quantum Measurements in the Presence of Symmetry* (University of York, UK: PhD thesis) `http://etheses.whiterose.ac.uk/2670/`
[15] Busch P and Loveridge L D 2013 Quantum measurements constrained by symmetries *Symmetries and Groups in Contemporary Physics* (Nankai Series in Pure, Applied Mathematics and Theoretical Physics vol 11) ed C Bai *et al* pp 587–92
[16] Busch P, Grabowski M and Lahti P J 1995 *Operational Quantum Physics* (Berlin: Springer)
[17] Busch P, Lahti P J and Mittelstaedt P 1996 *The Quantum Theory of Measurement* (Berlin, Springer)
[18] Busch P, Lahti P, Pellonpää J-P and Ylinen K 2016 *Quantum Measurement* ([Cham]: Springer)
[19] Busch P 2003 The role of entanglement in quantum measurement and information processing *Int. J. Theor. Phys.* **42** 937–41
[20] Mittelstaedt P 2004 *The Interpretation of Quantum Mechanics and the Measurement Process* (Cambridge: Cambridge University Press)
[21] Ozawa M 1984 Quantum measuring processes of continuous observables *J. Math. Phys.* **25** 79–87
[22] von Neumann J 1932 *Mathematische Grundlagen der Quantenmechanik* (Berlin: Springer) English translation 1955 *Mathematical Foundations of Quantum Mechanics* (Princeton: Princeton University Press)

[23] Bostelmann H, Fewster C J and Ruep M H 2020 Impossible measurements require impossible apparatus *Preprint* arXiv:2003.04660 [quant-ph]
[24] Wick G C, Wightman A S and Wigner E P 1952 The intrinsic parity of elementary particles *Phys. Rev.* **88** 101–5
[25] Wigner E P 1952 Die Messung quantenmechanischer Operatoren *Z. Phys.* **133** 101–8
[26] Busch P 2010 Translation of "Die Messung quantenmechanischer Operatoren" by E. P. Wigner *Preprint* arXiv:1012.4372 [quant-ph]
[27] Yanase M M 1961 Optimal measuring apparatus *Phys. Rev.* **123** 666–8
[28] Araki H and Yanase M M 1960 Measurement of quantum mechanical operators *Phys. Rev.* **120** 622–6
[29] Ozawa, M 2002 Conservation laws, uncertainty relations, and quantum limits of measurements *Phys. Rev. Lett.* **88** 050402
[30] Beltrametti E G, Cassinelli G and Lahti P J 1990 Unitary measurements of discrete quantities in quantum mechanics *J. Math. Phys.* **31** 91–8
[31] Ozawa M 1991 Does a conservation law limit position measurements? *Phys. Rev. Lett.* **67** 1956–9
[32] Busch P 2009 "No information without disturbance": quantum limitations of measurement *Quantum Reality, Relativistic Causality, and Closing the Epistemic Circle* (Western Ontario Series in Philosophy of Science vol 73) ed W C Myrvold and J Christian (Dordrecht: Springer) pp 229–56
[33] Shimony A and Stein H 1979 A problem in Hilbert space theory arising from the quantum theory of measurement *Amer. Math. Month.* **86** 292–3
[34] Shimony A 2009 Unfinished work: a bequest *Quantum Reality, Relativistic Causality, and Closing the Epistemic Circle* (Western Ontario Series in Philosophy of Science vol 73) ed W C Myrvold and J Christian (Dordrecht: Springer) pp 479–91
[35] Busch P, Lahti P and Werner R F 2014 Quantum root-mean-square error and measurement uncertainty relations *Rev. Mod. Phys.* **86** 1261–81
[36] Tukiainen M 2017 Wigner-Araki-Yanase theorem beyond conservation laws *Phys. Rev.* A **95** 012127
[37] Łuczak A 2016 Wigner-Araki-Yanase theorem in general theory of quantum measurement *Op. Sys. & Inf. Dyn.* **23** 1650013
[38] Navascués M and Popescu S 2014 How energy conservation limits our measurements *Phys. Rev. Lett.* **112** 140502
[39] Mohammady M H and Anders J 2017 A quantum Szilard engine without heat from a thermal reservoir *New J. Phys.* **19** 113026
[40] Mohammady M H and Romito A 2019 Conditional work statistics of quantum measurements *Quantum* **3** 175–92
[41] Ahmadi M, Jennings D and Rudolph T 2013 The Wigner–Araki–Yanase theorem and the quantum resource theory of asymmetry *New J. Phys.* **15** 013057
[42] Chitambar E and Gour G 2019 Quantum resource theories *Rev. Mod. Phys.* **91** 025001
[43] Marvian I and Spekkens R W 2012 An information-theoretic account of the Wigner-Araki-Yanase theorem *Preprint* arXiv:1212.3378 [quant-ph]
[44] Haag R 1996 *Local Quantum Physics* (Berlin: Springer)
[45] Loveridge L and Miyadera T 2019 Relative quantum time *Found. Phys.* **49** 549–60
[46] Einstein A 1920 *Relativity: The Special and the General Theory* (New York: Holt)
[47] Davies E B 1976 *Quantum Theory of Open Systems* (London: Academic Press)
[48] Pellonpää J-P 2002 *Covariant Phase Observables in Quantum Mechanics* (Annales Universitatis Turkuensis AI vol 288) (University of Turku: PhD thesis)
[49] Heinonen T, Lahti P and Pellonpää J-P 2003 Covariant phase difference observables in quantum mechanics *J. Math. Phys.* **44** 466–79
[50] Waldron J and Loveridge L (unpublished)
[51] Heinonen T, Lahti P, Pellonpää J.-P, Pulmannova S and Ylinen K 2003 The norm-1-property of a quantum observable *J. Math. Phys.* **44** 1998–2008
[52] Miyadera T, Loveridge L and Busch P 2016 Approximating relational observables by absolute quantities: a quantum accuracy-size trade-off *J. Phys. A: Math. Theor.* **49** 185301
[53] Busch P and Jaeger G 2010 Unsharp Quantum Reality *Found. Phys* **40** 1341–67
[54] De la Hamette A-C and Galley T 2020 Quantum reference frames for general symmetry groups *Preprint* arXiv:2004.14292 [quant-ph]
[55] Palmer M C, Girelli F and Bartlett S D 2014 Changing quantum reference frames *Phys. Rev.* A **89** 052121
[56] Vanrietvelde A, Höhn P A, Giacomini F and Castro-Ruiz E 2020 A change of perspective: switching quantum reference frames via a perspective-neutral framework *Quantum* **4** 225–60
[57] Giacomini F, Castro-Ruiz E and Brukner Č 2019 Quantum mechanics and the covariance of physical laws in quantum reference frames *Nat. Comm.* **10** 494

[58] Zych M, Costa F and Ralph T C 2018 Relativity of quantum superpositions *Preprint* arXiv:1809.04999 [quant-ph]
[59] Smith A R H 2019 *Detectors, Reference Frames, and Time* (Springer Theses) (Cham: Springer)
[60] Smith A R H and Ahmadi M 2019 Quantizing time: interacting clocks and systems *Quantum* **3** 160–71
[61] Höhn P A, Smith A R H and Lock M P E 2019 The trinity of relational quantum dynamics *Preprint* arXiv:1912.00033 [quant-ph]
[62] Castro-Ruiz E, Giacomini F, Belenchia A and Brukner Č 2020 Quantum clocks and the temporal localisability of events in the presence of gravitating quantum systems *Nat. Comm.* **11** 2672
[63] Chataignier L 2020 Relational observables, reference frames and conditional probabilities *Preprint* arXiv:2006.05526 [quant-ph]

Journal of Physics: Conference Series **1638** (2020) 012010 doi:10.1088/1742-6596/1638/1/012010

Localizability and elementary particles

Gregg Jaeger

Quantum Communication and Measurement Laboratory, Department of Electrical and Computer Engineering (ECE) & Division of Natural Science and Mathematics (CGS), Boston University, 871 Commonwealth Ave., Boston MA 02215, USA

E-mail: jaeger@bu.edu

Abstract. The well-definedness of particles of any kind depends on the limits, approximations, or other conditions that may or may not be involved, for example, whether there are interactions and whether ostensibly related energy is localizable. In particular, their theoretical status differs between its non-relativistic and relativistic versions: One can properly define interacting elementary particles in single-system non-relativistic quantum mechanics, at least in the case of non-zero mass systems; by contrast, one is severely challenged to define even these properly in the relativistic quantum field theories that now underlie the study of particle physics. Here, the impact of localizability on this status is reviewed in relation to the work of Paul Busch on positive operator valued measures that significantly probes the relevance of quantum unsharpness to it.

1. Introduction

In quantum theory, the definability of objects such as elementary particles when required to be localizable depends on the version of the theory under consideration and the interactions, limits, and other conditions imposed on the system via which they would be instantiated; cf. e.g. [1, Chapter 1]. Combining special relativity and quantum theory, as is needed for the study of elementary particles, requires that a field theory, relativistic quantum field theory (RQFT) be considered; cf. e.g. [2, Chapter 2]. Straightforward attempts within RQFT to describe particles—something typically done via energy-momentum eigenstates—that might subsequently be localized fail when interactions are allowed. This is so because adequate creation and annihilation operators needed for their accounting via corresponding number operators are definable only when no interaction can take place. Strictly speaking, this last condition is never satisfied in the absence of approximation for common interactions such as those of electromagnetism; cf. e.g. [3] and references therein. From a more operationalist perspective, there are also specific difficulties for the provision of the mathematical entities needed to define spatial localization consistently with relativistic causality requirements; it would appear that the operation of localization is impossible even though experiments are standardly made via what is assumed to be the local detection of particles. Yet, it might still be thought these latter difficulties could be resolved by appealing to unsharp localization operators because, before the work of Paul Busch, it was *sharp localization* that had been studied at the formal level in relativistic quantum theory. With localization defined via quantum observables, Busch considered whether these observables might be simply *unsharp*, so that the apparent issues with creation, annihilation, and number operators might eventually be finessed.

Whether unsharpness indeed provides a solution to this quandary was considered by Busch and coworkers specifically in the mathematical context of positive operator valued measures

Journal of Physics: Conference Series **1638** (2020) 012010 doi:10.1088/1742-6596/1638/1/012010

(POVMs), and significant results were obtained: (i) a generalization of Schlieder's theorem to show that the local commutativity of localization observables implies that they are, at best, *unsharp in a strong sense* and (ii) that, in important special cases, two (discrete) unsharp (as well as sharp) observables commute, for any given state, if and only if the statistics of a measurement of one is unaffected by a nonselective Lüders measurement of the other [4]. One is thus naturally led to the reconsideration of the assumptions that have been made in defining localization. Busch subsequently found that the local commutativity assumption is a necessary consequence of Einstein causality for local unsharp measurements. Accordingly, he indicated that it could be the specific requirement that localization operators *themselves be local* that is responsible for the difficulties encountered in regard to localization [4].

Here, a review of the above and related results, as well as their interplay is provided and further aspects of unsharp measurement in quantum theory that relate to the operators playing a role in localization are briefly discussed. It is emphasized that, in agreement with the position taken by Busch, the assumption that localization itself be local could be given up in order for quantum theory to accord with special relativity and that, in a relativistic physics with elementary particles, theories or models of measurement in the space-time context should include the quantum aspects of measurement devices to better illuminate the physics involved.

2. The Schlieder theorem

In their analysis of localization in relativistic quantum theory, Busch *et al.* considered the following Hilbert-space of states and group representation associated with what is sometimes called a localization system [4] which has been the standard setting for the subject.

(1) (State space) A complex Hilbert space $\mathcal{H}$, with rays representing the pure states

under the conditions

(2) (Spectral condition) A strongly continuous unitary representation (CUR) $a \mapsto U(a)$ in $\mathcal{H}$ of the Minkowski space (M) translation group;

(3) (Hamiltonian-boundedness condition) The generator $H(a)$ (Hamiltonian) is bounded below for future directed, timelike unit vectors a,

the latter being also sometimes called the energy-boundedness condition. In this setting, they examined the Schlieder theorem [5], with the goal of generalizing it, in the valuable form given it by Malament [6] which considers a spatial localization observable in relation to what he called the *localization-event structure*.

(i) (Localization event structure) With a foliation of Minkowski space M selected via a family $\mathcal{S}$ of parallel spacelike hyperplanes S, each S is equipped with a family $\mathcal{F}(S)$ of subsets (spatial sets), including a covering family of bounded subsets, such that $\mathcal{F}(S+a)$ is the translation by a of the sets from $\mathcal{F}(S)$, and there is a map $\Delta \mapsto E_\Delta$ from $\mathcal{F}(S)$ to effects of $\mathcal{H}$ for each S.

The Malament theorem assumes satisfaction of the following explicit conditions for localization and locality in an inertial frame of reference:

(ii) (Translational covariance) For all $a \in M$,

$$U(a)E_\Delta U(a)^* = E_{\Delta+a}.$$

(iii) (Localizability) For each $S \in \mathcal{S}$ and $\Delta_1, \Delta_2 \in \mathcal{F}(S)$,

$$\text{if } \Delta_1 \cap \Delta_2 = \emptyset \quad \text{then} \quad E_{\Delta_1}E_{\Delta_2} = E_{\Delta_2}E_{\Delta_1} = 0.$$

Journal of Physics: Conference Series **1638** (2020) 012010 doi:10.1088/1742-6596/1638/1/012010

(iv) (Local Commutativity) For $S_1, S_2 \in \mathcal{S}$, $\Delta_1 \in \mathcal{F}(S_1), \Delta_2 \in \mathcal{F}(S_2)$,

$$\text{if } \Delta_1, \Delta_2 \text{ are spacelike separated, then } \mathrm{E}_{\Delta_1}\mathrm{E}_{\Delta_2} = \mathrm{E}_{\Delta_2}\mathrm{E}_{\Delta_1}.$$

The Schlieder theorem in Malament's form was then written [4]

> If $(\mathcal{H}, a \mapsto U(a), \Delta \mapsto P_\Delta)$, where the P_Δ are projections (special cases of E_Δ), satisfies (ii)–(iv), then $P_\Delta = 0$ for all bounded spatial sets Δ.

The relevant implication of this theorem which, again, regards only projection-valued measures (i.e. the sharp case) is that *there can be no physical adequate notion of localizability* in the above context, with the implication that there is no adequate standard notion of localization of particles in relativistic quantum theory; cf. [7]. The Schlieder theorem excludes local number operators from Wightman's quantum field theory [8] or algebraic quantum field theories; cf. [9] and references therein.

3. Generalizing the Schlieder theorem to the case of unsharp observables

To find out whether some alternative approach to achieving a well-defined sense of localization in relativistic quantum theory may be possible in light of this result, Busch looked to the intrinsically *unsharp* observables defined via POVMs rather than only the special case of projection-valued measures. He addressed the question of whether a theorem analogous to Malament's form of Schlieder's theorem holds for unsharp localization observables by considering *non-projective* effects E_Δ in addition to projections P_Δ, that is, with the localization event structure such that the maps $\Delta \mapsto E_\Delta$ are POVMs defined on spatial Borel sets. Busch found that to be the case, proving the following.

> **Theorem** (Busch). If the structure $(\mathcal{H}, a \mapsto U(a), \Delta \mapsto E_\Delta)$ satisfies conditions (ii)–(iv), then $E_\Delta = 0$ for all (bounded) spatial sets Δ.

He then noted that the physical basis for the algebraic condition imposed in the assumptions *changes* in the move from projective measures to positive operator valued measures, in that the localization, when imposed via $E_{\Delta_1}E_{\Delta_2} = 0$ for disjoint spatial sets, no longer achieves the *physical* goal of imposing the condition "If the system is in Δ_1, it certainly is not in Δ_2 whenever these sets are disjoint" [4]. Therefore, he introduced, instead of (iii), the condition

(iii′) For all states $\varphi \in \mathcal{H}$, $\|\varphi\| = 1$, $\Delta_1, \Delta_2 \in \mathcal{F}(S)$,

$$\text{if } \Delta_1 \cap \Delta_2 = \emptyset \quad \text{then} \quad \langle \varphi | E_{\Delta_1} \varphi \rangle = 1 \Longrightarrow \langle \varphi | E_{\Delta_2} \varphi \rangle = 0 \ ,$$

which is a direct consequence of the assumption that $\Delta \mapsto E_\Delta$ is a (not necessarily normalized) POVM, tantamount to $P^{(1)}_{\Delta_1} \leq P^{(0)}_{\Delta_2}$, where $P^{(1)}_\Delta$, $P^{(0)}_\Delta$ denote the spectral projections of E_Δ associated with the eigenvalues 1 and 0, respectively, which is equivalent to (iii) in the case of projections, but better captures what is required for localization via effects, in that (iii′) implies only $P^{(1)}_{\Delta_1} P^{(1)}_{\Delta_2} = 0$; $E_{\Delta_1}E_{\Delta_2} \neq 0$ in general.

Assuming the alternative algebraic condition (iii′), which is a strong causality assumption that entails localizability when one considers $\Delta_2 = \Delta_1 + ta$ with t a time (cf. [9, Sec. 2]), Busch proved the following.

> **Theorem** (Busch). If the structure $(\mathcal{H}, a \mapsto U(a), \Delta \mapsto E_\Delta)$ satisfies conditions (ii), (iii′) and (iv), then $P^{(1)}_\Delta = 0$ for all (bounded) spatial sets Δ .

He took this result to indicate that any localization observable satisfying the conditions (ii), (iii′), and (iv) would necessarily be *strongly unsharp*, in that its effects E_Δ do not have 1 as an eigenvalue (cf. [4]), and posed the associated question of "whether among the strongly unsharp, covariant localization observables there exist any that satisfy local commutativity" (i.e. iv), pointing out that "an indication to the negative is provided by a recent theorem stating that *spacetime* localization observables cannot belong to any quasilocal algebra" [10].[1]

4. The strongly unsharp

After Busch proved the above theorems, Halvorson and Clifton provided an additional one on the basis of which they argued that even an appeal to strongly unsharp localization does not mitigate the force of the problems with localization indicated by the theorems of Schlieder and Malament, namely, that under the same assumptions of Busch's generalization, but with the addition of the condition

(NAV) (No absolute velocity) Let a be a spacelike translation of M. Then there is a pair (b, c) of timelike translations of M such that $a = b - c$,

the following holds [9].

> **Theorem** (Halvorson–Clifton). Suppose that the unsharp localization system $(\mathcal{H}, a \mapsto U(a), \Delta \mapsto E_\Delta)$ satisfies the conditions (ii),(iii′), (iv), and NAV. Then $E_\Delta = 0$ for all Δ.

This result was seen by its authors as further evidence in support of the need for a field-based ontology in that it shows more strongly that a relativistic quantum mechanics of individual localizable systems, such as particles, cannot be constructed along recognizable lines.

To strengthen the argument against a particle ontology within quantum field theory, Halvorson and Clifton produced another result even more squarely focused on particles by addressing the issue of *number operators* rather than position operators, that is, operators the eigenvalues of which would be considered the number of particles within a given region of spacetime. That is, they consider a number structure $(\mathcal{H}, a \mapsto U(a), \Delta \mapsto N_\Delta)$, where the E_Δ, that is, a system of local operators over spacetime M, which is a generalization of the event-localization structure where the operator eigenvalues is not restricted to 0 and 1, and not discrete as would be intuitive, but also even non-discrete non-negative, to reach what they identify as the *minimal requirement* for a sensible notion of localizable particles [9, Sec. 6].

In addition to the assumptions of Busch's theorem, the Halvorson–Clifton theorem makes two additional ones: The first is (particle-)number conservation, namely that if $\{\Delta_n\}$ where n is a natural number is a disjoint covering of Σ, then the sum $\Sigma_n N_{\Delta_n}$ converges to a densely defined, self-adjoint operator N on the Hilbert state space $\mathcal{H}$ and $U(a)NU^*(a) = N$ for any timelike a so that the number operator is well defined (a condition satisfied in non-interacting QFT); the second is the additivity of the number operator, i.e. if Δ and Δ' are disjoint subsets of a single spacetime hyperplane then $N_\Delta + N_{\Delta'} = N_{\Delta+\Delta'}$. The conclusion of this additional theorem of Halvorson and Clifton is that, under with all these assumptions $N_\Delta = 0$ for all Δ, that is, in every state there will be no particles in any local region.

5. Relation to causality

In quantum theory, two forms of causality may be relevant: weak Einstein causality, the requirement that any changes of observables' *expectation values* take place with at most light speed, and strong Einstein causality, the requirement that individual property values change with at most light speed; cf. [4]. Strong causality implies localizability and is natural for relativistic

[1] Beneduci and Schroeck Jr have made steps toward answering this by introducing a different space localization observable for the photon [11].

theories. However, Malament's original theorem does not require it, but assumes only the weak form; cf. [9]. Busch demonstrated that such *weak* Einstein causality is logically equivalent to the local commutativity condition [12]; as a consequence of Lüders theorem [13], two (discrete) observables represented by self-adjoint operators commute if and only if, for any state, the statistics of measurement of one are unaffected by nonselective *Lüders measurements* of the other, which is extendable to observables having non-discrete spectra [4]. One then notes that unsharp phase-space observables can be jointly measurable without commuting with each other. Thus, even weak Einstein causality appears open to violation in the face of the generalized Lüders theorem (applying to POVMs beyond the projectors) demonstrated by Busch and Singh [14], in two cases that Busch argued are the only ones of genuine physical significance [4]. However, the above results are obtained with the assumption that arbitrary measurement accuracy can be obtained as in classical physics.

Denying arbitrary accuracy provides a possible way out of an apparent difficulty for Einstein causality in the context of POVMs, in that it is operationally suspect, at least in quantum mechanics: Busch argued that localizing a quantum system entirely within a sharp, finite boundary can only be done using an infinite potential well or, in the context of quantum field theory, with enough energy to make significant particle-pair creation possible, leading to further complications which clearly take one beyond simple, single-particle analyses [4].

In addition, as Busch also pointed out, the concept of localization involves globality in that the collection of all bounded spatial subsets of S and *translational covariance* itself are global in character; cf. [15]. But even if localization observables allowing sharply localized states aren't required to be local, so as not to menace weak causality, one might still be concerned that there could be an instantaneous spread of wave functions which might be thought support superluminal signaling. Hegerfeldt's theorem is understood to show that superluminal spreading of wave functions occurs in any quantum theory in which there are states of localized systems. To understand this, one can consider, in particular, the following conditions—in addition to translation covariance and a Hamiltonian bounded from below—imposed by Hegerfeldt in the following forms [9, 16]:

(NIWS) If $\Delta \subseteq \Delta'$ and the boundaries of Δ and Δ' have finite distance, then there is an $\epsilon > 0$ such that $P_\Delta \leq P_{\Delta'+t}$ whenever $0 \leq t < \epsilon$;

(Monotonicity) If $\{\Delta_n\}$, where n is a natural number, is a downward nested family of subsets of Σ such that $\cap_n \Delta_n = \Delta$, then $\wedge_n P_{\Delta_n} = P_\Delta$.

Here each operator of the form P_Δ is a projection onto the corresponding spatial set, Δ, with the interpretation that it represents the proposition that the particle in question is certainly localized in Δ. The conclusion of Hegerfeldt is that a particle localized at some initial time in any finite region Δ must remain so localized; the assumptions of monotonicity, translational covariance, Hamiltonian boundedness, and NIWS can be satisfied only for entirely trivial dynamics; otherwise, instantaneous wave function spreading will occur [9].

The natural states in which to consider quantum systems would seem therefore to be unlocalized ones, such as those of momentum. Nonetheless, as Busch further argued, an instantaneous wave function spread would not in itself underwrite superluminal signaling: Any signal sender could not *control* the particle in question by releasing the particle or not to communicate a bit of information as this would required an infinite amount of energy, as it could only be accomplished via an infinite potential well (or the equivalent) and, if the particle is free, it cannot be under sufficient control to enable any quasi-signal to be received at space-like separation, even though it had been instantaneously localized. Thus, the notion of local measurement and local observable algebra would be subject to reformulation under some new, truly operationally significant notion of causality [4].

The results considered above bring into question the significance of the quantum-field theoretical postulates of local commutativity and strong and weak causality, the last being introduced via a local commutativity requirement [4]. The situation thus clarified by the work of Busch *et al.* has motivated recent analyses of Barat and Kimball [18] and Terno *et al.* [17, 19] focusing on energy density, which is directly related to photodetection, as the quantity best indicating the presence of any putative particle. Terno *et al.* argued that causality is not endangered by the requirements of localization, concluding that "energy density cannot be 'localized' enough to violate causality".

Terno *et al.* constructed an operator density POVM element with the probability of detecting a particle in a volume Δ assumed proportional to the integral of the energy density within it. The probability of finding a particle in the volume, given that a detection has occurred anywhere at all, was taken to be

$$\Pi(t,\mathbf{x}) = H^{-1/2}\hat{T}_{00}(t,\mathbf{x})H^{-1/2}, \tag{1}$$

the first moment of which is proportional to the Born-Infeld position operator; cf. [20]. Their conclusion was then reached based on previous results of Barat and Kimball [18] showing that physical states of a real scalar field described by configuration-space wave-functions $\psi(t,\mathbf{x})$ are such that the requirement that the probability of a particle be outside a sphere of radius R be bounded by

$$\mathrm{Prob}_{\notin R} < C^2 \exp(-2\gamma R) \ , \tag{2}$$

C being a constant and $\gamma > m$ cannot be satisfied for the one-particle state $|\Psi\rangle$ having positive normal-ordered Hamiltonian energy density $T_{00}(t,\mathbf{x})$ for the Hamiltonian H [18]. Their POVM element is allowed to be non-local, as is evidenced by the presence of the Hamiltonian in its definition. In taking this operational approach, inconsistency with the causality requirements of quantum field theory appears to be avoided.

6. Conclusion

The work on localizability of Busch and others for elementary particle physics in a well-rounded world picture shows that the *theory of measurement* in relativistic quantum mechanics is in need of greater study. The approach that has been followed so far is one where the constitution of the instruments used for carrying out the operation of localization—the possible significance of which was also noted by Busch [4]—has not come under sufficiently explicit consideration. If the discussion is carried out fully in the realm of particle physics, as would serve the consistency of application of relativistic physical theory, the measuring instruments taken to localize particles should be considered themselves to be constituted by fields or elementary particles, something which thus far has not been carefully done. Thus, detailed theoretical descriptions at the *atomic-* and *subatomic-levels* of the instruments that would be involved in the operation of localization appear necessary for progress in exact treatments of the phenomena of particle physics. It could also serve to enrich notions such as Heisenberg's actualization of quantum potentiality, cf. [21], by providing greater physical detail to the measurement operations in RQFT, which goes beyond the quantum mechanics of systems fixed particle number in relation to which the former was first introduced. It is, therefore, suggested that this be done in future investigations of localizability in quantum theory.

References

[1] Jaeger G 2014 *Quantum Objects* (Berlin: Springer)
[2] Weinberg S 1995 *The Quantum Theory of Fields I* (Cambridge: Cambridge University Press)
[3] Jaeger G 2019 *Entropy* **21** 141
[4] Busch P 1999 *J. Phys. A: Math. Gen.* **32** 6535–46
[5] Schlieder S 1971 Zum kausalen Verhalten eines relativistischen quantenmechanischen Systems *Quanten und Felder, W. Heisenberg zum 70. Geburtstag* ed H P Dürr (Braunschweig: Vieweg) pp 145–60

[6] Malament D 1996 In defense of dogma: why there cannot be a relativistic quantum mechanics of (localizable) particles *Perspectives on Quantum Reality* ed R Clifton (Dordrecht: Kluwer) pp 1–10
[7] Castrigiano D P L and Mutze U 1982 *J. Math. Phys.* **26** 3499–505
[8] Wightman A S 1962 *Rev. Mod. Phys.* **34** 845–72
[9] Halvorson H and Clifton R 2002 *Phil. Sci.* **69** 1–28
[10] Giannitrapani R 1998 *J. Math. Phys.* **39** 5180–2
[11] Beneduci R and Schroeck Jr F E 2019 *Found. Phys.* **49** 561–76
[12] Schlieder S 1968 *Comm. Math. Phys.* **7** 305–31
[13] Lüders G 1951 *Ann. Phys.* **8** 322–8
[14] Busch P and Singh J 1998 *Phys. Lett.* A **249** 10–2
[15] Fleming G and Butterfield J 1999 Strange positions *From Physics to Philosophy* ed J Butterfield and C Pagonis (Cambridge: Cambridge University Press) pp 108–68
[16] Hegerfeldt G C 1998 *Ann. Phys.* **7** 716–25
[17] Terno D R 2014 *Phys. Rev.* A **89** 042111
[18] Barat N and Kimball J C 2003 *Phys. Lett.* A **308** 110–5
[19] Céleri L C, Kiosses V and Terno D R 2016 *Phys. Rev.* A **94** 062115
[20] Białynicki-Birula I and Białynicki-Birula Z 2012 *Phys. Rev.* A **86** 02218
[21] Jaeger G 2017 *Philos. Trans. Royal Soc.* A **375** 20160390

Journal of Physics: Conference Series **1638** (2020) 012011 doi:10.1088/1742-6596/1638/1/012011

What is unsharp quantum reality? A conceptual analysis

Brigitte Falkenburg
Department of Philosophy and Political Science, Technische Universität Dortmund, Germany
E-mail: brigitte.falkenburg@tu-dortmund.de

Abstract. In their paper *Unsharp Quantum Reality*, Paul Busch and Gregg Jaeger explain the conception of unsharp reality in terms of Positive Operator Valued (POV) measures, potentiality, and single-case probabilities. The present paper discusses their conception from a philosophical point of view, focusing on the ontological interpretation of POV measures employed in it and on its relation to Peter Mittelstaedt's work on the semantics and ontology of quantum mechanics. Busch and Jaeger extend Mittelstaedt's approach by adding a realistic interpretation of unsharp quantum properties to it. Their account of unsharp quantum reality generalizes Heisenberg's potentiality interpretation of quantum states. In order to explain the transition from potential to actual quantum properties, they suggest a Ghirardi-Rimini-Weber (GRW) collapse theory. Their approach comes close to recent work on quantum propensities. However, the question remains what unsharp quantum reality means when no measurement takes place, as for example claimed in the quark model of the subatomic constitution of matter.

1. Introduction

In 1994, I first met Paul Busch as a member of Peter Mittelstaedt's research group in Cologne. Since then, I had intriguing discussions with them on the semantics and ontology of quantum mechanics and their conception of unsharp reality. Beyond the formal quantum theory of measurement and its extension in terms of Positive Operator Valued (POV) measures, the question remained in which philosophical terms one may conceive of unsharp quantum reality. Busch's joint paper with Gregg Jaeger *Unsharp Quantum Reality* [1] provides an answer in terms of potentiality, the causal power of unsharp quantum properties, and the transition from potential to actual quantum properties by a spontaneous collapse as described in the Ghirardi-Rimini-Weber (GRW) approach. Busch and Jaeger's conception of unsharp quantum reality is a promising step towards a realistic interpretation of quantum theory without recourse to many-worlds or hidden variables. However, due to the operational approach to quantum theory on which it is based [2] it has certain limitations, as concerns a realistic account of the constitution of matter.

2. The question

The background of Busch's and Jaeger's account of unsharp quantum reality is the quantum theory of measurement [3, 4] and Mittelstaedt's work on the semantics and ontology of quantum theory [5, 6]. Mittelstaedt always claimed that the principles of quantum theory are fundamental and universally valid, whereas the theories of classical physics have a restricted domain in which they only approximately hold. The assumption that quantum theory is universally valid gives rise to the quest for a quantum ontology. The unresolved measurement problem, however, precludes to establish it in a straightforward way. In order to spell it out, Mittelstaedt attributed an object semantics to the world of classical physics, but a (weaker) process semantics to the quantum domain [5]. In face of the quantum measurement problem, the crucial question is then: To what extent are both kinds of semantics compatible?

In order to clarify the relation of both kinds of theory and their respective ontologies, Mittelstaedt followed Busch, taking up the conception of *unsharp properties* which may be attributed to single quantum systems [7]. It is based on Heisenberg's uncertainty relation and the possibility of joint unsharp measurements of position and momentum [8] or other pairs of non-commuting quantum observables (which Bohr called complementary [9]). According to the probabilistic interpretation of quantum mechanics, the uncertainty relation $\Delta x \Delta p \sim \hbar$ expresses the widths of the probability distributions of the observables position and momentum in a given state [10]. Hence, the corresponding unsharp properties of position and momentum have probabilistic meaning. They are properties only attributed with a certain probability to a quantum system. However, no ignorance of probability is tenable for them, according to the standard interpretation of quantum mechanics. Therefore, the question arises: In which sense may such unsharp properties give rise to a quantum ontology? A convincing ontology should not describe mere possibilities but the actual world. So, what is the reality of these probabilistic properties?

Mittelstaedt attempted to spell the unsharp quantum ontology out [6] relying on the formal approach to unsharp observables in terms of Positive Operator Valued (POV) measures [2, 3]. Based on his earlier work on the semantics and ontology of quantum theory [5], he showed on the basis of Quantum Logic how the classical ontology has to be relaxed to a Quantum Ontology $O(Q)$ restricted to the value definite propositions that still remain in the quantum domain. Then, he emphasized that this is not sufficient, given that the resulting assumptions about value definite propositions are still too strong:

> *In order to re-establish consistency* [...], $O(Q)$ *must be further relaxed, in particular with respect to the value definiteness of elementary properties.* [...] *As the envisaged result we expect self-consistency between quantum mechanics and a new weak ontology* $O(Q^U)$ *of unsharp properties.* [6, p 82]

According to his previous investigations concerning universality, quantum mechanics is semantically consistent at the probabilistic level, but not at the level of single quantum systems [11, 12]. Hence the need for a further ontological relaxation. The resulting unsharp quantum ontology $O(Q^U)$ is obviously expressed in terms of probabilities. It deals with possibilities rather than with the actual facts of physical reality. Is this *really* an ontology? Or, to put it in other words, what is unsharp reality?

3. Potentiality

To shed light on this intriguing question, Busch and Jaeger wrote the paper entitled *Unsharp Quantum Reality* as a contribution to the *Festschrift* for P Mittelstaedt on his 80th birthday [1]. In it, they address Mittelstaedt's claim that quantum mechanics is semantically consistent at the probabilistic level. The proof of this claim is based on considering an ensemble of N identically prepared quantum systems as one composite system [4, 11, 12]. Busch and Jaeger comment on this approach as follows [1]:

> *However, the entity that carries the probability value as an (approximately) actual property is an* ensemble *of equally prepared quantum systems. Since this system is composed of independent subsystems, the question remains how one could then explain that such independent objects act together to cause the emergence of an approximately*

> *actual eigenvalue of the respective frequency operators of the ensemble. After all, the description of the ensemble as a compound system is hardly more than a conceptual construct.* (p 1350)

In spite of this conceptual construct, the ensemble exactly behaves as to be expected:

> *The answer is therefore to be sought in the fact that each constituent has been treated—*prepared*—in an identical way.* [...] *Through its preparation, each individual system is constrained to 'respond' to a measurement so as to induce a specific outcome in accordance with the probability law specified by its quantum state.* (pp 1350–1)

Then, Busch and Jaeger add their own claim concerning unsharp reality:

> *Each possible value of the observable to be measured has a limited degree of reality, quantified by the associated quantum probability. Accordingly, the individual has a limited capacity to cause the actualization of each of the corresponding pointer values (and, in a repeatable measurement, the values of the observable). On many repetitions (or in many parallel runs) of the measurement, the degrees of reality of the different values are then manifested as the frequencies of the outcomes.* (p 1351)

Sharp properties of quantum systems correspond to spectral projections on eigenstates, and hence to definite values of the respective observable, according to the eigenstate-eigenvalue link of standard quantum theory. Sharp properties are actual, they have probability 1 and belong to physical reality. In contrast, unsharp properties of quantum systems correspond to effects of POV measures which are no projections and give rise to fuzzy values. Unsharp properties obey uncertainty relations in Heisenberg's sense and express probabilities < 1. Busch and Jaeger suggest to interpret them in terms of *single case probabilities as potential properties* with a *degree of reality* < 1. In particular, they consider the effects E of an isolated quantum system in a pure state ψ, represented by a projection $P = P_\psi$. The time evolution of ψ is determined by the Schrödinger equation, resulting in a continuous evolution of single case probabilities or degrees of reality $f_E(P) = \mathrm{tr}[PE] = \langle \psi \,|\, E\psi \rangle$. They conclude that "all the effects E of the system [...] are simultaneously real to a degree (actualization tendency) given by $f_E(P)$." [1, p 1363]

The background of their approach is Heisenberg's potentiality interpretation of quantum mechanics, according to which the wave function of a quantum state expresses 'an objective tendency or possibility, a 'potentia' in the sense of Aristotelian philosophy' [13]. The wave function expresses the tendency to become actual by the measurement of an observable. The measurement transforms this potentiality into actuality. The result of this process is described by the collapse of the wave function to an eigenstate of the respective observable. Busch and Jaeger generalize Heisenberg's approach to all states and effects of quantum systems, ascribing physical reality with a degree $f_E(P) < 1$ to the quantum states ψ and their effects E. And they add an important causal aspect to it, namely that the quantum states, respectively their degrees of reality, are probabilistic causes for the actualization of certain measurement results.

4. Propensities, dispositions and GRW

Busch and Jaeger's account of unsharp reality stands in the tradition of the propensity interpretations of quantum theory proposed by Margenau [14] and Popper [15] in the 1950s. These approaches first provoked substantial criticism, but now there are new prospects for propensities [16, 17].

A propensity is a quantified tendency to become actual. According to this definition, the degrees of reality < 1 of unsharp properties determined by $f_E(P) = \mathrm{tr}[PE] = \langle \psi \,|\, E\psi \rangle$ are propensities. In philosophy of science, propensities or probabilistic causes are also discussed in terms of dispositions. Busch and Jaeger emphasize that their conception of unsharp reality comes close to Esfeld's account of quantum dispositions with causal powers [18].

One of the problems of Popper's approach was to apply his propensity conception to probability in general as well as to quantum theory [16]. It has indeed been argued 'that any broadly dispositional analysis of probability will either fail to give an adequate explication of probability, or else will fail to provide an explication that can be gainfully employed elsewhere (for instance, in empirical science or in the regulation of credence)', with the conclusion [19] that this 'does place quite strong constraints on what type of propensity interpretation can be maintained.'

The recent approaches to quantum propensities and dispositions are more successful. On the one hand, they do not identify propensities and probabilities. In particular, it has be shown that quantum propensities substantially differ from (classical) conditional Kolmogorov probabilities [20]. On the other hand, the recent approaches emphasize that quantum propensities or dispositions require a GRW type collapse theory in order to contribute to a genuinely metaphysical, or realistic, interpretation of quantum mechanics, beyond instrumentalist views [21]. Busch and Jaeger indeed claim that in order to explain the objectification of measurement results (that is, the transition from potential to actual quantum properties), a GRW type theory is needed [1, p 1359-61]. Vice versa, it has been claimed that the GRW approach is a good candidate for an ontology of quantum dispositions [22]. As to be expected, such an ontology of quantum propensities or dispositions and the associated GRW collapse theory also take the non-locality of quantum states into account [1, 22].

In recent papers, Jaeger enhanced the conception of unsharp quantum reality [1] to clarify the relation between Heisenberg's potentiality interpretation of quantum mechanics and the Aristotelian concept of *potentia* [23], and attempted to make it more precise in terms of the distinction of actuality vs. reality [24]. He argues that Heisenberg's account of Aristotle's distinction of *potentiality* ($\delta\upsilon\nu\alpha\mu\iota\varsigma$) and *actuality* ($\varepsilon\nu\varepsilon\rho\gamma\varepsilon\iota\alpha$) does not fall short of Aristotle's concepts, given that the spontaneous transition from potential to actual properties occurring in quantum measurements correspond to the Aristotelian concept of chance, which means a spontaneous event or process in nature devoid of any teleological connotation [23]. Indeed Aristotle's concept of chance refers to events occurring with objective probability. Applied to probabilistic quantum transitions, Aristotelian chances hence amount to quantum propensities and the associated spontaneous transitions from potential to actual quantum properties. In addition, Jaeger refines Heisenberg's distinctions of potentiality and actuality by distinguishing furthermore reality from actuality, and he suggests that unsharp quantum properties with an actualization tendency or degree of reality $\frac{1}{2} < f_E(P) < 1$ may be called *real*, in spite of not being *actual* [24].

Jaeger emphasizes that Heisenberg himself did not distinguish the concepts of reality and actuality. However, this matter of fact may partially be due to a translation problem raised by the English original version of *Physics and Philosophy.* In German, and in particular in the Kantian tradition to which Heisenberg belonged, one may distinguish *reality* (Realität) and *actuality* (*Wirklichkeit*). In the German translation of *Physics and Philosophy*, Heisenberg accordingly uses the term *Wirklichkeit* (*actuality*) as opposed to *potentiality* (*Möglichkeit*) [25]. But in the English original version, he had used the term *reality* rather than the term *actuality*, which would have been the correct opposite of *potentiality.*

Jaeger's way of refining Heisenberg's distinction of potentiality and actuality in terms of a non-Heisenbergian distinction of reality and actuality is in two regards problematic. On the one hand, it does not conform to Kant's distinction of reality and actuality [26], a conceptual distinction that is however more important for philosophers than for physicists. On the other hand, it introduces an arbitrary distinction between (approximately) real quantum properties with probability $\frac{1}{2} < f_E(P) < 1$ and unsharp quantum properties with a degree of reality $0 < f_E(P) \leq \frac{1}{2}$, referring to an early paper of Busch [26]. But unsharp quantum reality comes in continuous degrees. There is no reason to attribute significantly more reality to quantum properties with probability $\frac{1}{2} + \epsilon$ than to properties with $0 < f_E(P) \leq \frac{1}{2}$. Quantum events with a very low probability $0 < f_E(P) \ll 1$ happen. I resist to say that they happen *quite often*, but they *happen*. They do in the case of large angle Rutherford scattering, to mention just a famous example. The methods of data analysis in particle physics show that in the quantum field theory of scattering such examples indeed are ubiquitous. They cannot be neglected in the reconstruction of particle tracks and quantum data probability distributions [27, 28]. Hence, a convincing account of unsharp quantum reality should include them, and with them, any degrees of reality $f_E(P) \ll 1$. As long as the quantum probabilities are non-zero, the corresponding quantum states belong to unsharp quantum reality, as Jaeger indeed claims, attributing causal efficacy to them. But his approach still misses to emphasize and discuss the physical significance of unsharp quantum reality in the range of $0 < f_E(P)) < \frac{1}{2}$.

5. Conclusions

Busch's and Jaeger's account of unsharp quantum reality [1] offers a realistic interpretation of quantum states in terms of quantum actualization tendencies, or quantum propensities. Their approach is based on a formal generalization of quantum mechanics in Hilbert space by POV measures which make it possible to express joint measurements of non-commuting (complementary) quantum observables and their joint probabilities, expressed by (generalized) Heisenberg uncertainty relations. They express unsharp quantum reality in terms of the degrees of reality $0 < f_E(P) < 1$ of quantum effects $f_E(P)$, including quantum effects of low but non-zero probabilities. In order to explain the transition from quantum superpositions, or potential quantum properties, to actual measurement outcomes, they suggest that the realistic interpretation of unsharp quantum properties requires a GRW type collapse theory.

The advantage of this new realistic interpretation of quantum mechanics is that it makes no recourse to speculative metaphysical assumptions such as many-world or hidden variables, which have to be added to quantum mechanics and its probabilistic standard interpretation. It is in accordance with the recent discussion of quantum propensities, dispositions, and the corresponding ontology of a GRW approach to the quantum measurement problem. And it even admits degrees of reality $0 < f_E(P) \ll 1$.

However, due to its foundation in the operational account of quantum physics [2] Busch's and Jaeger's account of unsharp quantum reality is limited. In order to be in accordance with the practice of physics, any realistic interpretation of quantum theory should be able to explain the current constituent model of matter which is based on the standard model of particle physics. According to this model, quarks and leptons are the parts of matter. In particle physics, the existence of these quantum parts of matter is established in terms of sum rules for momentum-energy, spin, parity, and so on, which are based on the scattering experiments of high energy physics [2]. The quarks within matter make up the protons and neutrons, which in turn make up the atomic nuclei. Within the protons and neutrons, however, the quarks only come in superpositions, given that their mass eigenstates differ from their flavour eigenstates. In addition, due to quark confinement no single quarks can be measured. Hence, any merely operational interpretation of the unsharp properties of quarks misses the role of quarks as actual matter constituents, which are well-established parts of physical reality. In particular with regard to compound quantum systems, additional concepts are needed in order to conceive of unsharp quantum reality beyond an operational quantum physics.

Acknowledgments

I would like to thank an anonymous referee for helpful comments on an earlier version of this paper.

Journal of Physics: Conference Series **1638** (2020) 012011 doi:10.1088/1742-6596/1638/1/012011

References

[1] Busch P and Jaeger G 2010 Unsharp quantum reality *Found. Phys.* **40** 1341–67
[2] Busch P, Grabowski M and Lahti P J 1995 *Operational Quantum Physics* (Berlin: Springer)
[3] Busch P, Lahti P J and Mittelstaedt P 1991 *The Quantum Theory of Measurement* (Berlin: Springer)
[4] Mittelstaedt P 1998 *The Interpretation of Quantum Mechanics and the Measurement Process* (Cambridge: Cambridge University Press)
[5] Mittelstaedt P 1986 *Sprache und Realität in der modernen Physik.* (Darmstadt: Bibliographisches Institut)
[6] Mittelstaedt P 2013 *Rational Reconstruction of Modern Physics* (Dordrecht: Springer) 2nd ed.
[7] Busch P 1982 Indeterminacy relations and simultaneous measurements in quantum theory *Int. J. Theor. Phys.* **24** 63–92
[8] Busch P, Lahti P J and Mittelstaedt P 1992 Weak objectification, joint probabilities, and Bell inequalities in quantum mechanics *Found. Phys.* **22** 949–62
[9] Bohr N 1927 The quantum postulate and the recent development of atomic theory (Como lecture) revised version: 1928 *Nature* **121** 580–90
[10] Busch P and Falkenburg B 2009 Heisenberg uncertainty relation (indeterminacy relations) *Compendium of Quantum Physics* ed D Greenberger *et al* (Dordrecht: Springer) pp 281–3
[11] Mittelstaedt P 1997 Is quantum mechanics a probabilistic theory? *Potentiality, Entanglement and Passion-at-a-distance* ed R S Cohen *et al* (Dordrecht: Kluwer) pp 159–75
[12] Mittelstaedt P 1998 Unifying quantum mechanics and its interpretation *Philosophia Naturalis* **35** 127–51
[13] Heisenberg W 1958 *Physics and Philosophy* (New York: Harper and Row)
[14] Margenau H 1954 Advantages and disadvantages of various interpretations of quantum theory *Phys. Today* **7** 6–13
[15] Popper K 1957 The propensity interpretation of the calculus of probability, and the quantum theory *Observation and Interpretation in the Philosophy of Physics* ed S Körner (London: Butterworths) pp 65–70
[16] Suárez M 2009 Propensities in quantum mechanics *Compendium of Quantum Physics* ed D Greenberger *et al* (Dordrecht: Springer) pp 500–5
[17] Suárez M (ed) 2011 *Probabilities, Causes and Propensities in Physics* (Dordrecht: Springer)
[18] Esfeld M 2010 Physics and causation *Found. Phys.* **40** 1597–610
[19] Eagle A 2004 Twenty-one arguments against propensity analyses of probability *Erkenntnis* **60** 371–416
[20] Suárez M 2017 Propensities, probabilities, and experimental statistics *EPSA15: Selected Papers* ed M Massimi *et al* (Cham: Springer)
[21] Dorato M 2011 Do dispositions and propensities have a role in the ontology of quantum mechanics? *Probabilities, Causes and Propensities in Physics* ed M Suárez M (Dordrecht: Springer) pp 192–219
[22] Dorato M and Esfeld M 2010 GRW as an ontology of dispositions *Studies in Philosophy of Modern Physics* **41** 41–9
[23] Jaeger G 2017 Quantum potentiality revisited *Philos. Trans. Royal Soc.* A **375** 20160390
[24] Jaeger G 2019 Quantum unsharpness, potentiality, and reality *Found. Phys.* **49** 663–76
[25] Heisenberg W 1959 *Physik und Philosophie* (Stuttgart: Hirzel) p 175 (German translation of [13])
[26] Busch P 1985 Can quantum theoretical reality be considered sharp? *Recent Developments in Quantum Logic* ed P Mittelstaedt and E W Stachow (Mannheim: Bibliographisches Institut) pp 81–101
[27] Falkenburg B 2007 *Particle Metaphysics* (Berlin: Springer)
[28] Falkenburg B 2012 From waves to particle tracks and quantum probabilities *From Ultrarays to Astroparticles* ed B Falkenburg and W Rhode pp 265–95

Mathematical Foundations of Quantum Mechanics - *in memoriam* Paul Busch IOP Publishing
Journal of Physics: Conference Series **1638** (2020) 012012 doi:10.1088/1742-6596/1638/1/012012

Two entanglement measures

Stan Gudder

Department of Mathematics, University of Denver, Denver, Colorado 80208, USA

E-mail: sgudder@du.edu

Abstract. In a previous work we introduced an entanglement number $e(\psi)$ for a vector state ψ. This number relied on the Schmidt decomposition of ψ which may be difficult to compute. We now present a method for finding $e(\psi)$ that is simple and efficient. We show that $e(\psi)$ is continuous in the vector norm. We next extend $e(\psi)$ to an entanglement number $e(\rho)$ for a general mixed state ρ and show that ρ is separable (not entangled) if and only if $e(\rho) = 0$. We next define a related quantity $e_s(\rho) \geq e(\rho)$ which we call the spectral entanglement number. We argue that $e_s(\rho)$ is easier to compute than $e(\rho)$ and that the physical motivation for $e_s(\rho)$ is superior to that of $e(\rho)$. It is shown that $e_s(\rho)$ is continuous in the operator norm. Although $e(\rho)$ is a convex function, we show that $e_s(\rho)$ need not be convex. An open problem is to characterize the states ρ such that $e(\rho) = e_s(\rho)$. Many illustrative examples are presented.

1. Introduction

Entangled states are an important resource for quantum computation and information theory [1–4]. With this application in mind, we shall only consider finite-dimensional physical systems. This assumption also provides mathematical simplifications that make the theory more manageable. Also, for simplicity, we only consider bipartite systems and leave the study of multipartite systems to later work. Because of this significance of entanglement, it is useful to have the ability to compare the entanglement of different states. We thus require a measure of entanglement for quantum states and various measures have been proposed [5–10]. In a recent work, we introduced an entanglement measure for a vector state ψ that we called an entanglement number $e(\psi)$ [11]. This number relied on the Schmidt decomposition of ψ which may be difficult to compute [2, 3]. We now present a method for finding $e(\psi)$ that applies to any orthonormal bases of the component Hilbert spaces. This method is simple and efficient. It is shown that $e(\psi)$ is continuous in the norm topology.

We next extend $e(\psi)$ to an entanglement number $e(\rho)$ for a general mixed state ρ and show that ρ is separable (not entangled) if and only if $e(\rho) = 0$. We also show that $e(\rho)$ is a convex function. A related quantity $e_s(\rho) \geq e(\rho)$, which we call the spectral entanglement is introduced. We argue that $e_s(\rho)$ is easier to compute than $e(\rho)$ and that the physical motivation for $e_s(\rho)$ is superior to that of $e(\rho)$. One of the reasons for this latter argument is that $e_s(\rho) = 0$ if and only if the nonzero eigenvalues of ρ (including multiplicity) have factorized eigenvectors. It is shown that $e_s(\rho)$ is continuous in the operator norm topology. Although $e(\rho)$ is a convex function, we show that $e_s(\rho)$ need not also be convex.

An open problem is to characterize the states ρ such that $e(\rho) = e_s(\rho)$. Eight examples that illustrate various concepts of the theory are presented.

2. Entanglement measure for pure states

In this work, we consider a bipartite physical system described by a complex finite-dimensional Hilbert space $H = K_1 \otimes K_2$ where K_1 and K_2 are its composite parts. We can enlarge one of the spaces if necessary so we can assume without loss of generality that $\dim K_1 = \dim K_2 = n$. We then have that $\dim H = n^2$. Let $\psi \in H$ be a unit vector that we call a *vector state*. The vector ψ corresponds to a pure state in H given by the one-dimensional projection operator $P_\psi = |\psi\rangle\langle\psi|$. We say that ψ (or P_ψ) is *factorized* if $\psi = \alpha \otimes \beta$ for unit vectors $\alpha \in K_1$, $\beta \in K_2$. Otherwise, we say that ψ is *entangled.*

According to the Schmidt Decomposition Theorem [2, 3], there exist numbers $\lambda_i \geq 0$ and orthonormal bases $\{\alpha_i\}$, $\{\beta_i\}$ for K_1 and K_2, respectively, such that

$$\psi = \sum \sqrt{\lambda_i}\, \alpha_i \otimes \beta_i. \tag{2.1}$$

The real numbers λ_i are unique, although the bases are not and, of course, $\sum \lambda_i = 1$. In a previous paper [11], we defined a measure of entanglement for ψ (or P_ψ) given by the *entanglement number*

$$e(\psi) = e(P_\psi) = \left[1 - \sum \lambda_i^2\right]^{1/2}. \tag{2.2}$$

Moreover, we showed in [11] that ψ is factorized if and only if $e(\psi) = 0$. It is clear that $e(a\psi) = e(\psi)$ for all $a \in \mathbb{C}$ with $|a| = 1$. A disadvantage of (2.2) is that the singular-values $\sqrt{\lambda_i}$ in (2.1) may be difficult to compute. (It's similar to finding the eigenvalues of a matrix.) We now give a method for finding $e(\psi)$ that applies to any orthonormal bases $\{\phi_i\}$, $\{\psi_i\}$ for K_1 and K_2, respectively. Let $\psi = \sum c_{ij}\phi_i \otimes \psi_j$ and define the matrix $C = [c_{ij}]$. The matrix $|C|$ is defined to be the positive semidefinite matrix given by $|C| = (C^*C)^{1/2}$.

Theorem 2.1. (a) $e(\psi) = \left[1 - \operatorname{tr}\left(|C|^4\right)\right]^{1/2}$. (b) *$\psi$ is factorized if and only if* $\operatorname{tr}\left(|C|^4\right) = 1$. (c)

$$\operatorname{tr}\left(|C|^4\right) = \sum_{r,s}\left[\left|\sum_i \overline{c}_{ri}c_{si}\right|^2\right].$$

Proof. (a) By the Singular-Value Theorem [3], we can write

$$C = [c_{ij}] = UDV,$$

where U and V are unitary matrices, $D = \operatorname{diag}(\lambda_1^{1/2}, \dots, \lambda_n^{1/2})$, $\lambda_i \geq 0$, and $\sqrt{\lambda_i}$ are the singular values of C. These singular values coincide with those given in (2.1). Now

$$|C|^2 = C^*C = V^*DU^*UDV = V^*D^2V,$$

and hence, $|C|^4 = V^*D^4V$. Therefore,

$$\sum \lambda_i^2 = \operatorname{tr}(D^4) = \operatorname{tr}\left(|C|^4\right).$$

We conclude from (2.2) that $e(\psi) = \left[1 - \operatorname{tr}\left(|C|^4\right)\right]^{1/2}$. (b) follows from (a). To verify (c) we have that

$$|C|^4_{ij} = (C^*C)(C^*C)_{ij} = \sum_k (C^*C)_{ik}(C^*C)_{kj}$$

$$= \sum_{k,r} C_{ir}^* C_{rk} \sum_s C_{ks}^* C_{sj} = \sum_{r,s,k} \overline{c}_{ri} c_{rk} \overline{c}_{sk} c_{sj}.$$

Hence,

$$\operatorname{tr}\left[|C|^4\right] = \sum_i |C|_{ii}^4 = \sum_{i,k,r,s} \overline{c}_{ri} c_{rk} \overline{c}_{sk} c_{si}$$
$$= \sum_{r,s}\left[\sum_i \overline{c}_{ri} c_{si} \sum_k \overline{\overline{c}_{rk} c_{sk}}\right] = \sum_{r,s}\left[\left|\sum_i \overline{c}_{ri} c_{si}\right|^2\right]. \qquad \square$$

Example 1. Let $\psi \in \mathbb{C}^2 \otimes \mathbb{C}^2$ be the bipartite qubit state given by

$$\psi = \tfrac{1}{\sqrt{10}}(\phi_1 \otimes \phi_1 - 2i\phi_1 \otimes \phi_2 + \phi_2 \otimes \phi_1 - 2i\phi_2 \otimes \phi_2),$$

where $\{\phi_1, \phi_2\}$ is an orthonormal basis for $\mathbb{C}^2$. We then have that

$$C = \tfrac{1}{\sqrt{10}}\begin{bmatrix} 1 & -2i \\ 1 & -2i \end{bmatrix}.$$

Applying Theorem 2.1(c) gives

$$\operatorname{tr}\left(|C|^4\right) = \sum_{r,s=1}^{2}\left[\left|\sum_{i=1}^{2} \overline{c}_{ri} c_{si}\right|^2\right] = \sum_{r,s=1}^{2}\left[\left|\overline{c}_{r1} c_{s1} + \overline{c}_{r2} c_{s2}\right|^2\right]$$
$$= \left(|c_{11}|^2 + |c_{12}|^2\right)^2 + \left(|c_{21}|^2 + |c_{22}|^2\right)^2 + 2\left|\overline{c}_{11} c_{21} + \overline{c}_{12} c_{22}\right|^2$$
$$= \tfrac{1}{100}\left[(1+4)^2 + (1+4)^2 + 2(1+4)^2\right] = 1.$$

It follows from Theorem 2.1(b) that ψ is factorized. In fact, $\psi = \alpha \otimes \beta$ where

$$\alpha = \tfrac{1}{\sqrt{2}}(\phi_1 + \phi_2), \qquad \beta = \tfrac{1}{\sqrt{5}}(\phi_1 - 2i\phi_2). \qquad \square$$

Example 2. We change Example 1 slightly and let

$$\psi = \tfrac{1}{\sqrt{10}}(\phi_1 \otimes \phi_1 - 2i\phi_1 \otimes \phi_2 + \phi_2 \otimes \phi_1 + 2i\phi_2 \otimes \phi_2)$$

We then have that

$$C = \tfrac{1}{\sqrt{10}}\begin{bmatrix} 1 & -2i \\ 1 & 2i \end{bmatrix},$$

and Theorem 2.1(c) gives

$$\operatorname{tr}\left(|C|^4\right) = \tfrac{1}{100}\left[(1+4)^2 + (1+4)^2 + 2(1-4)^2\right] = \tfrac{17}{25}.$$

We conclude that ψ is entangled and

$$e(\psi) = \left(1 - \tfrac{17}{25}\right)^{1/2} = \tfrac{2\sqrt{2}}{5}. \qquad \square$$

The following two examples will be needed in Sections 3, 4.

Example 3. Let $\{\phi_1, \phi_2\}$ be an orthonormal basis for $\mathbb{C}^2$ and define

$$\psi_1 = \tfrac{1}{2}\left[(\phi_2 - \phi_1)\otimes\phi_1 + (\phi_1+\phi_2)\otimes\phi_2\right] \in \mathbb{C}^2\otimes\mathbb{C}^2.$$

It is clear that ψ_1 is entangled with Schmidt decomposition

$$\psi_1 = \tfrac{1}{\sqrt{2}}\left[\tfrac{1}{\sqrt{2}}(\phi_2-\phi_1)\otimes\phi_1\right] + \tfrac{1}{\sqrt{2}}\left[\tfrac{1}{\sqrt{2}}(\phi_1+\phi_2)\otimes\phi_2\right].$$

The Schmidt coefficients are $\lambda_1 = \lambda_2 = 1/2$. Hence,

$$e(\psi_1) = \left(1 - \sum \lambda_i^2\right)^{1/2} = \left(1-\tfrac{1}{2}\right)^{1/2} = 1/\sqrt{2}.$$

We can apply Theorem 2.1 to check this result by writing

$$\psi_1 = -\tfrac{1}{2}\,\phi_1\otimes\phi_1 + \tfrac{1}{2}\,\phi_1\otimes\phi_2 + \tfrac{1}{2}\,\phi_2\otimes\phi_1 + \tfrac{1}{2}\,\phi_2\otimes\phi_2.$$

The corresponding matrix becomes

$$C = \tfrac{1}{2}\begin{bmatrix} -1 & 1 \\ 1 & 1 \end{bmatrix}.$$

Hence,

$$|C|^2 = C^*C = \tfrac{1}{2}\,I,$$

and we have that $|C|^4 = \frac{1}{4}\,I$. Therefore,

$$e(\psi_1) = \left[1 - \operatorname{tr}\left(|C|^4\right)\right]^{1/2} = \left(1-\tfrac{1}{2}\right)^{1/2} = 1/\sqrt{2}.$$

□

Example 4. Define the vector

$$\psi_2 = \tfrac{1}{2\sqrt{3}}\left[(3\phi_1+\phi_2)\otimes\phi_1 + (\phi_1+\phi_2)\otimes\phi_2\right].$$

Unlike Example 3, the Schmidt decomposition of ψ_2 is not so clear. To find $e(\psi_2)$ we write

$$\psi_2 = \tfrac{\sqrt{3}}{2}\,\phi_1\otimes\phi_1 + \tfrac{1}{2\sqrt{3}}\,\phi_1\otimes\phi_2 + \tfrac{1}{2\sqrt{3}}\,\phi_2\otimes\phi_1 + \tfrac{1}{2\sqrt{3}}\,\phi_2\otimes\phi_2.$$

The corresponding matrix becomes

$$C = \tfrac{1}{2\sqrt{3}}\begin{bmatrix} 3 & 1 \\ 1 & 1 \end{bmatrix}.$$

Hence,

$$|C|^2 = \tfrac{1}{6}\begin{bmatrix} 5 & 2 \\ 2 & 1 \end{bmatrix}, \qquad |C|^4 = \tfrac{1}{36}\begin{bmatrix} 29 & 12 \\ 12 & 5 \end{bmatrix}.$$

We conclude that $\operatorname{tr}\left(|C|^4\right) = 17/18$ and hence,

$$e(\psi_2) = \left(1 - \tfrac{17}{18}\right)^{1/2} = 1/3\sqrt{2}.$$

Thus, ψ_2 is less entangled than ψ_1. □

Theorem 2.2. *The function $\psi \mapsto e(\psi)$ is continuous in the norm topology on H.*

Proof. Let $\varepsilon > 0$ and suppose that ψ and ϕ are unit vectors on H with $||\psi - \phi|| < \varepsilon$. Let $\{\phi_i\}$, $\{\psi_i\}$ be orthonormal bases for K_1 and K_2, respectively. Define the matrices $C_\psi = [c_{ij}]$ and $C_\phi = [d_{ij}]$ where $c_{ij} = \langle \psi, \phi_i \otimes \psi_j \rangle$ and $d_{ij} = \langle \phi, \phi_i \otimes \psi_j \rangle$. We then have that $\psi = \sum c_{ij}\phi_i \otimes \psi_j$ and $\phi = \sum d_{ij}\phi_i \otimes \psi_j$. By Schwarz's inequality we obtain

$$|c_{ij} - d_{ij}| = |\langle \phi - \psi, \phi_i \otimes \psi_j \rangle| \le ||\psi - \phi|| < \varepsilon.$$

Letting α_k be the vector $\alpha_k(j) = \delta_{jk}$ we have that

$$(C_\psi - C_\phi)\alpha_k = [c_{ij} - d_{ij}]\,\alpha_k = \beta,$$

where β is the vector given by $\beta(j) = c_{jk} - d_{jk}$. Hence,

$$||(C_\psi - C_\phi)\alpha_k||^2 = \sum_j |\beta(j)|^2 = \sum_j |c_{jk} - d_{jk}|^2 \le n\varepsilon^2,$$

so that $||(C_\psi - C_\phi)\alpha_k|| \le n^{1/2}\varepsilon$. If $\delta = \sum a_k\alpha_k$ is a unit vector we have that $\sum |a_k|^2 = ||\delta||^2 = 1$. Applying Schwarz's inequality gives

$$\begin{aligned} ||(C_\psi - C_\phi)\delta|| &= \left|\left|\sum_k a_k(C_\psi - C_\phi)\alpha_k\right|\right| \le \sum_k ||a_k(C_\psi - C_\phi)\alpha_k|| \\ &= \sum_k |a_k|\; ||(C_\psi - C_\phi)\alpha_k|| \\ &\le \left(\sum_k |a_k|^2\right)^{1/2} \left(\sum_k ||(C_\psi - C_\phi)\alpha_k||^2\right)^{1/2} \le n\varepsilon. \end{aligned}$$

It then follows that the operator norm $||C_\psi - C_\phi|| \le n\varepsilon$. We also have that $\left|\left|C_\psi^* - C_\phi^*\right|\right| \le n\varepsilon$ and $||C_\psi||, \left|\left|C_\psi^*\right|\right|, ||C_\phi||, \left|\left|C_\phi^*\right|\right| \le 1$. Applying the triangle inequality we conclude that

$$\begin{aligned} \left|\left||C_\psi|^2 - |C_\phi|^2\right|\right| &= ||C_\psi^* C_\psi - C_\phi^* C_\phi|| \\ &\le ||C_\psi^* C_\psi - C_\psi^* C_\phi|| + ||C_\psi^* C_\phi - C_\phi^* C_\phi|| \\ &\le ||C_\psi^*||\; ||C_\psi - C_\phi|| + ||C_\psi^* - C_\phi||\; ||C_\phi|| \le 2n\varepsilon. \end{aligned}$$

Applying the triangle inequality again gives

$$\begin{aligned} \left|\left||C_\psi|^4 - |C_\phi|^4\right|\right| &\le \left|\left||C_\psi|^4 - |C_\psi|^2|C_\phi|^2\right|\right| + \left|\left||C_\psi|^2|C_\phi|^2 - |C_\phi|^4\right|\right| \\ &\le \left(\left|\left||C_\psi|^2\right|\right| + \left|\left||C_\phi|^2\right|\right|\right) \left|\left||C_\psi|^2 - |C_\phi|^2\right|\right| \le 4n\varepsilon. \end{aligned}$$

It follows that

$$\begin{aligned} \left|\text{tr}\left(|C_\psi|^4\right) - \text{tr}\left(|C_\phi|^4\right)\right| &= \left|\text{tr}\left(|C_\psi|^4 - |C_\phi|^4\right)\right| \\ &\le n\left|\left||C_\psi|^4 - |C_\phi|^4\right|\right| \le 4n^2\varepsilon. \end{aligned}$$

Therefore,

$$|e(\psi)^2 - e(\phi)^2| = \left|\left(1 - \text{tr}\left(|C_\psi|^4\right)\right) - \left(1 - \text{tr}\left(|C_\phi|^4\right)\right)\right| \le 4n^2\varepsilon.$$

If ϕ_n are unit vectors with $\lim_{n\to\infty}\phi_n = \psi$ in norm, then $\lim_{n\to\infty} e(\phi_n)^2 = e(\psi)^2$ and hence $\lim_{n\to\infty} e(\phi_n) = e(\psi)$. □

Journal of Physics: Conference Series **1638** (2020) 012012 doi:10.1088/1742-6596/1638/1/012012

3. Entanglement measure for mixed states

We denote the set of states on H by $\mathcal{S}(H)$. The elements of $\mathcal{S}(H)$ are the positive operators ρ on H satisfying $\mathrm{tr}\,(\rho) = 1$. If ρ is not a pure state, we call ρ a *mixed state.* Every $\rho \in \mathcal{S}(H)$ has the form $\rho = \sum \lambda_i P_i$ where $\lambda_i > 0$, $\sum \lambda_i = 1$ and P_i are pure states [2]. We can and will assume there are at most n^2 terms in this decomposition of ρ. In general, ρ has uncountably many representations as such convex combinations [2]. If ρ has a representation $\rho = \sum \lambda_i P_{\phi_i}$ where ϕ_i are factorized, then ρ is *separable.* Otherwise, ρ is *entangled.* We say that ρ is *factorized* if $\rho = \rho_1 \otimes \rho_2$ for some $\rho_i = \mathcal{S}(K_i)$, $i = 1, 2$. If ρ is factorized, of course, it is separable. Notice that ρ is separable if and only if $\rho = \sum \lambda_{ij} P_{\phi_i \otimes \psi_j}$ for arbitrary unit vectors $\phi_i \in K_1$, $\psi_j \in K_2$, $\lambda_{ij} > 0$, $\sum \lambda_{ij} = 1$. We call $\{\lambda_{ij}\}$ where $\lambda_{ij} > 0$, $\sum \lambda_{ij} = 1$ a *product measure* if $\lambda_{ij} = \alpha_i \beta_j$ where $\alpha_i, \beta_j > 0$, $\sum \alpha_i = \sum \beta_j = 1$. We first consider the simple case of the factorized state.

Lemma 3.1. (a) *ρ is factorized if and only if ρ is separable and $\{\lambda_{ij}\}$ is a product measure.*
(b) *If $\lambda_{ij} > 0$, $\sum \lambda_{ij} = 1$, then λ_{ij} is a product measure if and only if for all i, j we have that*

$$\lambda_{ij} = \left(\sum_j \lambda_{ij}\right)\left(\sum_i \lambda_{ij}\right).$$

Proof. (a) If ρ is separable and $\{\lambda_{ij}\}$ is a product measure, then

$$\rho = \sum \lambda_{ij} P_{\phi_i \otimes \psi_j} = \sum \alpha_i \beta_j P_{\phi_i} \otimes P_{\psi_j} = \left(\sum \alpha_i P_{\phi_i}\right) \otimes \left(\sum \beta_j P_{\psi_j}\right) = \rho_1 \otimes \rho_2.$$

where clearly, $\rho_1 \in \mathcal{S}(K_1)$, $\rho_2 \in \mathcal{S}(K_2)$. Conversely, if ρ is factorized then $\rho = \rho_1 \otimes \rho_2$ where $\rho_1 = \sum \alpha_i P_{\phi_i}$ and $\rho_2 = \sum \beta_j P_{\psi_j}$ are spectral decompositions. Hence,

$$\rho = \rho_1 \otimes \rho_2 = \left(\sum \alpha_i P_{\phi_i}\right) \otimes \left(\sum \beta_j P_{\psi_j}\right) = \sum \alpha_i \beta_j P_{\phi_i \otimes \psi_j}.$$

Letting $\lambda_{ij} = \alpha_i \beta_j$we have that ρ is separable and $\{\lambda_{ij}\}$ is a product measure.
(b) If $\{\lambda_{ij}\}$ is a product measure, then $\lambda_{ij} = \alpha_i \beta_j$, $\alpha_i, \beta_j > 0$ and $\sum \alpha_i = \sum \beta_j = 1$. Hence, for all i, j we have that

$$\left(\sum_j \lambda_{ij}\right)\left(\sum_i \lambda_{ij}\right) = \left(\sum_j \alpha_i \beta_j\right)\left(\sum_i \alpha_i \beta_j\right) = \alpha_i \beta_j = \lambda_{ij}.$$

Conversely, if the condition in (b) holds, then letting $\alpha_i = \sum_j \lambda_{ij}$, $\beta_j = \sum_i \lambda_{ij}$ we have that $\alpha_i, \beta_j > 0$, $\sum_i \alpha_i = \sum_{i,j} \lambda_{ij} = 1$ and $\sum_j \beta_j = \sum_{i,j} \lambda_{ij} = 1$ and $\lambda_{ij} = \alpha_i \beta_j$. □

Example 5. Let $H = \mathbb{C}^2 \otimes \mathbb{C}^2$ and let $\{\phi_1, \phi_2\}$ be an orthonormal basis for $\mathbb{C}^2$. Define $\phi = \frac{1}{\sqrt{2}}(\phi_1 + \phi_2)$ and consider the separable state

$$\rho = \tfrac{1}{2} |\phi \otimes \phi\rangle\langle\phi \otimes \phi| + \tfrac{1}{2} |\phi_1 \otimes \phi_1\rangle\langle\phi_1 \otimes \phi_1|.$$

The eigenvalues of ρ are 0 (multiplicity 2), 3/4 and 1/4. The eigenvectors for 3/4 and 1/4 are

$$\psi_1 = \tfrac{1}{2}\left[(\phi_2 - \phi_1) \otimes \phi_1 + (\phi_1 + \phi_2) \otimes \phi_2\right]$$
$$\psi_2 = \tfrac{1}{2\sqrt{3}}\left[(3\phi_1 + \phi_2) \otimes \phi_1 + (\phi_1 + \phi_2) \otimes \phi_2\right].$$

which were considered in Examples 3 and 4. We showed in these previous examples that ψ_1 and ψ_2 are entangled. This illustrates the interesting fact that the entanglement of a mixed state is not related to the entanglement of its eigenvectors [2]. □

Let $\rho = \sum \lambda_i P_i$, $\lambda_i > 0$, $\sum \lambda_i = 1$ be a decomposition of ρ into pure states P_i. We denote this decomposition by $\mathcal{A} = \{\lambda_i P_i\}$ and let Γ_ρ be the set of all such decompositions. We then let $e_{\mathcal{A}}(\rho) = \sum \lambda_i e(P_i)$ and define the *entanglement number* of ρ to be

$$e(\rho) = \inf_{\mathcal{A} \in \Gamma_\rho} \left[e_{\mathcal{A}}(\rho) \right]. \tag{3.1}$$

Since a pure state has the unique nontrivial decomposition $P = P$, (3.1) reduces to the usual definition of entanglement number for pure states. We say that the infimum in (3.1) is *attained* if there exists an $\mathcal{A} \in \Gamma_\rho$ such that $e(\rho) = e_{\mathcal{A}}(\rho)$. We could produce a complete proof of the next result as we did in Theorem 2.2, but this would be rather long and involved so we just give a sketch.

Theorem 3.2. *$e(\rho)$ is attained for every $\rho \in \mathcal{S}(H)$.*

Sketch of Proof. Fix $\rho \in \mathcal{S}(H)$ and for $\{\lambda_i P_i\}, \{\mu_i Q_i\} \in \Gamma_\rho$ define

$$\sigma\left(\{\lambda_i P_i\}, \{\mu_i Q_i\}\right) = \min\left\{\sum \left|\left|\lambda_{i_j} P_{i_j} - \mu_{i_k} Q_{i_k}\right|\right|\right\},$$

where the minimum is taken over all orderings of the indices in $\{\lambda_i P_i\}$ and $\{\mu_i Q_i\}$. We first show that σ is a metric on Γ_ρ. Clearly, $\sigma(x,x) \geq 0$, $\sigma(x,y) = 0$ if and only if $x = y$ and $\sigma(x,y) = \sigma(y,x)$. To verify the triangle inequality, let $\{\lambda_i P_i\}, \{\mu_i Q_i\}, \{\delta_i R_i\} \in \Gamma_\rho$. Then there exist indices such that

$$\begin{aligned} &\sigma\left(\{\lambda_i P_i\}, \{\delta_i R_i\}\right) + \sigma\left(\{\delta_i R_i\}, \{\mu_i Q_i\}\right) \\ &\qquad = \sum \left|\left|\lambda_{i_j} P_{i_j} - \delta_{i_r} R_{i_r}\right|\right| + \sum \left|\left|\delta_{i_s} R_{i_s} - \mu_{i_t} Q_{i_t}\right|\right|. \end{aligned} \tag{3.2}$$

Consider a term $\left|\left|\lambda_{i_j} P_{i_j} - \delta_{i_r} R_{i_r}\right|\right|$ in the left summation of (3.2). Then $\left|\left|\delta_{i_r} R_{i_r} - \mu_{i_k} Q_{i_k}\right|\right|$ appears in a term in the right summation of (3.2). Since

$$\left|\left|\lambda_{i_j} P_{i_j} - \delta_{i_r} R_{i_r}\right|\right| + \left|\left|\delta_{i_r} R_{i_r} - \mu_{i_k} Q_{i_k}\right|\right| \geq \left|\left|\lambda_{i_j} P_{i_j} - \mu_{i_k} Q_{i_k}\right|\right|,$$

we conclude that

$$\begin{aligned} \sigma\left(\{\lambda_i P_i\}, \{\delta_i R_i\}\right) + \sigma\left(\{\delta_i R_i\}, \{\mu_i Q_i\}\right) &\geq \sum \left|\left|\lambda_{i_j} P_{i_j} - \mu_{i_k} Q_{i_k}\right|\right| \\ &\geq \min\left\{\sum \left|\left|\lambda_{i_j} P_{i_j} - \mu_{i_k} Q_{i_k}\right|\right|\right\} = \sigma\left(\{\lambda_i P_i\}, \{\mu_i Q_i\}\right). \end{aligned}$$

It follows that σ is a metric on Γ_ρ. Now suppose that $\lambda, \mu \in (0,1)$ and $\lambda P_\phi \approx \mu P_\psi$ in the operator norm topology (where $x \approx y$ means that x and y are within a multiple of ε in the appropriate metric topology). Then

$$\left|\lambda - \mu \left|\langle \psi, \phi \rangle\right|^2\right| = \left|\langle (\lambda P_\phi - \mu P_\psi)\phi, \phi \rangle\right| \approx 0,$$

so that $\lambda \approx \mu \left|\langle \psi, \phi \rangle\right|^2$. In a similar way, we have that $\mu \approx \lambda \left|\langle \psi, \phi \rangle\right|^2$. Hence, $\lambda \approx \lambda \left|\langle \psi, \phi \rangle\right|^4$ and $\left|\langle \psi, \phi \rangle\right| \approx 1$. Now there exist $\theta \in [0, 2\pi)$ such that

$$\langle \psi, \phi \rangle = \left|\langle \psi, \phi \rangle\right| e^{i\theta} \approx e^{i\theta}.$$

Letting $\phi' = e^{-i\theta}\phi$ we have that

$$\mathrm{Re}\left\langle \psi, \phi' \right\rangle = \mathrm{Re}\, e^{-i\theta} \langle \psi, \phi \rangle \approx 1.$$

Therefore
$$||\psi-\phi'||^2=2\left(1-\operatorname{Re}\langle\psi,\phi'\rangle\right)\approx 0,$$
and we conclude that $\psi\approx\phi'$. Hence, $\lambda\approx\mu$ and of course, $P_\phi=P_{\phi'}\approx P_\psi$. We now show that $e(\rho)$ is continuous relative to the metric σ. If $\sigma\left(\{\lambda_iP_i\},\{\mu_iQ_i\}\right)\approx 0$, then reordering the indices if necessary, we have that
$$\sum||\lambda_iP_i-\mu_iQ_i||\approx 0.$$
Hence, $\lambda_iP_i\approx\mu_iP_i$ and by our previous discussion, we have that $\lambda_i\approx\mu_i$ and $P_i\approx Q_i$. Applying Theorem 2.2 we obtain $e(P_i)\approx e(Q_i)$. Defining $e\left(\{\lambda_iP_i\}\right)=\sum\lambda_ie(P_i)$ we conclude that
$$e\left(\{\lambda_iP_i\}\right)=\sum\lambda_ie(P_i)\approx\sum\mu_ie(Q_i)=e\left(\{\mu_iQ_i\}\right).$$

Thus, e is a continuous function on the metric space Γ_ρ. Now Γ_ρ can be considered as a closed, bounded (and, hence, compact) subset of a finite-dimensional real linear metric space. Since e is continuous on Γ_ρ, e attains a minimum on Γ_ρ. □

The next result improves a theorem in [11].

Corollary 3.3. *$e\in\mathcal{S}(H)$ is separable if and only if $e(\rho)=0$.*

Proof. If ρ is separable, then ρ has a decomposition $\rho=\sum\lambda_iP_i$ where the pure states P_i are factorized. Then for $\mathcal{A}=\{\lambda_iP_i\}\in\Gamma_\rho$ we have that $e(P_i)=0$ so that
$$e_{\mathcal{A}}(\rho)=\sum\lambda_ie(P_i)=0.$$

Hence, $e(\rho)=e_{\mathcal{A}}(\rho)=0$. Conversely, suppose that $e(\rho)=0$. By Theorem 3.2, $e(\rho)$ is attained. Thus, there exists $\{\lambda_iP_i\}\in\Gamma_\rho$ such that
$$\sum\lambda_ie(P_i)=e_{\mathcal{A}}(\rho)=e(\rho)=0.$$

Therefore, $e(P_i)=0$ so P_i is factorized for every i. Thus, $\rho=\sum\lambda_iP_i$ is a decomposition of ρ where P_i is factorized. We conclude that ρ is separable. □

We conjecture that $e(\rho)$ is continuous in the operator norm topology, but this remains an open problem.

Corollary 3.4. *The function $e\colon\mathcal{S}(H)\to[0,1)$ is convex.*

Proof. Suppose $\rho_1,\rho_2\in\mathcal{S}(H)$, $\lambda\in[1,0]$ and $\rho=\lambda\rho_1+(1-\lambda)\rho_2$. By Theorem 3.2 there exist decompositions $\rho_1=\sum a_iP_i$, $\rho_2=\sum b_iQ_i$ such that $e(\rho_1)=\sum a_ie(P_i)$ and $e(\rho_2)=\sum b_ie(Q_i)$. Now
$$\rho=\sum\left[\lambda a_iP_i+(1-\lambda)b_iQ_i\right],$$
and we denote this decomposition by $\mathcal{A}$. We then have that
$$\begin{aligned}e\left(\lambda\rho_1+(1-\lambda)\rho_2\right)=e(\rho)\le e_{\mathcal{A}}(\rho)&=\lambda\sum a_ie(P_i)+(1-\lambda)\sum b_ie(Q_i)\\&=\lambda e(\rho_1)+(1-\lambda)e(\rho_2),\end{aligned}$$

so that e is convex. □

4. Spectral entanglement

The next example illustrates the difficulty of finding the entanglement number for an arbitrary $\rho \in \mathcal{S}(H)$.

Example 6. Let $\{\phi_1, \phi_2\}$ be an orthonormal basis for $\mathbb{C}^2$ and let $\rho \in \mathcal{S}(\mathbb{C}^2 \otimes \mathbb{C}^2)$ be given by

$$\rho = \tfrac{1}{2} P_\alpha + \tfrac{1}{2} P_\beta, \tag{4.1}$$

where $\alpha = \frac{1}{\sqrt{2}}\phi_1 \otimes \phi_1 + \frac{1}{\sqrt{2}}\phi_2 \otimes \phi_2$ and $\beta = \phi_1 \otimes \phi_1$. It is clear that ρ is entangled so $e(\rho) \neq 0$. Now $e(\beta) = 0$ and $e(\alpha) = 1/\sqrt{2}$. Letting $\mathcal{A} = \left\{\frac{1}{2} P_\alpha, \frac{1}{2} P_\beta\right\}$ we have that

$$e_{\mathcal{A}}(\rho) = \tfrac{1}{2} e(P_\alpha) + \tfrac{1}{2} e(P_\beta) = \tfrac{1}{2\sqrt{2}}.$$

Now the eigenvalues of ρ are 0 (multiplicity 2), $\mu_1 = \frac{1}{2}\left(1 + \frac{1}{\sqrt{2}}\right)$ and $\mu_2 = \frac{1}{2}\left(1 - \frac{1}{\sqrt{2}}\right)$. The corresponding eigenvectors for μ_1, μ_2, respectively are

$$v_1 = \frac{1}{\sqrt{4+2\sqrt{2}}}\left[\left(1+\sqrt{2}\right)\phi_1 \otimes \phi_1 + \phi_2 \otimes \phi_2\right]$$
$$v_2 = \frac{1}{\sqrt{4-2\sqrt{2}}}\left[\left(\sqrt{2}-1\right)(-\phi_1) \otimes \phi_1 + \phi_2 \otimes \phi_2\right].$$

The spectral decomposition of ρ is $\rho = \mu_1 P_{v_1} + \mu_2 P_{v_2}$. Letting $\mathcal{B} = \{\mu_1 P_{v_1}, \mu_2 P_{v_2}\}$ we have that

$$e_{\mathcal{B}}(\rho) = \mu_1 e(v_1) + \mu_2 e(v_2).$$

The Schmidt coefficients of v_1 are $\lambda_1 = \frac{2+\sqrt{2}}{4}$, $\lambda_2 = \frac{2-\sqrt{2}}{4}$. Hence,

$$e(v_1) = \left[1 - \left(\lambda_1^2 + \lambda_2^2\right)\right]^{1/2} = \tfrac{1}{2}.$$

Similarly, $e(v_2) = 1/2$ and we obtain

$$e_{\mathcal{B}}(\rho) = \tfrac{1}{2}\mu_1 + \tfrac{1}{2}\mu_2 = \tfrac{1}{2}.$$

We conclude that $e_{\mathcal{A}}(\rho) < e_{\mathcal{B}}(\rho)$ and $e(\rho) \leq e_{\mathcal{A}}(\rho) = 1/2\sqrt{2}$. However, it is not clear how to find $e(\rho)$. □

We have seen that it can be difficult to find $e(\rho)$ precisely. We now consider a weaker type of entanglement (or stronger type of separability) for which an entanglement measure is easier to compute.

A state $\rho \in \mathcal{S}(K_1 \otimes K_2)$ is *spectrally separable* if its nonzero eigenvalues (including multiplicity) have factorized eigenvectors. If the multiplicity of λ is greater than one, then λ may also have entangled eigenvectors. The important point is that if λ has multiplicity m, then there exists an orthonormal set of m factorized eigenvectors with corresponding eigenvalue λ. If ρ is spectrally separable, then the spectral decomposition of ρ has the form

$$\rho = \sum \lambda_{ij} P_{\phi_i \otimes \psi_j}, \tag{4.2}$$

where

$$\phi_i \otimes \psi_j \perp \phi_r \otimes \psi_s \quad \text{if } (i,j) \neq (r,s). \tag{4.3}$$

Since (4.2) characterizes separable states, it follows that ρ is separable. However, a general separable state need not satisfy the orthogonality condition (4.3). In summary, a spectrally

Journal of Physics: Conference Series **1638** (2020) 012012 doi:10.1088/1742-6596/1638/1/012012

separable state is separable. However, as shown in Example 5, a separable state need not be spectrally separable. If a state is not spectrally separable it is *spectrally entangled.* We conclude that if ρ is entangled, then ρ is spectrally entangled but the converse need not hold. Now the physical properties of a state are determined by its eigenvalues and eigenvectors. For this reason, we see that spectral entanglement is a more physical concept than entanglement. Moreover, as we shall see, spectral entanglement has a more manageable measure than entanglement.

Lemma 4.1. *A state $\rho \in \mathcal{S}(K_1 \otimes K_2)$ is spectrally separable if and only if there exist orthonormal bases $\{\phi_i\}$, $\{\psi_i\}$ for K_1, K_2, respectively, such that*

$$\rho = \sum \lambda_{ij} P_{\phi_i \otimes \psi_j}. \tag{4.4}$$

In this case, λ_{ij} are eigenvalues of ρ with corresponding eigenvectors $\phi_i \otimes \psi_j$.

Proof. If (4.4) holds, then $\phi_i \otimes \psi_j$ are factorized eigenvectors of ρ so ρ is spectrally separable. Conversely, suppose that ρ is spectrally separable. Then ρ has the form (4.4) where ϕ_i, ψ_j are unit vectors that need not form bases but $\phi_i \otimes \psi_j \perp \phi_r \otimes \psi_s$ when $(i,j) \neq (r,s)$. We can extend the sets $\{\phi_i\}$, $\{\psi_j\}$ to obtain

$$I = \sum_{i,j} P_{\phi_i \otimes \psi_j} = \sum_{i,j} P_{\phi_i} \otimes P_{\psi_j} = \left(\sum_i P_{\phi_i}\right) \otimes \left(\sum_j P_{\psi_j}\right).$$

This implies that $\sum P_{\phi_i} = I_1$, $\sum_{\psi_j} = I_2$ where I_1, I_2 are the identity operators on K_1, K_2, respectively. But then $\{\phi_i\}$, $\{\psi_j\}$ are orthonormal bases for K_1, K_2, respectively. □

Notice that if ρ is factorized, then $\rho = \rho_1 \otimes \rho_2$ and the eigenvectors of ρ have the form $\phi_i \otimes \psi_j$ where ϕ_i, ψ_j are eigenvectors of ρ_1, ρ_2, respectively. Hence, ρ is spectrally separable. Moreover, if ρ has the form (4.4), then ρ is factorized if and only if $\{\lambda_{ij}\}$ is a product measure.

If $\rho \in \mathcal{S}(K_1 \otimes K_2)$ has spectral decomposition $\rho = \sum \lambda_i P_{\phi_i}$, $\lambda_i > 0$, $\sum \lambda_i = 1$, we call $\mathcal{A} = \{(\lambda_i, P_{\phi_i})\}$ the set of *eigenpairs* for ρ and denote the set of eigenpairs by $\mathcal{E}_\rho$. We define the *spectral entanglement number* of ρ by

$$e_s(\rho) = \inf_{\mathcal{E}_\rho} \left[\sum \lambda_i e(\phi_i)\right].$$

We say that ρ is *nondegenerate* if its nonzero eigenvalues are distinct. If ρ is nondegenerate, then $\mathcal{E}_\rho$ is unique so the infimum is not needed. Of course, $e(\rho) \le e_s(\rho)$ and $e_s(\rho) = 0$ implies $e(\rho) = 0$. However, we can have $e(\rho) = 0$ and $e_s(\rho) > 0$.

Theorem 4.2. $e_s(\rho) = 0$ *if and only if ρ is spectrally separable.*

Proof. If ρ is spectrally separable then ρ has the form (4.2) so

$$e_s(\rho) = \sum \lambda_{ij} e(\phi_i \otimes \psi_j) = 0.$$

Conversely, suppose $e_s(\rho) = 0$. As in Theorem 3.2, e_s is attained. Hence, ρ has a spectral representation $\rho = \sum \lambda_i P_{\phi_i}$, $\lambda_i > 0$ with

$$\sum \lambda_i e(\phi_i) = e_s(\rho) = 0.$$

We conclude that $e(\phi_i) = 0$ so the eigenvectors ϕ_i are factorized. Therefore, ρ is spectrally separable. □

Example 7. In Example 5, we considered the separable state

$$\rho = \tfrac{3}{4}\,P_{\psi_1} + \tfrac{1}{4}\,P_{\psi_2}.$$

This state is not spectrally separable because the eigenvectors ψ_1 and ψ_2 are entangled and these are the only eigenvectors (up to a scalar multiple) for nonzero eigenvalues. The spectral entanglement number for ρ becomes

$$e_s(\rho) = \tfrac{3}{4}\,e(\psi_1) + \tfrac{1}{4}\,e(\psi_2).$$

We have found $e(\psi_1)$, $e(\psi_2)$ in Examples 3 and 4 to obtain

$$e_s(\rho) = \tfrac{3}{4}\tfrac{\sqrt{2}}{2} + \tfrac{1}{2}\tfrac{\sqrt{6}}{2} = \tfrac{5}{12}\sqrt{2}.$$

Of course, $e(\rho) = 0$. □

Let $H = K_1 \otimes K_2$ and let $\{\phi_i\}$, $\{\psi_i\}$ be orthonormal bases for K_1, K_2, respectively. A factorized vector state has the form

$$\eta = \phi \otimes \psi = \left(\sum c_j\phi_j\right) \otimes \left(\sum d_k\psi_k\right) = \sum c_jd_k\phi_j \otimes \psi_k.$$

If ϕ is not a scalar multiple of a ϕ_j and ψ is not a scalar multiple of a ψ_k, then at least two of the c_j and at least two of d_k are nonzero. Hence, $c_jd_k \neq 0$ for $j \neq k$. Hence, any unit vector of the form $\eta = \sum a_j\phi_j \otimes \psi_j$ where $a_j \neq 0$ for at least two j is entangled. This can also be seen by writing $a_j = |a_j|\, e^{i\theta j}$ and hence

$$\eta = \sum |a_j|\,(e^{i\theta j}\phi_i) \otimes \psi_j$$

gives a Schmidt decomposition of η. We also conclude that

$$e(\eta) = \left(1 - \sum |a_i|^4\right)^{1/2}.$$

The next example is based on an example in [2].

Example 8. Let $\{\phi_1, \phi_2\}$ be an orthonormal basis for $\mathbb{C}^2$ and let $a, b \in \mathbb{C}$ with $|a|^2 + |b|^2 = 1$, $a, b \neq 0$. We conclude from our previous discussion that the vector states in $H = \mathbb{C}^2 \otimes \mathbb{C}^2$ given by

$$\eta_\pm = a\phi_1 \otimes \phi_1 \pm b\phi_2 \otimes \phi_2$$

are entangled with $e(\eta_\pm) = \left[1 - |a|^4 - |b|^4\right]^{1/2}$. Notice that $\eta_+ \perp \eta_-$ if and only if $|a|^2 = |b|^2$ which is equivalent to $|a| = 1/\sqrt{2}$. The mixed state

$$\begin{aligned} T &= \tfrac{1}{2}\,|\eta_+\rangle\langle\eta_+| + \tfrac{1}{2}\,|\eta_-\rangle\langle\eta_-| \\ &= |a|^2\,|\phi_1 \otimes \phi_1\rangle\langle\phi_1 \otimes \phi_1| + |b|^2\,|\phi_2 \otimes \phi_2\rangle\langle\phi_2 \otimes \phi_2| \end{aligned}$$

is spectrally separable and hence separable and, of course, $e(T) = e_s(T) = 0$. Moreover, $\phi_1 \otimes \phi_1$, $\phi_2 \otimes \phi_2$ are factorized eigenvectors of T with corresponding eigenvalues, $|a|^2$, $|b|^2$, respectively. If $|a| = |b| = 1/\sqrt{2}$, then $\eta_\pm$ are also eigenvectors of T that are entangled with $e(\eta_\pm) = 1/\sqrt{2}$. However, if $\lambda \neq 1/2$, then

$$T_\lambda = \lambda|\eta_+\rangle\langle\eta_+| + (1 - \lambda)|\eta_-\rangle\langle\eta_-| \tag{4.5}$$

are entangled and hence spectrally entangled.
We now compute $e_s(T_\lambda)$. Expanding (4.5) gives

$$\begin{aligned} T_\lambda &= \lambda|a\phi_1\otimes\phi_1 + b\phi_2\otimes\phi_2\rangle\langle a\phi_1\otimes\phi_1 + b\phi_2\otimes\phi_2| \\ &\quad + (1-\lambda)|a\phi_1\otimes\phi_1 - b\phi_2\otimes\phi_2\rangle\langle a\phi_1\otimes\phi_1 - b\phi_2\otimes\phi_2| \\ &= |a|^2 P_{\phi_1\otimes\phi_1} + |b|^2 P_{\phi_2\otimes\phi_2} + (2\lambda-1)\left[a\bar{b}|\phi_1\otimes\phi_1\rangle\langle\phi_2\otimes\phi_2| + a\bar{b}|\phi_2\otimes\phi_2\rangle\langle\phi_1\otimes\phi_1|\right]. \end{aligned}$$

We can represent T_λ by the matrix

$$T_\lambda = \begin{bmatrix} |a|^2 & 0 & 0 & (2\lambda-1)a\bar{b} \\ 0 & 0 & 0 & 0 \\ 0 & 0 & 0 & 0 \\ (2\lambda-1)\bar{a}b & 0 & 0 & |b|^2 \end{bmatrix}.$$

The eigenvalues of T_λ are

$$\mu_\pm = \frac{1 \pm \sqrt{1-16\lambda(1-\lambda)\,|ab|^2}}{2}. \tag{4.6}$$

Letting

$$c_\pm = \frac{\mu_\pm - |a|^2}{(2\lambda-1)ab}$$

and $N_\pm = \sqrt{1+|C_\pm|^2}$ the corresponding normalized eigenvectors become

$$\psi_\pm = \tfrac{1}{N_\pm}\,(\phi_1\otimes\phi_1 + c_\pm\phi_2\otimes\phi_2).$$

This gives a Schmidt decomposition

$$\psi_\pm = \tfrac{1}{N_\pm}\,\phi_1\otimes\phi_1 + \tfrac{|c_\pm|}{N_\pm}\,(e^{i\theta_\pm}\phi_2)\otimes\phi_2,$$

where $c_\pm = |c_\pm|\,e^{i\theta_\pm}$. We conclude that

$$e(\psi_\pm) = \left[1 - \frac{1+|c_\pm|^4}{\left(1+|c_\pm|^2\right)^2}\right]^{1/2} = \frac{\sqrt{2}\,|c_\pm|}{1+|c_\pm|^2}. \tag{4.7}$$

Hence,

$$e_s(T_\lambda) = \mu_+ e(\psi_+) + \mu_- e(\psi_-),$$

where the values of $\mu_\pm$, $e(\psi_\pm)$ are given by (4.6), (4.7), respectively. For example, suppose $a = b = 1/\sqrt{2}$. We then have that

$$\mu_\pm = \frac{1 \pm \sqrt{1-4\lambda(1-\lambda)}}{2}.$$

Hence,

$$e_s(T_\lambda) = e(\psi_+) = e(\psi_-) = \frac{|2\lambda-1|}{\sqrt{2}\,\sqrt{1-4\lambda(1-\lambda)}}.$$

In particular, if $\lambda = 1/4$, then $\mu_+ = 3/4$, $\mu_- = 1/4$ and $e_s(T_\lambda) = 1/\sqrt{2}$. We have not found $e(T_\lambda)$ and this illustrates why $e_s(T_\lambda)$ is easier to compute than $e(T_\lambda)$. Notice that $\lim_{\lambda\to 1/2} T_\lambda = T$ in the operator norm topology. By L'Hospital's Rule we have that $\lim_{\lambda\to 1/2} e_s(T_\lambda) = 0 = e_s(T)$ so in this case, e_s is continuous in the operator norm topology. We now indicate why this holds in general. □

Theorem 4.3. *e_s is continuous on $\mathcal{S}(H)$ in the operator norm topology.*

Sketch of Proof. If $\rho_1 \approx \rho_2$, then it is well known that the eigenvalues and eigenvectors of ρ_1 and ρ_2 are close. As in Theorem 3.2, $e_s(\rho_1)$ and $e_s(\rho_2)$ are attained so there exist spectral decompositions $\rho_1 = \sum \lambda_i P_{\phi_i}$ and $\rho_2 = \sum \mu_i P_{\psi_i}$ such that $e_s(\rho_1) = \sum \lambda_i e(\phi_i)$ and $e_s(\rho_2) = \sum \mu_i \phi(\psi_i)$. We can reorder the elements of the second sum so that $\lambda_i \approx \mu_i$ and $\phi_i \approx \psi_i$. Since by Theorem 2.2, $\psi \mapsto e(\psi)$ is continuous in the norm topology, we have that $e(\phi_i) \approx e(\psi_i)$. We conclude that $e_s(\rho_1) \approx e_s(\rho_2)$ and hence, e_s is continuous. □

Example 6 can be employed to show that e_s is not a convex function, in general. Consider the state $\rho = \frac{1}{2} P_\alpha + \frac{1}{2} P_\beta$ of Example 6. In that example we showed that

$$e_s(\rho) = e_{\mathcal{B}}(\rho) = 1/2.$$

We also showed that $e_s(P_\alpha) = e_s(\alpha) = 1/\sqrt{2}$ and $e_s(P_\beta) = 0$. But then

$$e_s\left(\tfrac{1}{2} P_\alpha + \tfrac{1}{2} P_\beta\right) = e_s(\rho) = \tfrac{1}{2} > \tfrac{1}{2\sqrt{2}} = \tfrac{1}{2} e_s(P_\alpha) + \tfrac{1}{2} e_s(P_\beta),$$

so e_s is not convex.

An open problem is to characterize the $\rho \in \mathcal{S}(H)$ for which $e(\rho) = e_s(\rho)$. One possibility is that $e(\rho) = e_s(\rho)$ if and only if ρ is spectrally separable in which case $e_s(\rho) = 0$. However, it is not at all clear whether this holds.

References

[1] Busch P 2003 The role of entanglement in quantum measurement and information processing *Int. J. Theor. Phys.* **42** 937–41
[2] Heinosaari T and Ziman M 2012 *The Mathematical Language of Quantum Theory* (Cambridge: Cambridge University Press)
[3] Nielson K and Chuang I 2000 *Quantum Computation and Quantum Information* (Cambridge: Cambridge University Press)
[4] Ziman M and Bužek V 2006 Entanglement-induced state ordering under local operations *Phys. Rev. A* **73** 012312
[5] Horodecki R, Horodecki P, Horodecki M and Horodecki K 2009 Quantum entanglement *Rev. Mod. Phys.* **81** 865–942
[6] Plenio M and Virmani S 2007 An introduction to entanglement measures *Quant. Inf. Comp.* **7** 1–15
[7] Steiner M 2003 Generalized robustness of entanglement *Phys. Rev.* A **67** 054305
[8] Vedral V and Plenio M 1998 Entanglement measures and purification procedures *Phys. Rev.* A **57** 1619–33
[9] Vedral V and Tarrach R 1999 Robustness of entanglement *Phys. Rev.* A **59** 141–55
[10] Wei T and Goldbart P 2003 Geometric measure of entanglement and applications to bipartite and multipartite quantum states *Phys. Rev.* A **68** 042307
[11] Gudder S 2019 A theory of entanglement *Preprint* arXiv:1904.06589 [quant-ph]

Comparison of entangling protocols in ABC-type spin chains

Marta P Estarellas[1], Timothy P Spiller[2] and Irene D'Amico[2]

[1] National Institute of Informatics, 2-1-2 Hitotsubashi, Chiyoda-ku, Tokyo 101-8430, Japan
[2] Department of Physics, University of York, YO10 5DD York, United Kingdom

E-mail: mpestarellas@nii.ac.jp, timothy.spiller@york.ac.uk, irene.damico@york.ac.uk

Abstract. In this contribution we consider an advantageous building block with potential for various quantum applications: a device based on coupled spins capable of generating and sharing out an entangled pair of qubits. Our model device is a dimerized spin chain with three weakly coupled embedded sites (defects). Three different entangling protocols were proposed for this chain in [1] and [2], one producing a Cluster state building block and two generating a Bell state, depending on the initial state injection. Here we compare the robustness of such protocols as quantum entangling gates against different types of fabrication (static energy fluctuations) and operation (timing injection delays) errors.

1. Introduction

Paul Busch was a real aficionado of the foundations and fundamentals of quantum theory. We all miss his deep thinking and wise words. Over recent decades, the work of Paul and many other quantum gurus has not only broadened our understanding and perspectives of quantum theory, but it has also opened up new potential for quantum applications in the real world. New quantum technologies are emerging, where the most fundamental features of quantum physics play centre stage, and provide advantage over current conventional technologies [3, 4]. Our contribution to this collection is about one such application.

Entanglement is perhaps the most fundamental "non-classical" feature of quantum physics. Entanglement underpins numerous new quantum applications. It is the resource for teleportation [5]; it powers measurement-based quantum processing and computing [6, 7]; it can be used for secure communications [8]; and it can be employed to sense and measure things better than we can do with conventional instruments [9, 10]. Generation and distribution of entanglement is therefore a very useful primitive, that can facilitate a whole range of quantum applications.

In this work we consider a model device, based on coupled spin-1/2 systems—or qubits, the building blocks of many new quantum technologies. One-dimensional systems of coupled spins, usually called chains, form versatile quantum devices. Spin chains can be employed to efficiently transport quantum information [11, 12], even perfectly if the couplings between the spins can be tuned [13–16], and such systems can be used to create, distribute and store entanglement [17, 18].

A very appealing reason for studying spin chains theoretically is that these investigations can apply to a range of physical systems, because qubits and their coupling mechanisms take many different forms. In the laboratory, the spins in a spin chain could be actual spins in a chain molecule, or nanoscale magnetic particles [19], or spins in a string of coupled fullerenes [20]. Or they could be atoms held in optical lattices [21, 22], ions in an ion trap [23, 24], or electrons or excitons confined in coupled nanostructures [25–27]. Optical realisations, through coupled micro-pillars [28] or waveguides [29], and graphene-based systems [30] are also relevant. Thus spin chain modelling has applicability across a broad spectrum of qubit systems, and thus potential future technology platforms.

2. ABC-type spin chains and entanglement generation protocols

The model system we consider for our entanglement generation protocols is a spin chain with alternating weak (δ) and strong (Δ) couplings, distributed such that there are three -defect-sites (labelled A, B and C) weakly coupled to the rest of the chain as shown in Fig. 1. For protocol demonstration, here we focus on a spin chain configuration with $N = 7$ sites. However, longer chains could be considered for implementation of the same protocols [1, 2].

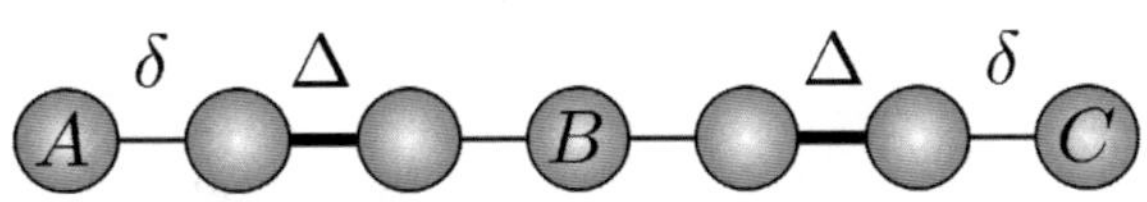

Figure 1: ABC-type spin chain configuration.

This spin chain can be described by the following time-independent Hamiltonian

$$\mathcal{H} = \sum_{i=1}^{N} \epsilon_i |1\rangle\langle 1|_i + \sum_{i=1}^{N-1} J_{i,i+1}[|1\rangle\langle 0|_i \otimes |0\rangle\langle 1|_{i+1} + \text{h.c.}], \tag{1}$$

with the coupling $J_{i,i+1}$ equal to either Δ or δ depending on the site (see Fig. 1) and the on-site energy $\epsilon_i = 0$ unless stated otherwise (i.e. when diagonal random disorder is added). For each spin qubit, the standard computational basis states are represented by spin up, $|1\rangle$, and spin down, $|0\rangle$, with the zero (excitation) state of the system being all spins down. We note that this zero state could be prepared as the ground state of the system (brought into contact with a dissipative environment at temperature T) in a suitable external magnetic field that makes all the on-site energies sufficiently positive, so $\epsilon_i \gg kT$ for all i. Alternatively, it could be prepared through projective measurement of every spin, with local spin flips applied to each that is measured up, although this second method would clearly require local addressing and measurements at each site. We note also that the Hamiltonian (1) preserves excitation (spin up) number, so if amplitudes of different excitation (spin up) numbers are prepared from the zero state, these then evolve separately according to the dynamics for that excitation number sector.

It has been demonstrated [31–34] that related dimerized chains have high fidelity quantum state transfer (QST) properties. In general QST properties in spin chains can be exploited to generate entanglement, for example as discussed in [37–41]. This is something that we exploit in our protocols in order to generate the desired entangled states in dimerized chains [1, 2]. The operation of these spin chain devices relies on their natural dynamics. A simple initial

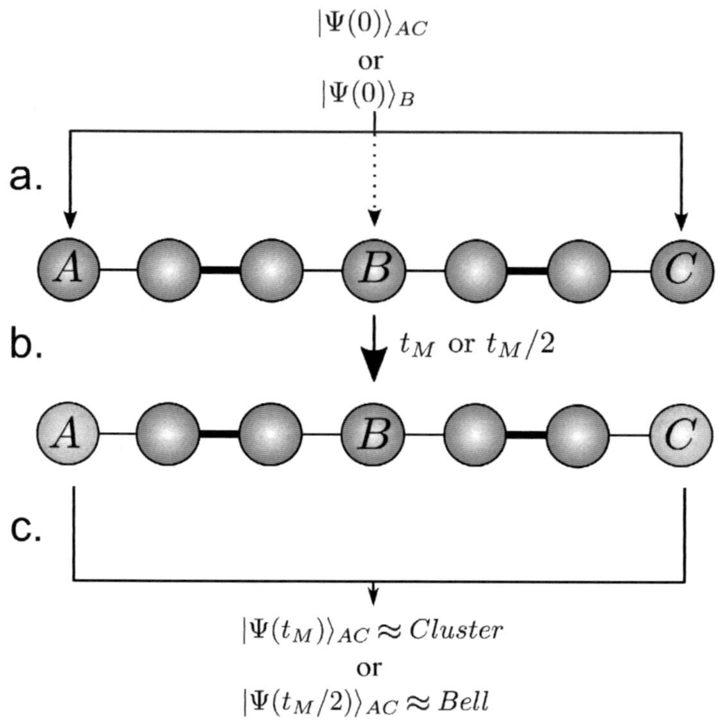

Figure 2: Entangling protocols: (a) Initial injection (red) at sites A and C (solid arrows) or at site B (dashed arrow); (b) Evolution of the system up to time $t_E \approx t_M$ or $t_E \approx t_M/2$; (c) Generation of a highly entangled Cluster state or Bell state, respectively, between sites A and C (green).

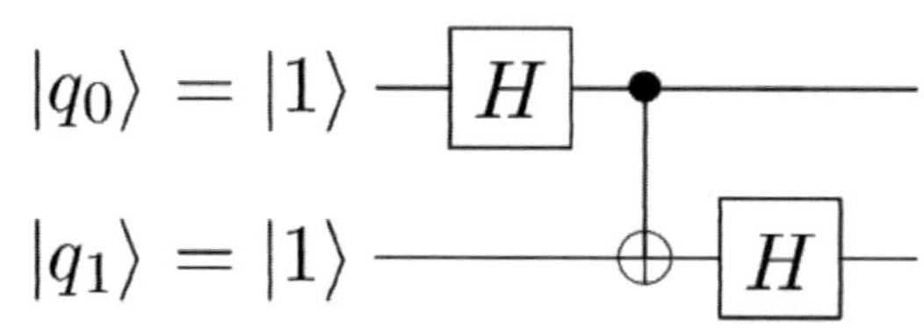

Figure 3: Cluster state creation written as a circuit. H is the single-qubit Hadamard gate and the two-qubit gate is a CNOT [3].

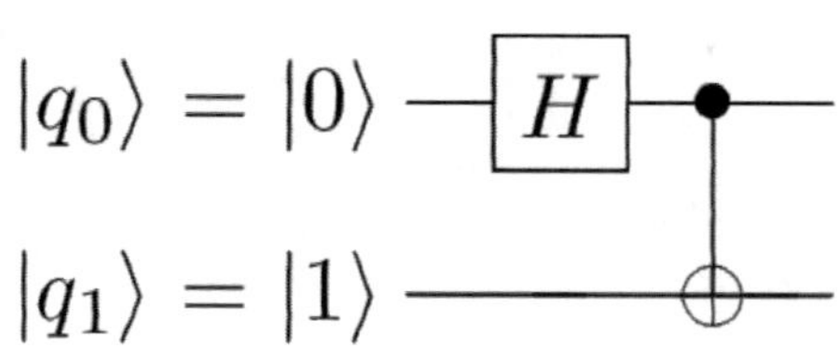

Figure 4: Bell state creation written as a circuit. H is the single-qubit Hadamard gate and the two-qubit gate is a CNOT [3].

state of the system is prepared and the protocol is then effected by evolution under $\mathcal{H}$ for some pre-determined time.

The entanglement generation protocols are schematically illustrated in Fig. 2. The initial injection, applied to the zero state of the system, defines the overall initial state of the chain, and hence the final entangled state. We note that injection is one of only two interactions the user has to make with the system. Initialization is done by either simultaneously injecting a double excitation state at sites A and C, or by a single excitation state injection at site B. Depending on the chosen initial state, the entangled state generated will either be approximated to a Cluster state building block (a superposition across the zero-, one- and two-excitation sectors) or a Bell state. The second user interaction with the system involves either retrieving the entangled state at a specific pre-determined time later in the dynamics, or performing a measurement. The Cluster state building block is so named because repeated application of such a protocol can be used to knit larger cluster states [42]. From now on, with this understood, we simply refer to this two-qubit entanglement example as a Cluster state.

Let us now look in more detail at the injection for the generation of this two-qubit Cluster state, denoted approach (i). In the ideal case, this is a maximally entangled state formed by an equal superposition of all the involved site basis vectors. Fig. 3 shows the circuit equivalent to an ideal Cluster state-generation gate. In this case, the protocol is initiated at $t = 0$ with the injection of two initial $|+\rangle = \frac{1}{\sqrt{2}}(|0\rangle + |1\rangle)$ states at the chain ends (sites A and C). We can write the initial state in the standard basis as follows:

$$\begin{aligned} \text{i) } |\Psi(0)\rangle_{(\text{i})} &= \frac{1}{2}\Big(|+\rangle_A \otimes |+\rangle_C\Big) \otimes |0\rangle_{\text{rest-of-chain}} \\ &= \frac{1}{2}\Big(|0\rangle_A|0\rangle_C + |1\rangle_A|0\rangle_C + |0\rangle_A|1\rangle_C + |1\rangle_A|1\rangle_C\Big) \otimes |0\rangle_{\text{rest-of-chain}}. \end{aligned} \tag{2}$$

We turn now to the injection protocols for Bell state creation. The circuit equivalent to the Bell state-generation gate is presented in Fig. 4. There are two approaches: (ii) the simultaneous injection of two excitations at sites A and C; or (iii) single-excitation injection at site B. These possibilities present different features and thus distinct advantages. The single injection at the centre, (iii), allows generation and distribution of a Bell state with the convenience and ease of having to initially interact with one site only (site B). Alternatively, some applications such as modular quantum processor proposals may require generation of this same state with simultaneous compliance and contribution of two separate parties or quantum registers [34]. Then the appropriate injection approach is (ii), which involves the two parties A and C, who will need to agree a priori initiation of the protocol. For Bell state creation, we can write the state of the chain at initial time, $t = 0$, as follows:

$$\text{ii) } |\Psi(0)\rangle_{(\text{ii})} = (|1\rangle_A \otimes |1\rangle_C) \otimes |0\rangle_{\text{rest-of-chain}}, \tag{3}$$

$$\text{iii) } |\Psi(0)\rangle_{(\text{iii})} = |1\rangle_B \otimes |0\rangle_{\text{rest-of-chain}}. \tag{4}$$

After initialization, the system is allowed to evolve under its natural dynamics to a given time, the 'entangling time', t_E, which depends on the version of the protocol used and the required state to be generated (Cluster or Bell state). For the Cluster state generation [1], t_E is the mirroring time ($t_E \approx t_M$), i.e. the time needed for an arbitrary initial state to evolve to its mirror position with respect to the centre of the chain. The reason an initial product state 'mirrors' to an entangled state is because of an additional fundamental sign change that arises for (just) the two-excitation amplitude [37–39]. This sign change, which effectively arises from the exchange of two fermions, underpins the generation of a Cluster state building block [42].

Journal of Physics: Conference Series **1638** (2020) 012013 doi:10.1088/1742-6596/1638/1/012013

For the Bell state generation [2], the 'entangling time' is approximately half the mirroring time ($t_E \approx t_M/2$). In case (iii), at this time a single excitation injected at the centre effectively delocalizes to a superposition of being at both ends of the chain [2, 41]. In case (ii), at this time the state of the chain becomes a product state of such a delocalized single excitation with a state of the rest of the chain containing the other excitation [2, 40].

In all cases, at t_E the system state becomes highly entangled between sites A and C, from which the two entangled qubits can be extracted, if desired[1]. In Table 1 we summarize the three different initial states we employ and their corresponding injection sites, as well as the generated entangled states and the associated entangling times, t_E.

Table 1: Initialization states (injection), their corresponding entangled states, and their corresponding entangling times t_E. The $|+\rangle$ state corresponds to $\frac{1}{\sqrt{2}}(|0\rangle + |1\rangle)$. For simplicity, only the states of the relevant sites (A, C or B) are presented and a compact notation is used.

	Initial state	**Entangled State**	$\mathbf{t_E}$
i	$|\Psi(0)\rangle_{AC} = (|+\rangle_A \otimes |+\rangle_C)$	$|\Psi(t_E)\rangle_{AC} \approx \frac{1}{2}(|00\rangle_{AC} - |01\rangle_{AC} - |10\rangle_{AC} - |11\rangle_{AC})$	t_M
ii	$|\Psi(0)\rangle_{AC} = |1\rangle_A \otimes |1\rangle_C$	$|\Psi(t_E)\rangle_{AC} \approx \frac{-i}{\sqrt{2}}(|10\rangle_{AC} + |01\rangle_{AC})$	$t_M/2$
iii	$|\Psi(0)\rangle_B = |1\rangle_B$	$|\Psi(t_E)\rangle_{AC} \approx \frac{-i}{\sqrt{2}}(|10\rangle_{AC} + |01\rangle_{AC})$	$t_M/2$

3. Comparison of the entangling protocols

The chain mirroring time t_M, and hence the entangling time t_E, can be estimated analytically in terms of the coupling ratio δ/Δ, and it decreases as this ratio increases [1, 2]. Intuitively this can be understood by considering the smaller δ-couplings to form 'bottle-necks' for the excitation dynamics, with respect to those of a uniform chain with all Δ-couplings. Of course, due to quantum interference, we cannot think of the excitation dynamics as a simple flow between contiguous sites, so increasing δ/Δ also affects the level of entanglement and in a non-monotonic way. Fig. 5 presents the behaviour of the three protocols for different coupling ratios. Here the exact entangling times, t_E, for each ratio are calculated numerically by determining when the entanglement peaks at a maximum for that ratio value. The entanglement is measured as entanglement of formation (EOF) [35] of the reduced state for the sites of interest in the chain, with $EOF \in [0, 1]$ and $EOF = 1$ indicating a maximally entangled state. This EOF therefore implicitly includes any damage due to the sites of interest having residual entanglement with other sites. The exact entangling times t_E are still close to t_M for the Cluster state and $t_M/2$ for the Bell state protocols. As would be expected, with an increasing value of the coupling ratio, δ/Δ, the entangling protocols are indeed faster, as observed in the inset of Fig. 5. However the corresponding value for the entanglement has an oscillatory behaviour, with well-defined peaks whose maxima decrease with increasing coupling ratio δ/Δ for two protocols out of three. The (ii) protocol, corresponding to creation of Bell states from two initial injections, indeed continues to achieve (almost) maximal entanglement at every peak.

Ideally, one would desire maximal entanglement at a short t_E (to optimize entanglement delivery in the presence of decoherence processes occurring in the device), as well as some level

[1] An alternative to this is to effectively freeze and then protectively store the state using a slightly modified system configuration and the protocol, see [1].

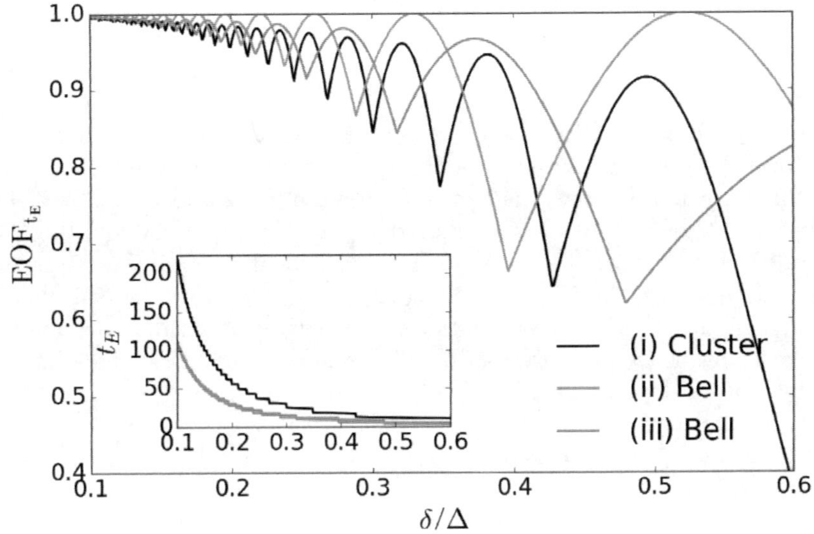

Figure 5: Entanglement of formation at t_E for the three different protocols, (i), (ii) and (iii), against varying coupling ratio, δ/Δ, for an ABC chain with $N = 7$. The inset shows how the entangling times, t_E, vary with the coupling ratio.

of robustness against static fabrication defects and/or injection errors. In the rest of this section, we will seek a compromise between the entangling times, t_E, the EOF of the state obtained at those times, and the robustness of the protocol.

Having already analysed the schemes and presented results for the ideal cases, we next turn to study the behaviour of the protocols against fabrication or injection errors with increasing coupling ratio. For these analyses we consider the overall protocols for a selection of three different coupling ratios, corresponding to different maxima of each profile of Fig. 5, as summarized in Table 2.

Table 2: Selected coupling ratios δ/Δ corresponding to some chosen maxima of the Fig. 5 profiles for protocols (i), (ii) and (iii).

i. Cluster state	**ii. Bell state**	**iii. Bell state**
0.205	0.260	0.204
0.382	0.330	0.280
0.490	0.523	0.373

We will test the protocols' robustness against static random perturbations of both on-site energies ϵ_i (diagonal disorder) and couplings $J_{i,i+1}$ (off-diagonal disorder) in the Hamiltonian (1). In addition, we will consider the effects of relative time delays between the injections of the two excitations in protocols (i) and (ii). Details of the methods and corresponding results are described below.

3.1. Diagonal and off-diagonal disorder

Local energy fluctuations may be induced by fabrication defects and/or local field fluctuations, linked to the specific physical realisation of the spin chain. We simulate static (compared to the timescale of the entanglement dynamics) energy fluctuations by adding random disorder to the diagonal terms of the Hamiltonian. Diagonal disorder has been shown to be one of the most damaging sources of static decoherence for the single excitation sector in spin chains [36].

The first term on the right hand side of the Hamiltonian (1) is set to have $\epsilon_i = Er_{i,i}\delta$, where $r_{i,i}$ is a random number from a uniform distribution between $-1/2$ and $1/2$, E is a dimensionless positive parameter that fixes the scale of the disorder, and δ is the weak coupling.

Our second approach to model fabrication errors and local defects is accounting for static (again, compared to the timescale of the entanglement dynamics) errors in the off-diagonal terms of the Hamiltonian (1). Such perturbations represent random disorder in the couplings of the chain, $J_{i,i+1}$. In order to do so we modify the second term on the right hand side of the Hamiltonian (1) by setting $J^{\text{eff}}_{i,i+1} = J_{i,i+1} + Er_{i,i+1}\delta$, where $r_{i,i+1}$ are again randomly generated from a uniform distribution.

For each type of disorder, each protocol, and each selected coupling ratio, we calculate 1000 random realisations and then average over the values of the EOF to estimate the robustness of the protocols. The various panels of Fig. 6 show the averaged EOF obtained at the unperturbed t_E, for each of the studied coupling ratios and against an increasing level of diagonal and off-diagonal disorder (up to 50% of the weak coupling), for protocols (i), (ii) and (iii). We observe that even for coupling ratios as high as $0.4 \sim 0.5$ our protocols are still considerably resilient against static perturbations. Importantly, for diagonal disorder—previously thought to be the most damaging—the performance of all protocols is dramatically improved by an increasing coupling ratio. So *in all three cases* (i), (ii) and (iii), for $\delta/\Delta > 0.37$ the averaged EOF at t_E is always close or even greater than 0.8, even with a diagonal disorder at 50% of the weak coupling value. Additionally, for both types of disorder and all protocols, as the coupling ratio is increased, each realisation for the EOF calculation deviates less from the averaged value, as shown by the narrower standard deviation shades from Fig. 6. All of these analyses offer clear evidence that our protocols can be optimized, in terms of making the entangling operation times faster without necessarily sacrificing the quality of the entangled state. Hence it is possible to offer optimal and useful EOF values but also with resilience against actual system disorder.

3.2. Time Delays

For protocols (i) and (ii), the ideal scenario is synchronous injection at sites A and C. We want to assess how damaging it is for the target entangled state when the excitations are injected in an asynchronous manner. This is modelled by adding a time delay $t_D = Dt_E$ between the two injections, where D is the dimensionless scale of the delay ($0 \leq D \leq 0.1$ in this work) and t_E is the unperturbed entangling time, different for each protocol and coupling ratio. The justification for investigating delays at a small fraction of t_E is that any viable physical implementation will always need to operate with accuracy at the timescale of its expected t_E. Therefore it is appropriate to investigate a window of injection (and extraction) timing errors covering a small fraction of the relevant t_E.

The initial injection is then performed in two parts. First, we initiate the time evolution with injection at site A, so that the initial state becomes $|+,0...0,0\rangle$ for protocol (i) or $|1,0...0,0\rangle$ for protocol (ii). Immediately before the second, delayed, injection time, so at t_D^-, we retrieve the overall state of the system $|\Psi(t_D^-)\rangle$. We can write $|\Psi(t_D^-)\rangle = \sum_i^N c_i|\Phi_i^{1\text{ex}}\rangle + c_0|0,0...0,0\rangle$, with N being the total number of sites, $c_0 = 1/\sqrt{2}$ for protocol (i) and $c_0 = 0$ for protocol (ii), and where $\sum_i^{0,N} |c_i|^2 = 1$ and $|\Phi_i^{1\text{ex}}\rangle = |0,0..1_i..0,0\rangle$. At time t_D^+ and for protocol (ii), injection at site C is simulated by mapping the coefficients c_i into the two-excitation vectors, $|\Phi_v^{2\text{ex}}\rangle = |00..1_i1_b..00\rangle$ with $i \neq b$, to find $|\Psi(t_D^+)\rangle_{\text{(ii)}} = \sum_v c_v|\Phi_v^{2\text{ex}}\rangle$. For protocol (i), the coefficients c_i must be mapped

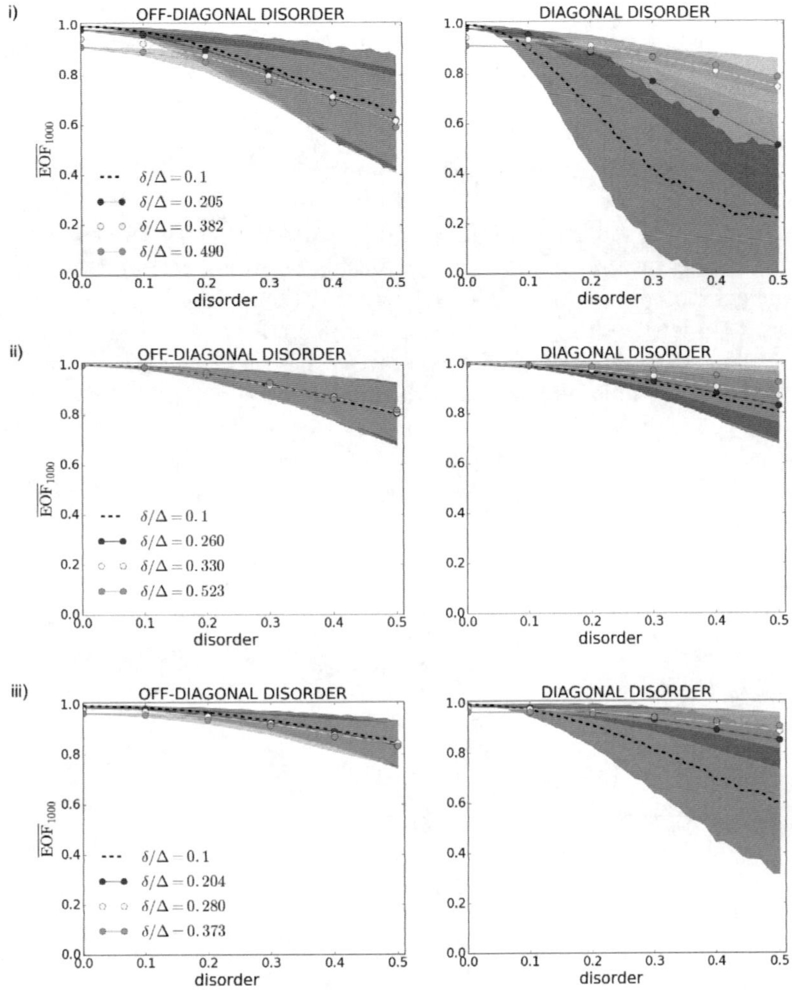

Figure 6: Averaged EOF at the unperturbed t_E for different levels of off-diagonal (l.h.s.) and diagonal (r.h.s.) disorders and coupling ratios from Table 2 for each of the three protocols (i), (ii) and (iii). Averages are calculated over 1000 random realizations. The black dashed line shows the averaged EOF for a ratio $\delta/\Delta = 0.1$ and orange, yellow and blue solid lines show the other three compared ratios, as labelled. The shadows represent the standard deviation of a single realization for each averaged profile.

into the zero, one and two excitation sectors. For the zero-excitation sector, it is enough to renormalize the whole state with an additional $1/\sqrt{2}$ factor; for the single-excitation sector, one needs to account for the state with a single excitation at site C; and for the two-excitation sector we proceed as in protocol (ii), such that $|\Psi(t_D^+)\rangle_{(\mathrm{i})} = 1/\sqrt{2}(|\Psi(t_D^-)\rangle + |0,0...0,1_C\rangle + \sum_v c_v|\Phi_v^{2\mathrm{ex}}\rangle)$. We note that in our simulations the time delay is short enough so that at t_D^- there is negligible probability for site C to be occupied[2], hence the aforementioned treatments are an accurate approximation and no state renormalisation is required. In cases when the occupation

[2] This probability is at most 2×10^{-8} and 9×10^{-12}, for the Cluster and Bell state generation respectively, corresponding to the scenarios with larger coupling ratios (red curves in Fig. 7) and a delay of 10%.

probability of an ideally empty site is non-negligible at the time of the second, delayed injection, it is clearly important to consider the actual injection mechanism. The actual state prepared will then clearly depend on if the mechanism is some sort of SWAP operation at the relevant site, or a spin flip operation, or some kind of projection. A detailed discussion of injection mechanisms is given in [36].

The effect of time delay errors on protocols (i) and (ii) is shown in Fig. 7 for the different chosen coupling ratios from Table 2. For $\delta/\Delta = 0.1$, we observe that even with an injection time delay of 10% t_E, the EOF is still very high for both protocols, with values of 0.91 and 0.95, respectively. This robustness decreases with an increasing coupling ratio for both protocols. As δ/Δ increases, the eigenstates peaking at sites A, B, and C are less localized [2]. This implies that when injecting at these sites, the excitation is not injected in a single eigenstate (as it would be for $\delta/\Delta \to 0$), but in a superposition of eigenstates with increased weight of the additional eigenstates as δ/Δ increases. This introduces additional phases in the wavefunction, and hence a more complicated dynamics, so that a delayed second injection affects the desired dynamics more, and the performance worsen with increasing δ/Δ. However, the EOF still remains high, with values of 0.70, and 0.77 for the larger ratios ($\delta/\Delta = 0.490$ and $\delta/\Delta = 0.523$), respectively, and a time delay of 10% t_E. We can conclude that our protocol is therefore also considerably robust against asynchronous state injections, which one would expect to be a small fraction of the relevant t_E for good physical implementations.

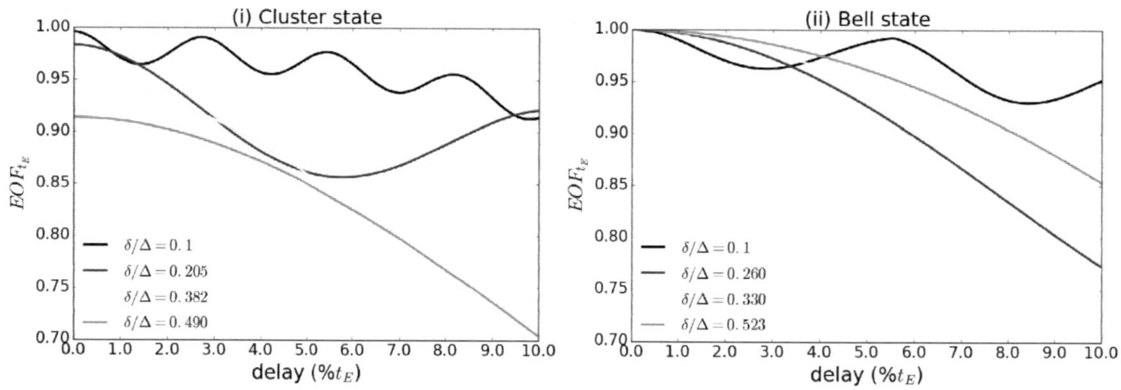

Figure 7: EOF at t_E against the input delay as a fraction of the entangling time (t_E) for protocols (i) (l.h.s.) and (ii) (r.h.s.) with asynchronous injections of $|+\rangle$ and $|1\rangle$ respectively at sites A and C, and the different chosen coupling ratios, δ/Δ, depicted in Table 2.

We finish this section by commenting on the scalability of the system. The chain length in this model can be increased by adding sets of two dimers, one either side of site B to preserve the symmetry. The resulting expanded systems will still support all the protocols discussed here. However, the time taken for entanglement creation would increase exponentially with chain length [2]. The important feature of the application discussed in this work is the robust creation of entanglement and its usefulness for modular quantum computation schemes. Scalability with chain length is less of an issue for this, in comparison to applications that utilize spin chains as quantum communication buses.

4. Conclusions

In this contribution we have compared the robustness of the entangling protocols proposed in [1] and [2] with respect to fabrication errors, slowly fluctuating fields and asynchronous injection of excitations during initialization. For all protocols and all forms of error considered, the resulting entanglement of formation has been shown to be very robust, even when the maximum percentage error considered is very large (e.g. up to 50% of the relevant Hamiltonian parameters for fabrication and slowly fluctuating fields). Energy fluctuation (diagonal) errors are generally more damaging than coupling (off-diagonal) disorder (against which there is excellent robustness) for small coupling ratios. So for practical implementations it is most important to focus on the reduction of fabrication errors that give rise to diagonal disorder. Timing injection errors, up to 10% of the time needed for producing entanglement, reduce the entanglement of formation by only a few % for $\delta/\Delta = 0.1$, with a more damaging effect when the coupling ratio is increased. However, good physical implementations would always be expected to have timing control errors limited to at most a small fraction of the actual device operation time.

We have provided a systematic comparison of the effect of imperfections on a large range of characteristic coupling ratios: this is important due to the exponential speedup of all entanglement protocols as the coupling ratio increases [1, 2]. We find that for all protocols the robustness to off-diagonal disorder is almost independent of the coupling ratio; even more exciting, the robustness to diagonal disorder increases significantly as the coupling ratio is increased. This result demonstrates a positive feedback between the speed and robustness to disorder of the protocols. The ability to maximize both of these crucial factors at little to no reduction in the maximum EOF allows us to propose ABC-type chains as rapid and reliable entanglement generation devices. ABC-type chains could therefore be of use in applications utilising various quantum computer architectures and across a variety of physical platforms, particularly where 'off-line' and robust entanglement creation and distribution between two parties is required as a resource.

References

[1] Estarellas M P, D'Amico I and Spiller T P 2017 *Phys. Rev.* A **95** 042335
[2] Wilkinson K N, Estarellas M P, D'Amico I and Spiller T P 2018 *Quant. Inf. Comp.* **18** (3&4) 0249
[3] Nielsen M A and Chuang I L 2010 *Quantum Computation and Quantum Information* (Cambridge: Cambridge University Press)
[4] Many countries now have major national programmes of applied research, technology development and commercialisation, e.g. `http://uknqt.epsrc.ac.uk/`
[5] Bennett C H, Brassard G, Crpeau C, Jozsa R, Peres A and Wootters W K 1993 *Phys. Rev. Lett.* **70** 1895–9
[6] Raussendorf R and Briegel H J 2001 *Phys. Rev. Lett.* **86** 5188–91
[7] Walther P, Resch K J, Rudolph T, Schenck E, Weinfurter H, Vedral V, Aspelmeyer M and Zeilinger A 2005 *Nature* **434** 169–76
[8] Ekert A 1991 *Phys. Rev. Lett.* **67** 661–3
[9] Giovannetti V, Lloyd S and Maccone L 2006 *Phys. Rev. Lett.* 2006 **96** 010401
[10] Dowling J P 2008 *Contemp. Phys.* **49** 125–43
[11] Bose S 2003 *Phys. Rev. Lett.* **91** 207901
[12] Bose S 2007 *Contemp. Phys.* **48** 13–30
[13] Christandl M, Datta N, Ekert A and Landahl A J 2004 *Phys. Rev. Lett.* **92** 187902
[14] Christandl M, Datta N, Dorlas T C, Ekert A, Kay A and Landahl A J 2005 *Phys. Rev.* A **71** 032312
[15] Plenio M B, Hartley J and Eisert J 2004 *New J. Phys.* **6** 36
[16] Kay A 2010 *Int. J. Quantum Inf.* **8** 641–76
[17] Spiller T P, D'Amico I and Lovett B W 2007 *New J. Phys.* **9** 20
[18] D'Amico I, Lovett B W and Spiller T P 2007 *Phys. Rev.* A **76** 030302
[19] Tejada J, Chudnovsky E M, del Barco F, Hernandez J M and Spiller T P 2001 *Nanotechnology* **12** 181
[20] Twamley J 2003 *Phys. Rev.* A **67** 052318
[21] Atala M, Aidelsburger M, Barreiro J T, Abanin D, Kitagawa T, Demler E and Bloch I 2013 *Nat. Phys.* **9** 795–800
[22] Meier E J, An F A and Gadway B 2016 *Nat. Commun.* **7** 13986
[23] Kielpinski D, Monroe C and Wineland D J 2002 *Nature* **417** 709–11
[24] Monroe C, Raussendorf R, Ruthven A, Brown K R, Maunz P, Duan L M and Kim J 2014 *Phys. Rev.* A **89** 022317
[25] D'Amico I 2007 *Semiconductor Research Trends* ed K G Sachs (New York: Nova Science) pp 281-300
[26] D'Amico I 2006 *Microelectron. J.* **37** 1440–1

[27] Nikolopoulos G M, Petrosyan D and Lambropoulos P 2004 *J. Phys. Condens. Matter* **16** 4991–5002
[28] Jacqmin T, Carusotto I, Sagnes I, Abbarchi M, Solnyshkov D D, Malpuech G, Galopin E, Lemaître A, Bloch J and Amo A 2014 *Phys. Rev. Lett.* **112** 116402
[29] Blanco-Redondo A, Andonegui I, Collins M J, Harari G, Lumer Y, Rechtsman M C, Eggleton B J and Segev M 2016 *Phys. Rev. Lett.* **116** 163901
[30] Delplace P, Ullmo D and Montambaux G 2011 *Phys. Rev.* B **84** 195452
[31] Estarellas M P, D'Amico I and Spiller T P 2017 *Sci. Rep.* **17** 42904
[32] Huo M X, Ying Li, Song Z and Sun C P 2008 *EPL* **84** 30004
[33] Almeida G M A, Ciccarello F, Apollaro T J G and Souza A M C 2016 *Phys. Rev.* A **93** 032310
[34] Gualdi G, Giampaolo S M and Illuminati F 2011 *Phys. Rev. Lett.* **106** 050501
[35] Wootters W K 1998 *Phys. Rev. Lett.* **80** 2245–8
[36] Ronke R, Spiller T P and D'Amico I 2011 *Phys. Rev.* A **83** 012325
[37] Yung M H and Bose S 2005 *Phys. Rev.* A **71** 032310
[38] Clark S R, Moura Alves C and Jaksch D 2005 *New J. Phys.* **7** 124
[39] Clark S R, Klein A, Bruderer M and Jaksch D 2007 *New J. Phys.* **9** 202
[40] Alkurtass B, Banchi L and Bose S 2014 *Phys. Rev.* A **90** 042304
[41] Banchi L, Compagno E and Bose S 2015 *Phys. Rev.* A **91** 052323
[42] Ronke R, D'Amico I and Spiller T P 2011 *Phys. Rev.* A **84** 032308

Journal of Physics: Conference Series **1638** (2020) 012014 doi:10.1088/1742-6596/1638/1/012014

Paul Busch: Contributions to Quantum Theory

Roger Colbeck[1], **Leon Loveridge**[2] **and Stefan Weigert**[1]
[1] Department of Mathematics, University of York, York YO10 5DD, United Kingdom
[2] Quantum Technology Group, Department of Science and Industry Systems, University of South-Eastern Norway, 3616 Kongsberg, Norway

E-mail: roger.colbeck@york.ac.uk, leon.d.loveridge@usn.no, stefan.weigert@york.ac.uk

Abstract. A bibliography of Paul Busch's scientific contributions is presented.

Over nearly forty years, Paul Busch made contributions to many fundamental aspects of quantum theory. He is well-known for adding mathematical and conceptual rigor to common lore, and his work also impinged on interpretational and philosophical questions. Much of Paul's research focused on exploring the behaviour of quantum systems in operational terms. He refined the formalism for describing quantum measurements and explored the properties of quantum observables, in particular through the lens of uncertainty relations and complementarity.

With this bibliography, we provide a comprehensive record of Paul's scientific contributions. The bibliography is based on a list in a CV, dated January 2018, that Paul made available on a personal website highlighting fundamental aspects of quantum mechanics. We have updated the list and included posthumously published work. Looking back at this body of work, we are struck by the consistency with which Paul interweaved mathematical rigour, conceptual precision and intellectual honesty.

The bulk of Paul's contributions to quantum theory can be found in *Journal articles* [1–92] and *Books* [93–96], both written and edited, as well as a number of *Chapters in books* [97–105] and refereed *Proceedings contributions* [106–121]. Next on the list are his *Editorials* [122–125] written for *Special issues* edited [126–129], and *Reviews of books* [130–135] published in the *Mathematical Reviews* (to which Paul contributed another 40 items, not listed here since they mainly summarize the papers under review). The list of *Unpublished contributions* [136–143] gives the titles of Paul's "Diplomarbeit", his PhD thesis and his "Habilitationsschrift", and a few items available at `arxiv.org`. The section *Varia* [144–147] contains the translation of a paper by Wigner and some software illustrating the properties of imcompatible observables. Paul also wrote three *Obituaries* [148–150], two of which are for his PhD supervisor, Peter Mittelstaedt.

Journal articles

[1] Busch P 1980 On the behavior of an oscillator clock near the singularity of a gravitational field *Gen. Relativ. Gravit.* **12** 483–92

[2] Busch P 1984 On joint lower bounds of position and momentum observables in quantum mechanics *J. Math. Phys.* **25** 1794–7

[3] Busch P and Lahti P J 1984 On various joint measurements of position and momentum observables in quantum theory *Phys. Rev.* D **29** 1634–46

[4] Busch P 1985 Indeterminacy relations and simultaneous measurements in quantum theory *Int. J. Theor. Phys.* **24** 63–92

[5] Busch P and Lahti P J 1985 A note on quantum theory, complementarity, and uncertainty *Philos. Sci.* **52** 64–77
[6] Busch P 1985 Momentum conservation forbids sharp localisation *J. Phys. A: Math. Gen.* **18** 3351–4
[7] Busch P 1986 Unsharp reality and joint measurements for spin observables *Phys. Rev.* D **33** 2253–61
[8] Busch P and Lahti P J 1986 To what extent do position and momentum commute? *Phys. Lett.* A **115** 259–64
[9] Busch P and Lahti P J 1987 Minimal uncertainty and maximal information for quantum position and momentum *J. Phys. A: Math. Gen.* **20** 899–906
[10] Busch P, Schonbek T P and Schroeck Jr F E 1987 Quantum observables: compatibility versus commutativity and maximal information *J. Math. Phys.* **28** 2866–72
[11] Busch P 1987 Some realizable joint measurements of complementary observables *Found. Phys.* **17** 905–37
[12] Ali S T, Brooke J A, Busch P, Gagnon R and Schroeck Jr F E 1988 Current conservation as a geometric property of spacetime *Can. J. Phys.* **66** 238–44
[13] Busch P 1988 Linearity versus symmetry? *Phys. Lett.* A **126** 300–2
[14] Busch P 1988 Surprising features of unsharp quantum measurements *Phys. Lett.* A **130** 323–9
[15] Busch P and Lahti P J 1989 The determination of the past and the future of a physical system in quantum mechanics *Found. Phys.* **19** 633–78
[16] Busch P, Grabowski M and Lahti P J 1989 Some remarks on effects, operations, and unsharp measurements *Found. Phys. Lett.* **2** 331–45
[17] Busch P and Schroeck Jr F E 1989 On the reality of spin and helicity *Found. Phys.* **19** 807–72
[18] Busch P, Cassinelli G and Lahti P J 1990 On the quantum theory of sequential measurements *Found. Phys.* **20** 757–75
[19] Busch P and Lahti P J 1990 Completely positive mappings in quantum dynamics and measurement theory *Found. Phys.* **20** 1429–39
[20] Busch P 1990 On the energy-time uncertainty relation. Part I: dynamical time and time indeterminacy *Found. Phys.* **20** 1–32
[21] Busch P 1990 On the energy-time uncertainty relation. Part II: pragmatic time versus energy indeterminacy *Found. Phys.* **20** 33–43
[22] Busch P and Lahti P J 1990 Some remarks on unsharp quantum measurements, quantum non-demolition, and all that *Ann. Phys.* **502** 369–82
[23] Busch P and Quadt R 1990 On Ruch's principle of decreasing mixing distance in classical statistical physics *J. Stat. Phys.* **61** 311–28
[24] Busch P 1991 Informationally complete sets of physical quantities *Int. J. Theor. Phys.* **30** 1217–27
[25] Busch P and Mittelstaedt P 1991 The problem of objectification in quantum mechanics *Found. Phys.* **21** 889–904
[26] Lahti P J, Busch P and Mittelstaedt P 1991 Some important classes of quantum measurements and their information gain *J. Math. Phys.* **32** 2770–5
[27] Busch P, Lahti P J and Mittelstaedt P 1992 Weak objectification, joint probabilities, and Bell inequalities in quantum mechanics *Found. Phys.* **22** 949–62
[28] Busch P and Ruch E 1992 The measure cone: irreversibility as a geometrical phenomenon *Int. J. Quantum Chem.* **41** 163–85
[29] Scherer H and Busch P 1993 Problem of signal transmission via quantum correlations and Einstein incompleteness in quantum mechanics *Phys. Rev.* A **47** 1647–51
[30] Busch P, Hellwig K-E and Stulpe W 1993 On classical representations of finite-dimensional quantum mechanics *Int. J. Theor. Phys.* **32** 399–405
[31] Busch P and Quadt R 1993 Concepts of coarse graining in quantum mechanics *Int. J. Theor. Phys.* **32** 2261–9
[32] Busch P, Kienzler P, Lahti P J and Mittelstaedt P 1993 Testing quantum mechanics against a full set of Bell inequalities *Phys. Rev.* A **47** 4627–31
[33] Quadt R and Busch P 1994 Coarse graining and the quantum-classical connection *Open Syst. Inf. Dyn.* **2** 129–55
[34] Hass K and Busch P 1994 Causality of superluminal barrier traversal *Phys. Lett.* A **185** 9–13
[35] Busch P, Grabowski M and Lahti P J 1994 Time observables in quantum theory *Phys. Lett.* A **191** 357–61
[36] Busch P, Grabowski M and Lahti P J 1994 Phase statistics and phase-space measurements *Phys. Rev.* A **50** 2881–4
[37] Busch P, Grabowski M and Lahti P J 1995 Who is afraid of POV measures? Unified approach to quantum phase observables *Ann. Phys.* **237** 1–11
[38] Busch P and Lahti P J 1995 The complementarity of quantum observables: theory and experiment *Riv. del Nuovo Cim.* **18** 1–27

[39] Busch P, Grabowski M and Lahti P J 1995 Repeatable measurements in quantum theory: their role and feasibility *Found. Phys.* **25** 1239–66
[40] Busch P, Cassinelli G and Lahti P J 1995 Probability structures for quantum state spaces *Rev. Math. Phys.* **7** 1105–21
[41] Busch P and Lahti P J 1996 The standard model of quantum measurement theory: history and applications *Found. Phys.* **26** 875–93
[42] Busch P and Shimony A 1996 Insolubility of the quantum measurement problem for unsharp observables *Studies in History and Philosophy of Modern Physics* **27** 397–404
[43] Busch P and Lahti P J 1996 Correlation properties of quantum measurements *J. Math. Phys.* **37** 2585–601
[44] Busch P and Lahti P J 1996 Individual aspects of quantum measurements *J. Phys. A: Math. Gen.* **29** 5899–907
[45] Busch P and Lahti P J 1997 Remarks on separability of compound quantum systems and time reversal *Found. Phys. Lett.* **10** 113–7
[46] Busch P 1998 Orthogonality and disjointness in spaces of measures *Lett. Math. Phys.* **44** 215–24
[47] Busch P 1998 Can 'unsharp objectification' solve the quantum measurement problem? *Int. J. Theor. Phys.* **37** 241–7
[48] Busch P and Singh J 1998 Lüders theorem for unsharp quantum measurements *Phys. Lett.* A **249** 10–2
[49] Busch P and Gudder S P 1999 Effects as functions on projective Hilbert space *Lett. Math. Phys.* **47** 329–37
[50] Busch P 1999 Stochastic isometries in quantum mechanics *Math. Phys. Anal. Geom.* **2** 83–106
[51] Busch P 1999 Unsharp localization and causality in relativistic quantum theory *J. Phys. A: Math. Gen.* **32** 6535–46
[52] Busch P 2000 Geometric and operational aspects of irreversibility *MATCH Commun. Math. Comput. Chem.* **42** 71–84
[53] Busch P 2001 Just how final are today's quantum structures? *Soft Comput.* **5** 137–8
[54] Busch P, Cassinelli G, De Vito E, Lahti P J and Levrero A 2001 Teleportation and measurement *Phys. Lett.* A **284** 141–5
[55] Busch P, Lahti P J, Pellonpää J-P and Ylinen K 2001 Are number and phase complementary observables? *J. Phys. A: Math. Gen.* **34** 5923–35
[56] Brooke J A, Busch P and Pearson D B 2002 Commutativity up to a factor of bounded operators in complex Hilbert space *Proc. Royal Soc.* A **458** 109–18
[57] Busch P 2003 The role of entanglement in quantum measurement and information processing *Int. J. Theor. Phys.* **42** 937–41
[58] Busch P and Shilladay C R 2003 Uncertainty reconciles complementarity with joint measurability *Phys. Rev.* A **68** 034102
[59] Busch P 2003 Quantum states and generalized observables: a simple proof of Gleason's theorem *Phys. Rev. Lett.* **91** 120403
[60] Weigert S and Busch P 2003 Lüders theorem for coherent-state POVMs *J. Math. Phys.* **44** 5474–86
[61] Busch P, Heinonen T and Lahti P J 2004 Noise and disturbance in quantum measurement *Phys. Lett.* A **320** 261–70
[62] Polachic C, Rangacharyulu C, van den Berg A, Hamieh S, Harakeh M, Hunyadi M, de Huu M, Wörtche H, Heyse J, Bäumer C, Frekers D, Rakers S, Brooke J and Busch P 2004 Polarization correlations of proton pairs as tests of hidden-variable theories *Phys. Lett.* A **323** 176–81
[63] Busch P and Shilladay C 2006 Complementarity and uncertainty in Mach-Zehnder interferometry and beyond *Phys. Rep.* **435** 1–31
[64] Busch P and Pearson D B 2007 Universal joint-measurement uncertainty relation for error bars *J. Math. Phys.* **48** 082103
[65] Busch P, Heinonen T and Lahti P J 2007 Heisenberg's uncertainty principle *Phys. Rep.* **452** 155–76
[66] Stulpe W and Busch P 2008 The structure of classical extensions of quantum probability theory *J. Math. Phys.* **49** 032104
[67] Busch P, Kiukas J and Lahti P J 2008 Measuring position and momentum together *Phys. Lett.* A **372** 4379–80
[68] Busch P and Heinosaari T 2008 Approximate joint measurements of qubit observables *Quantum Inf. Comput.* **8** 0797–0818
[69] Busch P 2009 On the sharpness and bias of quantum effects *Found. Phys.* **39** 712–30
[70] Busch P and Schmidt H-J 2010 Coexistence of qubit effects *Quantum Inf. Process.* **9** 143–69
[71] Busch P and Jaeger G 2010 Unsharp quantum reality *Found. Phys.* **40** 1341–67
[72] Busch P, Kiukas J and Lahti P J 2010 On the notion of coexistence in quantum mechanics *Math. Slovaca* **60** 665–80

[73] Busch P and Loveridge L 2011 Position measurements obeying momentum conservation *Phys. Rev. Lett.* **106** 110406
[74] Loveridge L and Busch P 2011 'Measurement of quantum mechanical operators' revisited *Eur. Phys. J.* D **62** 297–307
[75] Busch P, Heinosaari T, Schultz J and Stevens N 2013 Comparing the degrees of incompatibility inherent in probabilistic physical theories *EPL* **103** 10002
[76] Biniok J C G and Busch P 2013 Multislit interferometry and commuting functions of position and momentum *Phys. Rev.* A **87** 062116
[77] Beneduci R, Bullock T J, Busch P, Carmeli C, Heinosaari T and Toigo A 2013 Operational link between mutually unbiased bases and symmetric informationally complete positive operator-valued measures *Phys. Rev.* A **88** 032312
[78] Li Y and Busch P 2013 Von Neumann entropy and majorization *J. Math. Anal. Appl.* **408** 384–93
[79] Busch P, Lahti P J and Werner R F 2013 Proof of Heisenberg's error-disturbance relation *Phys. Rev. Lett.* **111** 160405
[80] Busch P, Lahti P J and Werner R F 2014 Heisenberg uncertainty for qubit measurements *Phys. Rev.* A **89** 012129
[81] Stevens N and Busch P 2014 Steering, incompatibility, and Bell-inequality violations in a class of probabilistic theories *Phys. Rev.* A **89** 022123
[82] Busch P, Lahti P J and Werner R F 2014 Measurement uncertainty relations *J. Math. Phys.* **55** 042111
[83] Busch P, Lahti P J and Werner R F 2014 Quantum root-mean-square error and measurement uncertainty relations *Rev. Mod. Phys.* **86** 1261–81
[84] Bullock T J and Busch P 2014 Focusing in Arthurs-Kelly-type joint measurements with correlated probes *Phys. Rev. Lett.* **113** 120401
[85] Biniok J C G, Busch P and Kiukas J 2014 Uncertainty in the context of multislit interferometry *Phys. Rev.* A **90** 022115
[86] Busch P and Stevens N 2015 Direct tests of measurement uncertainty relations: what it takes *Phys. Rev. Lett.* **114** 070402
[87] Miyadera T, Loveridge L and Busch P 2016 Approximating relational observables by absolute quantities: a quantum accuracy-size trade-off *J. Phys. A: Math. Theor.* **49** 185301
[88] Loveridge L, Busch P and Miyadera T 2017 Relativity of quantum states and observables *EPL* **117** 40004
[89] Xiong T P, Yan L L, Ma Z H, Zhou F, Chen L, Yang W L, Feng M and Busch P 2017 Optimal joint measurements of complementary observables by a single trapped ion *New J. Phys.* **19** 063032
[90] Loveridge L, Miyadera T and Busch P 2018 Symmetry, reference frames, and relational quantities in quantum mechanics *Found. Phys.* **48** 135–98
[91] Bullock T and Busch P 2018 Measurement uncertainty relations: characterising optimal error bounds for qubits *J. Phys. A: Math. Theor.* **51** 283001
[92] Busch P, Kiukas J and Werner R F 2018 Sharp uncertainty relations for number and angle *J. Math. Phys.* **59** 042102

Books

[93] P Busch, P J Lahti and P Mittelstaedt (eds) 1993 *Quantum Measurement, Irreversibility and the Physics of Information* (Singapore: World Scientific)
[94] Busch P, Grabowski M and Lahti P J 1995 *Operational Quantum Physics* (*Lecture Notes in Physics Monographs* vol 31) (Berlin: Springer)
[95] Busch P, Lahti P J and Mittelstaedt P 1996 *The Quantum Theory of Measurement* (*Lecture Notes in Physics Monographs* vol 2) (Berlin: Springer)
[96] Busch P, Lahti P J, Pellonpää J-P and Ylinen K 2016 *Quantum Measurement* Theoretical and Mathematical Physics (Cham: Springer)

Chapters in books

[97] Mittelstaedt P and Busch P 1994 Zur Quantentheorie der Messung *Quanten, Chaos und Dämonen: erkenntnistheoretische Aspekte der modernen Physik* ed K Mainzer and W Schirmacher (Mannheim: Bibliographisches Institut) pp 107–130
[98] Busch P 1997 Is the quantum state (an) observable? *Potentiality, Entanglement and Passion-at-a-Distance* vol 194 ed R S Cohen *et al* (Dordrecht: Springer) pp 61–70
[99] Busch P 2007 The time-energy uncertainty relation *Time in Quantum Mechanics* Lecture Notes in Physics vol 734 ed J Muga *et al* (Berlin: Springer) pp 73–105

[100] Busch P and Jaeger G 2009 Which-way or welcher-Weg-experiments *Compendium of Quantum Physics* ed D Greenberger *et al* (Berlin: Springer) pp 845–51

[101] Busch P and Falkenburg B 2009 Heisenberg uncertainty relation (indeterminacy relations) *Compendium of Quantum Physics* ed D Greenberger *et al* (Berlin: Springer) pp 281–3

[102] Busch P and Lahti P J 2009 Observable *Compendium of Quantum Physics* ed D Greenberger *et al* (Berlin: Springer) pp 425–8

[103] Busch P and Lahti P J 2009 Lüders rule *Compendium of Quantum Physics* ed D Greenberger *et al* (Berlin: Springer) pp 356–8

[104] Busch P and Lahti P J 2009 Measurement theory *Compendium of Quantum Physics* ed D Greenberger *et al* (Berlin: Springer) pp 374–9

[105] Busch P 2009 Effect *Compendium of Quantum Physics* ed D Greenberger *et al* (Berlin: Springer) pp 179–80

Proceedings contibutions

[106] Busch P 1985 Elements of unsharp reality in the EPR experiment *Symposium on the Foundations of Modern Physics 1985: 50 Years of the Einstein-Podolsky-Rosen Gedanken Experiment* ed P J Lahti and P Mittelstaedt (Singapore: World Scientific) pp 343–57

[107] Busch P 1985 Can quantum theoretical reality be considered sharp? *Recent Developments in Quantum Logic* ed P Mittelstaedt and E W Stachow (Mannheim: Bibliographisches Institut) pp 81–101

[108] Busch P 1987 Unsharp reality and the question of quantum systems *Symposium on the Foundations of Modern Physics 1987: The Copenhagen Interpretation 60 Years after the Como Lecture* ed P J Lahti and P Mittelstaedt (Singapore: World Scientific) pp 105–25

[109] Busch P 1990 A 'natural' explanation of the Aharonov et al 'surprising quantum effect' *Proc. 3rd International Symposium on Foundations of Quantum Mechanics in the Light of New Technology* ed N B Gakkai and S Kobayashi (Tokyo: Physical Society of Japan) pp 25–31

[110] Busch P 1991 Macroscopic quantum systems and the objectification problem *Symposium on the Foundations of Modern Physics: Quantum Theory of Measurement and related Philosophical Problems* ed P J Lahti and P Mittelstaedt (Singapore: World Scientific) pp 62–76

[111] Busch P 1993 The quantum theory of unsharp measurements *Classical and Quantum Systems–Foundations and Symmetries* Proc. 2nd International Wigner Symposium ed H D Doebner *et al* (Goslar: World Scientific) pp 19–28

[112] Busch P 1993 Quantum-to-classical transition for Galilei-covariant observables *Symposium on the Foundations of Modern Physics: Quantum Measurement, Irreveribility and the Physics of Information* ed P Busch *et al* (Singapore: World Scientific) pp 145–52

[113] Busch P 1994 The status of quantum mechanics in the light of the objectification problem *The Interpretation of Quantum Mechanics: Where do we stand?* ed L Accardi (Rome: Istituto della Enciclopedia Italiana) pp 205–17

[114] Scherer H and Busch P 1995 Weakly disturbing phase space measurements in quantum mechanics *Quantum Communications and Measurement* ed V P Belavkin *et al* (Boston, MA: Springer) pp 155–63

[115] Busch P 1998 Remarks on unsharp quantum observables, objectification, and modal interpretations *The Modal Interpretation of Quantum Mechanics* The Western Ontario Series in Philosophy of Science vol 60 ed W Demopoulos *et al* (Dordrecht: Springer) pp 279–88

[116] Busch P 2002 EPR-Bell tests with unsharp observables and relativistic quantum measurement *Non-locality and Modality* ed T Placek and J Butterfield (Dordrecht: Springer) pp 175–93

[117] Busch P 2003 Less (precision) is more (information): quantum information in fuzzy probability theory *Proc. International Conference Quantum Theory, Reconsideration of Foundations* ed A Khrennikov *et al* (Växjö: Växjö University Press) pp 113–28

[118] Busch P 2009 No information without disturbance: quantum limitations of measurement *Quantum Reality, Relativistic Causality, and Closing the Epistemic Circle* The Western Ontario Series in Philosophy of Science vol 73 ed Demopoulos W (Dordrecht: Springer) pp 229–56

[119] Busch P 2010 Quantum mechanics as a framework for dealing with uncertainty *Phys. Scr.* **2010** 014003

[120] Busch P and Loveridge L 2013 Quantum measurements constrained by symmetries *Symmetries and Groups in Contemporary Physics* Nankai Series in Pure, Applied Mathematics and Theoretical Physics vol 11 ed C Bai *et al* (Singapore: World Scientific) pp 587–92

[121] Busch P 2014 Quantum rms error and Heisenberg's error-disturbance relation *EPJ Web of Conferences* **78** 01002

Editorials

[122] Busch P and van der Merwe A 1989 Peter Mittelstaedt: philosopher-physicist *Found. Phys.* **19** 789–91

[123] Busch P, Dieks D and t'Hooft G 2009 Pekka Johannes Lahti – 60th birthday *Found. Phys.* **39** 519–20
[124] Weigert S, Busch P, Dodson M and Higuchi A 2010 Quantum groups, quantum foundations and quantum information: a Festschrift for Tony Sudbery *J. Phys.: Conf. Ser.* **254** 011001
[125] Busch P, Pfarr J, Ristig M L and Stachow E W 2010 Quantum–Matter–Spacetime: Peter Mittelstaedt's contributions to physics and its foundations *Found. Phys.* **40** 1163–70

Special issues edited

[126] P Busch and A van der Merwe (eds) 1989 Special issues in honor of Peter Mittelstaedt's 60th birthday *Found. Phys.* **19** (Issues 7, 8, 9)
[127] P Busch, D Dieks and G t'Hooft (eds) 2009 Festschrift for Pekka Johannes Lahti – 60th birthday *Found. Phys.* **39** (Issues 6 and 7)
[128] P Busch (ed) 2010 Between physics and philosophy – Festschrift for Peter Mittelstaedt on his 80th birthday *Found. Phys.* **40** (Issue 10)
[129] P Busch, T Heinosaari and T Miyadera (eds) 2016 Mathematics of quantum uncertainty *Mathematics* **4**

Book reviews

[130] Busch P 1990 Thomas S. Kuhn: Black-body theory and the quantum discontinuity Book review MR0957503 (90c:01040) American Mathematical Society
[131] Busch P 1999 Jeffrey Bub: Interpreting the quantum world Book review MR1464691 (99f:81006) American Mathematical Society
[132] Busch P 2002 Willem M. de Muynck: Foundations of quantum mechanics, an empiricist approach Book review MR2040793 (2005i:81004) American Mathematical Society
[133] Busch P 2002 Classical versus quantum ontology *Studies in History and Philosophy of Modern Physics* **33** 517–39
[134] Busch P 2009 Physical theory and its interpretation – essays in honor of Jeffrey Bub *Studies in History and Philosophy of Modern Physics* **40** 92–3
[135] Busch P 2013 The mathematical language of quantum theory Book review MR2896337 American Mathematical Society

Unpublished

[136] Busch P 1979 On the empirical relevance of proper time in a gravitational field with singularity Diploma Thesis University of Cologne
[137] Busch P 1982 Indeterminacy relations and simultaneous measurements in quantum theory PhD Thesis University of Cologne
[138] Busch P 1988 Physical aspects of a generalised observable concept in quantum theory Habilitation in Mathematical Physics University of Cologne
[139] Busch P, Lahti P J and Werner R F 2014 Measurement uncertainty: reply to critics *Preprint* arXiv:1402.3102 [quant-ph]
[140] Busch P, Lahti P J and Werner R F 2014 Comment on "Experimental test of error-disturbance uncertainty relations by weak measurement" *Preprint* arXiv:1403.0367 [quant-ph]
[141] Busch P, Lahti P J and Werner R F 2014 Comment on "Noise and disturbance in quantum measurements: an information-theoretic approach" *Preprint* arXiv:1403.0368 [quant-ph]
[142] Busch P and Pearson D B 2014 Error and unsharpness in approximate joint measurements of position and momentum *Preprint* arXiv:1405.6956 [quant-ph]
[143] Busch P and Reardon-Smith O 2019 On quantum uncertainty relations and uncertainty regions *Preprint* arXiv:1901.03695 [quant-ph]

Varia

[144] Busch P, Lahti P J and Mittelstaedt P 1993 Symposium on the foundations of modern physics 1993 *Found. Phys.* **23** 169–71
[145] Busch P 2010 Translation of "Die Messung quantenmechanischer Operatoren" by E. P. Wigner *Preprint* arXiv:1012.4372 [quant-ph]
[146] Busch P and Biniok J C G 2014 Heisenberg-type uncertainty relation for qubits – Wolfram demonstrations project `https://demonstrations.wolfram.com/HeisenbergTypeUncertaintyRelationForQubits/`
[147] Busch P and Bullock T 2017 Optimal joint measurements of qubit observables – Wolfram demonstrations project `https://demonstrations.wolfram.com/OptimalJointMeasurementsOfQubitObservables/`

Journal of Physics: Conference Series **1638** (2020) 012014 doi:10.1088/1742-6596/1638/1/012014

Obituaries

[148] Beltrametti E, Busch P and Lahti P J 2003 Sławomir Bugajski 1941–2003 *Int. J. Theor. Phys.* **42** 1133–7
[149] Busch P 2015 Nachruf auf Peter Mittelstaedt *Phys. J.* **14** 46
[150] Busch P 2015 Philosophical problems of modern physics: Peter Mittelstaedt 1929–2014 *Found. Phys.* **45** 483–95

AUTHOR INDEX

Institute of Physics
Dirac House, Temple Back
Bristol BS1 6BE UK

ISSN: 1742-6588
ISBN 978-1-7138-3914-9